T0325671

Introduction to Food Toxicology

Second Edition

Food Science and Technology International Series

A complete list of books in this series appears at the end of this volume.

Introduction to Food Toxicology

Second Edition

Takayuki Shibamoto
Department of Environmental Toxicology
University of California
Davis, CA

Leonard Bjeldanes
Department of Nutritional Sciences and Toxicology
University of California
Berkeley, CA

ELSEVIER

AMSTERDAM • BOSTON • HEIDELBERG • LONDON
NEW YORK • OXFORD • PARIS • SAN DIEGO
SAN FRANCISCO • SINGAPORE • SYDNEY • TOKYO

Academic Press is an imprint of Elsevier

Academic Press is an imprint of Elsevier
30 Corporate Drive, Suite 400, Burlington, MA 01803, USA
525 B Street, Suite 1900, San Diego, California 92101-4495, USA
84 Theobald's Road, London WC1X 8RR, UK

Library of Congress Cataloging-in-Publication Data
APPLICATION SUBMITTED

British Library Cataloguing-in-Publication Data
A catalogue record for this book is available from the British Library.

ISBN: 978-0-12-374286-5

For information on all Academic Press publications
visit our Web site at www.elsevierdirect.com

Printed in the United States of America

Transferred to Digital Printing in 2011

Table of Contents

Preface

Food is one of the most essential materials for the survival of living organisms, following perhaps only oxygen and water in importance. People have been learning how to identify and prepare appropriate foods since prehistoric times. However, there was probably a tremendous sacrifice of human lives before people learned to find and prepare safe foods. For thousands of years trial and error was the only method to detect the presence of poisons in the diet. Systematic data on poisons in foods have been recorded for only approximately 200 years or so. Moreover, food toxicology as a classroom discipline taught in universities has a relatively recent origin. The revolution in the last two decades in our knowledge of the sciences of chemistry and molecular biology that are the foundation of modern toxicology have enhanced to previously unimagined levels; our abilities to both detect extremely small amounts of toxic agents and to understand in great detail the mechanisms of action of these toxic substances.

This volume is a classroom reference for students who do not have strong backgrounds in either toxicology or food science, but who would like to be introduced to the exciting field of toxicology and its application to toxins in food and the environment. The format of the book is designed primarily to teach students basic toxicology of food and environmental toxins and to extend this knowledge to consider molecular targets and mechanisms of action of important toxic agents. The chemical identities of the toxicants and their fates in foods and in the human body are discussed, along with historical notes on the discoveries of the toxins and possible use in ancient times.

Student interest in toxicology has continued to grow since the publication of the first edition of this text. Issues related to toxic materials have received increased attention from the scientific community, regulatory agencies and the general public. The issues and potential problems are reported almost daily by the mass media, and are often the focus of attention in nightly newscasts. The major misunderstanding and confusion raised by many of these reports are almost always due to lack of basic knowledge about toxicology among most reporters and consumers. This volume presents basic principles of modern food toxicology and their application to topics of major interest for human health that will allow students of the subject to better identify and understand the significant problems of toxic materials in foods and the environment.

Takayuki Shibamoto
Leonard Bjeldanes

CHAPTER 1

Principles of Toxicology

CHAPTER CONTENTS

Toxicology is defined as the study of the adverse effects of chemicals on living organisms. Its origins may be traced to the time when our prehistoric ancestors first attempted to introduce substances into their diets that they had not encountered previously in their environments. By observing which substances could satisfy hunger without producing illness or death, ancient people developed dietary habits that improved survival and proliferation of the species in their traditional environment and allowed them to adapt to new environments. In its modern context, toxicology draws heavily on knowledge in chemical and biological fields and seeks a detailed understanding of toxic effects and means to prevent or reduce toxicity. In many instances, the original discoveries of toxins that caused devastating human illness and suffering have led to the development of the toxin as a probe of biological function that is used today to study basic mechanisms and to develop cures for human maladies as diverse as postpartum hemorrhage, psychosis, and cancer.

A brief history of documented uses of toxic agents serves to illustrate the importance of these substances since ancient cultures. The Ebers papyrus of about 1500 BCE, one of the oldest preserved medical documents, describes uses of many poisons such as hemlock, aconite arrow poison, opium, lead, and copper. By 399 BCE, death by hemlock poisoning was a well-established means of capital punishment in Greece, most notably in the forced suicide of Socrates. Around this same time, Hippocrates discussed bioavailability and overdosage of toxic agents, and intended poisonings—used mostly by aristocratic women as a means of dispatching unwanted husbands—were of common occurrence in Rome. By about 350 BCE, Theophrastus, a student of Aristotle, made many references to poisonous plants in his first *De Historia Plantarum*.

In about 75 BCE, King Mithridates VI of Pontus (in modern Turkey) was obsessed with poisons and, from a young age, took small amounts of as many as 50 poisons in the hopes of developing resistance to each of them. This practice apparently induced a considerable resistance to poisons, since according to legend, to avoid enemy capture, the standard poisonous mixture was not effective in a suicide attempt by the vanquished king and he had to fall on his sword instead. The term "mithridatic" refers to an antidotal or protective mixture of low but significant doses of toxins and has a firm scientific basis. However, the claim that vanishingly small doses of toxic agents also produce protective effects, which is the claimed basis for homeopathy, does not have scientific support.

In 82 BCE, *Lex Cornelia* (Law of Cornelius) was the first law to be enacted in Rome that included provisions against human poisonings. In approximately 60 CE, Dioscorides, a physician in the Roman armies of Emperors Nero, Caligula, and Claudius, authored a six to eight volume treatise that classified poisons on the basis of origin (plant, animal, mineral) and biological activity, while avoiding the common practice of classification based on fanciful theories of action that were considered important at the time, such as the theory of humors, which posed that body function is regulated by the proper balance of fluids called black bile, yellow bile, phlegm, and blood. This treatise often suggested effective therapies for poisonings such as the use of emetics, and was the standard source of such information for the next 1500 years.

Paracelsus (1493–1541) is considered to be the founder of toxicology as an objective science. Paracelsus, who changed his name from Phillip von Hohenheim, was an energetic, irascible, and iconoclastic thinker (Figure 1.1). He was trained in Switzerland as a physician and traveled widely in Europe and the Middle East to learn alchemy and medicine in other traditions of the day. Although astrology remained an important part of his philosophy, he eschewed magic in his medical practice. His introduction of the practice of keeping wounds clean and allowing them to drain to allow them to heal won him considerable acclaim in Europe. Most notably for toxicology, Paracelsus was the first person who attributed adverse effects of certain substances to the substance itself and not to its association with an evil or angered spirit or god. Paracelsus is accredited with conceiving the basic concept of toxicology, which often is stated as follows:

All substances are poisons; there is none that is not a poison. The right dose differentiates the poison from a remedy.

Although this and other concepts developed by Paracelsus were groundbreaking and major advances in thinking about disease for the time, they put him at odds with the major medical practitioners. As a result, he was forced to leave his home medical practice and spent several of his final years traveling. He was 48 when he died, and there are suspicions that his enemies caught up with him and ended his very fruitful life. How ironic it would be if the father of toxicology were murdered by poisoning!

It is useful to evaluate the significance of the Paracelsus axiom in our daily lives by considering examples of well-known substances with low and high toxicity. Water might be considered one of the least toxic of the substances that we commonly encounter. Can it be toxic? Indeed, there are many reports of water toxicity in the scientific literature. For example, in 2002 a student at California State University at Chico was undergoing a fraternity initiation ordeal in which he was required to drink up to five gallons

FIGURE 1.1 *Paracelsus (1493–1541).*

of water while engaged in rigorous calisthenics and being splashed with ice cold water. Consumption of this amount of water in a short span of time resulted in the dilution of the electrolytes in his blood to the point that normal neurological function was lost and tragically the young man died.

Let us now consider the converse concept that exposure to a small amount of a highly toxic agent can be of little consequence. For example, the bacterium that produces botulism, *Clostridium botulinum*, can produce deadly amounts of botulinum toxin in improperly sterilized canned goods. This bacterial toxin is one of the most toxic substances known. The same toxin, however, is used therapeutically, for example, to treat spastic colon and as a cosmetic to reduce wrinkles in skin.

BRANCHES OF TOXICOLOGY

The science of toxicology has flourished from its early origins in myth and superstition, and is of increasing importance to many aspects of modern life. Modern toxicology employs cutting-edge knowledge in chemistry, physiology, biochemistry and molecular biology, often aided by computational technology, to deal with problems of toxic agents in several fields of specialization.

The major traditional specialties of toxicology address several specific societal needs. Each specialty has its unique educational requirements, and employment in some areas may require professional certification. **Clinical toxicology** deals with the prevention, diagnosis, and management of poisoning, usually in a hospital or clinical environment. **Forensic toxicology** is the application of established techniques for the analysis of biological samples for the presence of drugs and other potentially toxic substances, and usually is practiced in association with law enforcement. **Occupational toxicology** seeks to identify the agents of concern in the workplace, define the conditions for their safe use, and prevent absorption of harmful amounts. **Environmental toxicology** deals with the potentially deleterious impact of man-made and natural environmental chemicals on living organisms, including wildlife and humans.

Regulatory toxicology encompasses the collection, processing, and evaluation of epidemiological and experimental toxicology data to permit scientifically based decisions directed toward the protection of humans from the harmful effects of chemical substances. Furthermore, this area of toxicology supports the development of standard protocols and new testing methods to continuously improve the scientific basis for decision-making processes. **Ecotoxicology** is concerned with the environmental distribution and toxic effects of chemical and physical agents on populations and communities of living organisms within defined ecosystems. Whereas traditional environmental toxicology is concerned with toxic effects on individual organisms, ecotoxicology is concerned with the impact on populations of living organisms or on ecosystems.

Food toxicology focuses on the analysis and toxic effects of bioactive substances as they occur in foods. Food toxicology is a distinct field that evaluates the effects of components of the complex chemical matrix of the diet on the activities of toxic agents that may be natural endogenous products or may be introduced from contaminating organisms, or from food production, processing, and preparation.

DOSE-RESPONSE

Since there are both toxic and nontoxic doses for any substance, we may also inquire about the effects of intermediate doses. In fact, the intensity of a biological response is proportional to the concentration of the substance in the body fluids of the exposed organism. The concentration of the substance in the body fluids, in turn, is usually proportional to the dose of the substance to which the organism is subjected. As the dose of a substance is increased, the severity of the toxic response will increase until at a high enough dose the substance will be lethal. This so-called individual dose-response can be represented as a plot of degree of severity of any quantifiable response, such as an enzyme activity, blood pressure, or respiratory rate, as a function of dose. The resulting plot of response against the $\log_{10}$ of concentration will provide a sigmoidal curve (as illustrated in Figure 1.2) that will be nearly linear within a mid-concentration range and will be asymptotic to the zero response and maximum response levels. This response behavior is called a **graded dose-response** since the severity of the response increases over a range of concentrations of the test substance.

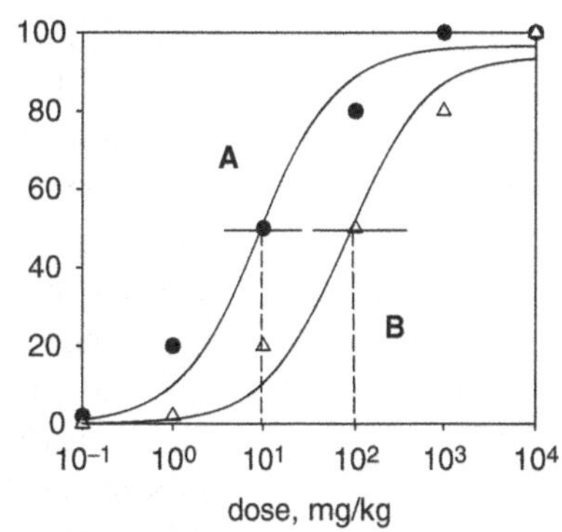

FIGURE 1.2 *Dose-response. The resulting plot of response against the $\log_{10}$ of concentration will provide a sigmoidal curve that will be nearly linear within a mid-concentration range and will be asymptotic to the zero response and maximum response levels.*

Toxicity evaluations with individual test organisms are not used often, however, because individual organisms, even inbred rodent species used in the laboratory, may vary from one another in their sensitivities to toxic agents. Indeed, in studies of groups of test organisms, as the dose is increased, there is not a dose at which all the organisms in the group will suddenly develop the same response.

Instead, there will be a range of doses over which the organisms respond in the same way to the test substance. In contrast to the graded individual dose-response, this type of evaluation of toxicity depends on whether or not the test subjects develop a specified response, and is called an **all-or-none** or **quantal population** response. To specify this group behavior, a plot of percent of individuals that respond in a specified manner against the log of the dose is generated.

Let us consider, for example, the generation of a dose-response curve for a hypothetical hypertensive agent. The test substance would be administered in increasing doses to groups of 10 subjects or test organisms. The percentage of individuals in each group that respond in a specific way to the substance (e.g., with blood pressure 140/100) then is determined. The data then are plotted as percent response in each group versus the log of the dose given to each group. Over a range of low doses, there will be no test subjects that develop the specified blood pressure. As the dose increases, there will be increased percentages of individuals in the groups that develop the required blood pressure, until a dose is reached for which a maximum number of individuals in the group respond with the specified blood pressure. This dose, determined statistically, is the mean dose for eliciting the defined response for the population. As the dose is further increased, the percentages of individuals that respond with the specified blood pressure will decrease, since the individuals that responded to the lower doses are now exhibiting blood pressures in excess of the specified level. Eventually, a dose will be reached at which all the test subjects develop blood pressures in excess of the specified level.

When the response has been properly defined, information from quantal dose-response experiments can be presented in several ways. A frequency-response plot (Figure 1.3) is generated by plotting the percentage of responding individuals in each dose group as a function of the dose.

The curve that is generated by these data has the form of the normal Gaussian distribution and, therefore, the data are subject to the statistical laws for such distributions. In this model, the numbers of individuals on either side of the mean are equal and the area under the curve represents the total population. The area under the curve bounded by the inflection points includes the number of individuals responding to the mean dose, plus or minus one standard deviation (SD) from the mean dose, or 95.5% of the population. This mean value is useful in specifying the dose range over which most individuals respond in the same way.

Frequency-response curves may be generated from any set of toxicological data where a quantifiable response is measured simply by recording the

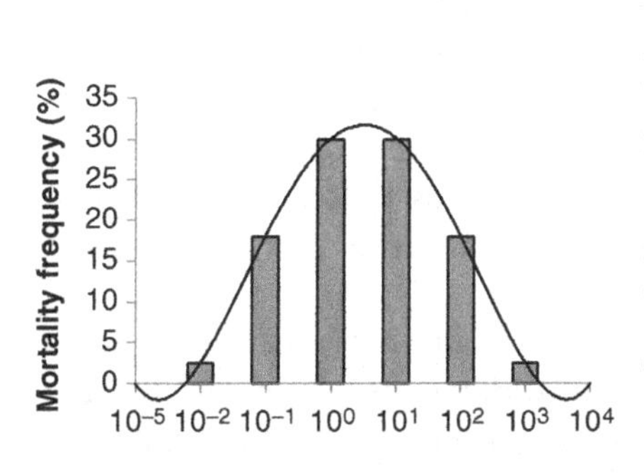

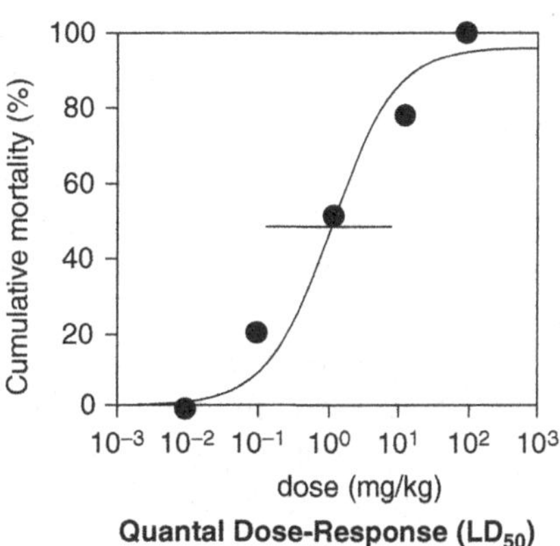

FIGURE 1.3

Comparison of shapes of the dose-response curves between Normal Frequency Distribution and Quantal Dose-Response.

percentage of subjects that respond at each dose minus the percentage that respond at the lower dose. Generally, the frequency-response curve obtained by experiment only approaches the shape of the true normal distribution. Such curves illustrate clearly, however, that there is a mean dose at which the greatest percentage of individuals will respond in a specific way. There will always be individuals who require either greater (hyposensitive) or smaller (hypersensitive) doses than the mean to elicit the same response.

Although the frequency-response distribution curves often are used for certain kinds of statistical analyses of dose-response data, the cumulative-response data presentation is employed more commonly, especially for representing lethal response data. The cumulative-response curve may be generated for nonlethal frequency-response data by plotting log dose versus percentage of individuals responding with at least a specified response. As illustrated in Figure 1.3, if the blood pressure responses used in the previous example are plotted as the percentage of individuals in each dosing group that respond with at least a level of 140/100, the resulting curve will be sigmoidal. Several important values used to characterize toxicity are obtained from this type of curve. The NOAEL (no observed adverse effect level) is the highest dose at which none of the specified toxicity was seen. The LOAEL (lowest observed adverse effect level) is the lowest dose at which toxicity was produced. The TD_{50} is the statistically determined dose that produced toxicity in 50% of the test organisms. If the toxic response of interest is lethality, then LD_{50} is the proper notation. At a high enough dose, 100% of the individuals will respond in the specified manner. Since the LD and TD values are determined statistically and based on results of multiple experiments with multiple test organisms, the values should be accompanied by some means of estimating the variability of the value. The probability range (or p value), which is commonly used, generally is accepted to be less than 0.05. This value indicates that the same LD or TD value would be obtained in 95 out of a hypothetical 100 repetitions of the experiment.

The cumulative-response curves can facilitate comparisons of toxic potencies between compounds or between different test populations. For example, for two substances with nonoverlapping cumulative dose-response curves, the substance with the curve that covers the lower dose range is clearly the more toxic of the two. If prior treatment of a test population with substance A results in a shift of the dose-response curve to the right for toxin B, then substance A exerts a protective effect against substance B. In the case where the dose-response curves for different toxins overlap, the comparison becomes a bit more complex. This can occur when the slopes of the dose-response curves are not the same, as shown in Figure 1.5. These hypothetical compounds have the same LD_{50}, and are said to be equally toxic at this dose. Below this dose, however, compound B produced the higher percentage of toxicity than compound A and, therefore, compound B is more toxic. At doses above the LD_{50}, compound A produces the higher percentage of lethality and, thus, is the more toxic substance. Based on their LD_{50} values only, compounds A and B have the same toxicity. Thus, in comparing the toxicities of two substances, the toxic response must be specified, the dose range of toxicity must be stated, and if the toxicities are similar, the slopes of the linear portions of the dose-response curves must be indicated.

POTENCY

Although all substances exhibit toxic and lethal dose-response behavior, there is a wide range of LD_{50} values for toxic substances. By convention, the toxic potencies fall into several categories. A list of LD_{50} values for several fairly common substances, along with a categorization of the toxicities from extreme to slight, are provided in Table 1.1.

Substances with LD_{50} values greater than about 2 g/kg body wt. generally are considered to be of slight toxicity and that relatively large amounts, in the range of at least one cup, are required to produce a lethal effect in an adult human and are easily avoided under most circumstances. However, exposure

Table 1.1 Potency of Common Toxins

Agent	LD_{50} (mg/kg)	Toxicity
Ethyl alcohol	9,000	
Sodium chloride	4,000	
BHA/BHT (antioxidants)	2,000	Slight
Morphine sulfate	900	
Caffeine	200	Moderate
Nicotine	1	High
Curare	0.5	
Shellfish toxin	0.01	
Dioxin	0.001	
Botulinum toxin	0.00001	Extreme

to substances in the extreme category with $LD_{50} < 1$ mg/kg requires only a few drops or less to be lethal and may be a considerable hazard.

HORMESIS

Hormesis is a dose-response phenomenon characterized by a low dose beneficial effect and a high dose toxic effect, resulting in either a J-shaped or an inverted U-shaped dose-response curve. A hormetic substance, therefore, instead of having no effect at low doses, as is the case for most toxins, produces a positive effect compared to the untreated subjects. A representative dose-response curve of such activity is presented in Figure 1.4.

Substances required for normal physiological function and survival exhibit hormetic dose-response behavior. At very low doses, there is an adverse effect (deficiency), and with increasing dose beneficial effects are produced (homeostasis). At very high doses, an adverse response appears from toxicity. For example, high doses of vitamin A can cause liver toxicity and birth defects while vitamin A deficiency contributes to blindness and increases the risk of disease and death from severe infections. Nonnutritional substances may also impart beneficial or stimulatory effects at low doses but produce toxicity at higher doses. Thus, chronic alcohol consumption at high doses causes esophageal and liver cancer, whereas low doses can reduce coronary heart disease. Another example is radiation, which at low levels induces beneficial adaptive responses and at high levels causes tissue destruction and cancer.

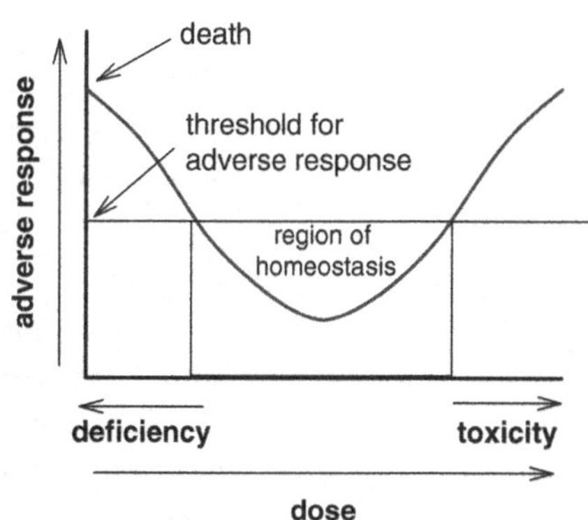

FIGURE 1.4 *Hormesis dose-response curve.*

MARGIN OF SAFETY

Safety is defined as freedom from danger, injury, or damage. Absolute safety of a substance cannot be proven since proof of safety is based on negative evidence, or the lack of harm or damage caused by the substance. A large number of experiments can be run that may build confidence that the substance will not cause an adverse effect, but these experiments will not prove the safety of the substance. There is always the chance that the next experiment might show that the substance produces an adverse effect in standard or new testing protocols. In addition, our concept of safety continues to evolve and we are now aware that even minute changes, for example, in the activity of an important enzyme, could portend a highly negative effect in the future. Indeed, our concept of safety in regard to toxic exposure continues to develop as our knowledge of biochemical and molecular effects of toxins, and our ability to measure them, grow.

Since absolute safety cannot be proven, we must evaluate relative safety, which requires a comparison of toxic effects between different substances or of the same substance under different conditions. When the experimental conditions for toxicity testing in a species have been carefully defined, and the slopes of the dose-response curves are nearly the same, the toxicities of two substances can often be calculated simply by determining the ratio of the TD_{50}s or LD_{50}s. Often, however, a more useful concept is the comparison of doses of a substance that elicit desired and undesired effects. The **margin of safety** of a substance is the range of doses between the toxic and beneficial effects; to allow for possible differences in the slopes of the effective and toxic dose-response curves, it is computed as follows:

$$\text{Margin of Safety (MS)} = LD_1/ED_{99}$$

LD_1 is the 1% lethal dose level and ED_{99} is the 99% effective dose level. A less desirable measure of the relative safety of a substance is the **Therapeutic Index**, which is defined as follows:

$$\text{Therapeutic Index (TI)} = LD_{50}/ED_{50}$$

TI may provide a misleading indication of the degree of safety of a substance because this computation does not take into account differences in the slopes of the LD and ED response curves. Nevertheless, this method has been used traditionally for estimations of relative safety. The dose-response data presented in Figure 1.5 serves to illustrate how the use of TI can provide misleading comparisons of the relative toxicities of substances.

In this example, drug A and drug B have the same LD_{50} = 100 mg/kg and ED_{50} = 2 mg/kg. The comparison of toxicities, therefore, provides the same TI = 100/2 = 50. Therapeutic index does not take into account the slope of the dose-response curves. Margin of safety, however, can overcome this deficiency by using ED_{99} for the desired effect and LD_1 for the undesired effect. Thus,

$$MS = LD_1/ED_{99} = 10/10 = 1 \text{ for drug A, and for drug B} = 0.002/10 = 0.0002$$

Therefore, according to the MS comparison, drug B is much less safe than drug A.

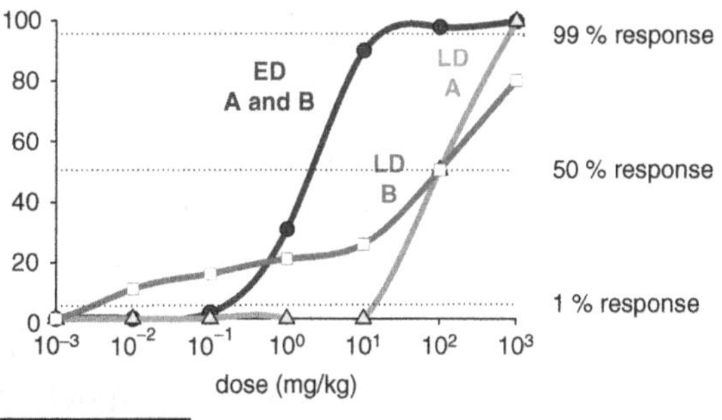

FIGURE 1.5 *The dose-response data serves to illustrate how the use of TI can provide misleading comparisons of the relative toxicities of substances.*

For substances without a relevant beneficial biological response, the concepts of MS and TI have little meaning. Many substances as diverse as environmental contaminants and food additives fall into this category. For these substances, safety of exposures is estimated based on the NOAEL adjusted by a series of population susceptibility factors to provide a value for the **Acceptable Daily Intake** (ADI). The ADI is an estimate of the level of daily exposure to an agent that is projected to be without adverse health impact on the human population. For pesticides and food additives, it is the daily intake of a chemical, which during an entire lifetime appears to be without appreciable risk on the basis of all known facts at the time, with the inclusion of additional safety factors. The ADI is computed as follows:

$$ADI = NOAEL/(UF \times MF)$$

where UF is the uncertainty factor and MF is the modifying factor.

UF and MF provide adjustments to ADI that are presumed to ensure safety by accounting for uncertainty in dose extrapolation, uncertainty in duration extrapolation, differential sensitivities between humans and animals, and differential sensitivities among humans (e.g., the presumed increased sensitivity for children compared to adults). The common default value for each uncertainty factor is 10, but the degree of safety provided by factors of 10 has not been quantified satisfactorily and is the subject of continuing experimentation and debate. Thus, for a substance that triggers all four of the uncertainty factors indicated previously, the calculation would be ADI = NOAEL/10,000. In some cases, for example, if the metabolism of the substance is known to provide greater sensitivity in the test organism compared to humans, an MF of less than 1 may be applied in the ADI calculation.

BIOLOGIC FACTORS THAT INFLUENCE TOXICITY

It is clear from the foregoing discussion that all substances can exhibit toxicity at sufficiently high doses and there will be a range of sensitivities among individuals to the toxic effects. We will now consider physiologic and anatomical factors that can influence this sensitivity. The scheme presented in Figure 1.6 summarizes biological processes that can modulate responses, both beneficial and adverse, to an administered chemical.

Tissue absorption generally is required for most substances to exhibit their toxic effects. This absorption, for example, can result in a limited distribution of the substance at or near the point of contact, or it can lead to the entry of the substance into the blood or lymph circulation and distribution into the entire body. When a substance enters the biologic fluid, it can exist in a free form or a form in which it is bound, most often to proteins in the blood. Also, while in the body fluid, the substance can be translocated in bound or free form to distant sites in the body. Storage sites are body compartments in which the compound is bound with sufficiently high affinity to reduce its free concentration in the general circulation. Bone, lipid tissue, and liver are common sites of storage of xenobiotics. The residence time for the substance in the storage site can be as long as decades and depends on the binding affinity for the site and the concentration of the substance in the circulating fluid. As the concentration of the substance in the body fluid drops due to a cessation in exposure, the substance will be released to the circulation at a rate that depends on its binding affinity for a component of the binding tissue.

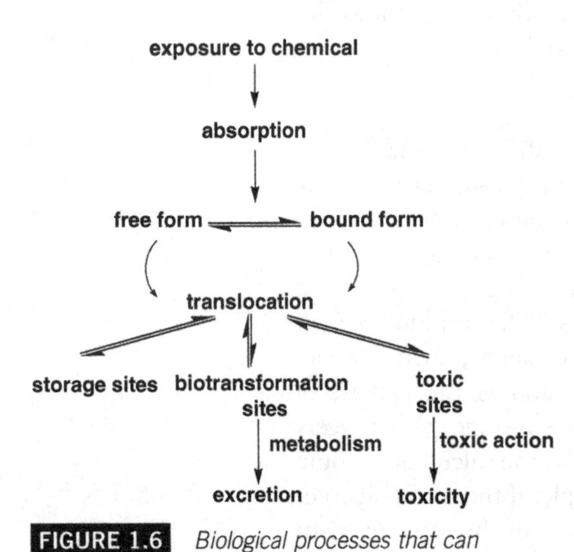

FIGURE 1.6 *Biological processes that can modulate responses, both beneficial and adverse, to an administered chemical.*

Biotransformation sites are locations in cells of certain organs that mediate the metabolism of xenobiotics. The most active tissues for this biotransformation are the portals of entry in the liver and small intestine. In most cases, biotransformation converts the substance to an oxidized and conjugated form that is water soluble and more readily excreted via the urine or the bile. In some cases, however, intermediates in the biotransformation process are responsible for the toxic effects of the administered substance.

Finally, the xenobiotic or its activated metabolite will encounter its site of action and toxicity. The molecular target is a component of a metabolic or signaling pathway that is important to the normal function or development of the organ.

Although a toxic agent may adversely affect the functions of many tissue macromolecules and the cells that contain them, these effects may not be important to the well being of the organ and the organism, and are not considered to be central sites of toxic action. Each of these factors that influence toxicity will be discussed in the following sections.

ABSORPTION

For a substance to gain access to a specific effector site within an organelle of a complex organism, the substance must often pass through a series of membranes. Although the membranes in various cells of the organism—such as the skin keratinocytes, intestinal enterocytes, vascular endothelial cells, liver hepatocytes, and the nuclear membrane—have certain characteristics that distinguish them from one another, the basic compositions of the membranes are very similar. An accepted general membrane model is illustrated in Figure 1.7.

In this model, the membrane is represented as a phospholipid bilayer with hydrophilic outer portions and a hydrophobic interior. Proteins are dispersed throughout the membrane with some proteins traversing the entire width and projecting beyond the surfaces of the membrane. The basic cell membrane is approximately 7.5 to 10 nanometers thick and is elastic. It is composed almost entirely of phospholipids and proteins with small quantities of carbohydrates on the surface.

A closer look at the chemical structure of the phospholipid component of the membrane provides insight into the effect of its composition on the function of the membrane. As represented in Figure 1.8, the polar head of the phospholipid is composed of a phosphate moiety bound to other small molecules such as choline, serine, ethanolamine, and inositol that can increase the polarity of the phospholipid or serve as sites for further modifications that control cell function, for example, by addition of carbohydrate or phosphate groups.

The composition of the lipid components of the phospholipid contributes to the fluidity of the membrane and, thereby, can affect cell function. For example, adequate fluidity of the membrane is maintained, in part, by incorporation of cis-unsaturated fatty acids. The cis-double bonds decrease the strength of interactions between adjacent lipid chains compared to lipids that are saturated or that contain trans-double bonds. Modification in fluidity can affect many cellular functions, including carrier-mediated transport, properties of certain membrane-bound enzymes and receptors, membrane transporters, immunological and chemotherapeutic cytotoxicity, and cell growth.

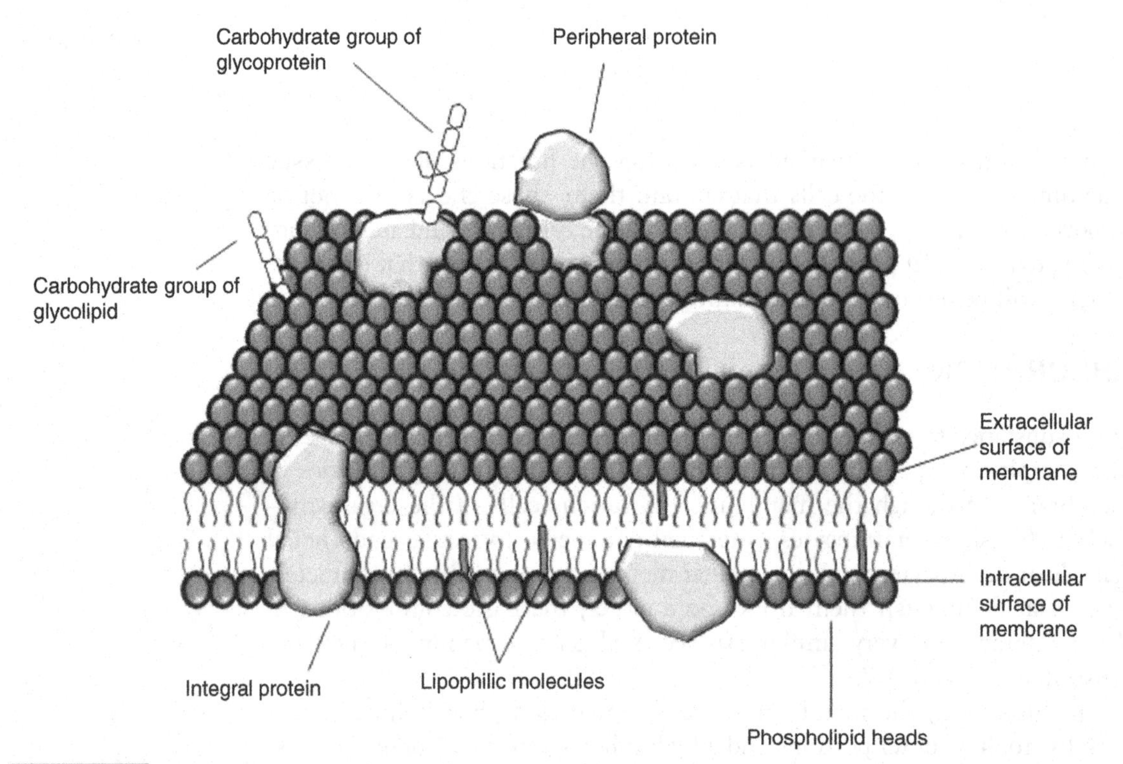

FIGURE 1.7 *General membrane model of animal cell.*

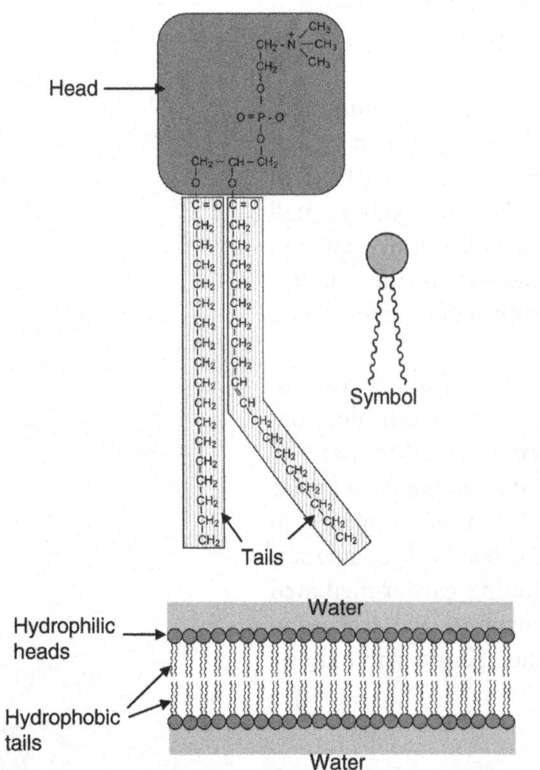

FIGURE 1.8

Chemical structure of the phospholipid component of the membrane.

Another important feature of the cell membranes is the presence of aqueous channels or pores. Although water can diffuse passively at a low rate through the continuous phospholipid bilayer of the membrane, some cell types exhibit much higher rates of water transport than others due to the presence of pores in the membrane. The transmembrane proteins that form these pores comprise a family of about 12 members called either **aquaporins**, if they allow passage of only water, or **aquaglyceroporins**, if they allow passage of glycerol and other small neutral solutes. The conformation of the most studied aquaporins, aquaporin 1 (AQP1) from red blood cells, is indicated in Figure 1.9. As visualized from the extracellular surface, AQP1 forms an elegant and highly symmetrical tetramer in the pore. Water passes through channels in each of the AQP1 molecules in the pore. The rates of passage of water and solutes through these pores depend on the size of the pore, which can be tissue specific and may be hormonally regulated. For example, channels in most cell types are less than 4 nm in diameter and allow passage of molecules with a molecular weight of only a few hundred Daltons. In contrast, the pores in the kidney glomerulus are much larger at approximately 70 nm, and allow passage of some small proteins (< 60,000 Da).

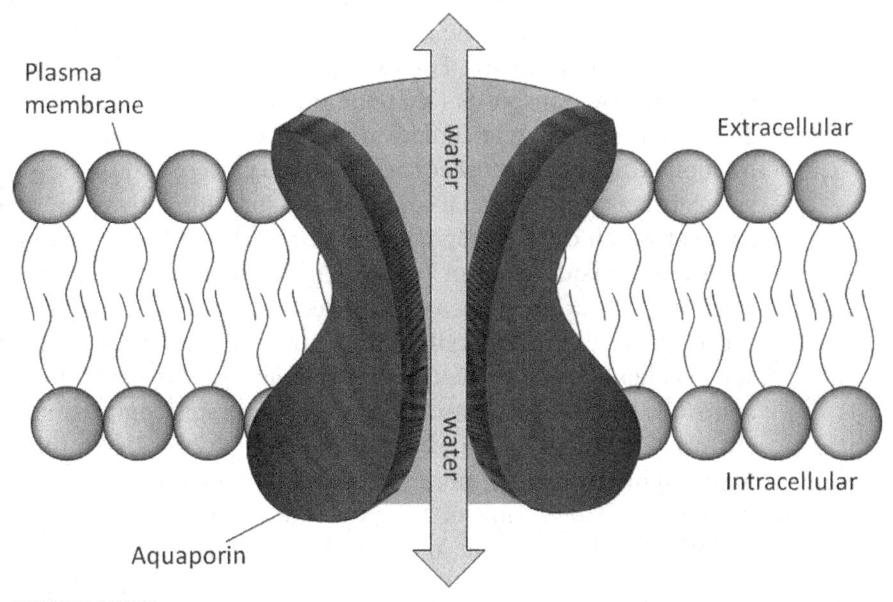

FIGURE 1.9 *Aquaporins, aquaporin 1 (AQP1) from red blood cells.*

TYPES OF MEMBRANE TRANSPORT

The processes of movement of substances across biological membranes are classified as passive diffusion and active transport. **Passive diffusion** processes, which include simple diffusion and facilitated diffusion, do not require energy in the form of ATP and are driven by concentration gradients.

Simple diffusion is characterized by the passive diffusion of hydrophobic molecules across the lipid membranes or of small hydrophilic molecules through aqueous pores. The rate of transport of lipophilic xenobiotics across the standard membrane depends on molecular size, hydrogen bonding, and polar surface area. Lipophilic xenobiotics with molecular weights of greater than about 500 Da tend not to pass readily across the lipid membrane. Such is also the case for substances that have very high hydrogen bonding characteristics. In general, the rate of passage of lipophilic xenobiotics across the membrane is proportional to the octanol/water partition coefficient or logP. Indeed, substances with a relatively high logP such as the pesticide DDT and environmental contaminant TCDD (logP approx. 7) are readily absorbed into the lipid membrane. Conversely, substances with relatively low logP values such as the pesticide paraquat, the antibiotic cephalosporin, and other charged molecules (logP < -4.5) are not appreciably absorbed by a passive mechanism into the lipid membrane.

Facilitated diffusion is a carrier-mediated transport of water soluble substances that mimic the structures of endogenous substances that normally exist in the body. For example, glucose normally enters cells via special glucose transporters that do not require ATP. Certain drugs such as 2-deoxyglucose and related derivatives have been designed to enter the cell by competing with glucose for access to the glucose transporters. Other passive carriers exist for metal ions such as sodium, potassium, and calcium, some of which can be coopted by toxic metals such as cadmium and lead.

Although both types of passive diffusion processes are driven by substrate concentration gradients, under certain circumstances these processes can result in the differential accumulation of substances in the absence of such gradients. Such is the case, for example, in the diffusion of weak acids and weak bases across a membrane that separates compartments of different pH. Since ionized molecules are poorly absorbed into the lipid membrane compared to the unionized forms, the rate of transport depends on the degree of ionization of the molecule. Weak acids are protonated and unionized at low pH and are more lipophilic than at high pH. Similarly, weak bases are deprotonated and unionized at high pH and are more lipophilic than at low pH. The degree of ionization, in turn, depends on the pK of the molecule and the pH of the medium as specified by the Henderson-Hasselbalch equation:

$$\text{For acids: pKa} - \text{pH} = \log[\text{nonionized}]/[\text{ionized}]$$

$$\text{For bases: pKa} - \text{pH} = \log[\text{ionized}]/[\text{nonionized}]$$

By these equations, the pH at which a weak organic acid or base is 50% ionized is the pKa. As the pH is changed by 2 units from the pK level, the degree of ionization of a weak acid or weak base either rises to 99% or declines to 1.0%. Thus, for a weak carboxylic acid such as benzoic with pKa = 4.2, on passage from the stomach, with a pH near 2, to the blood, with pH near 7, the degree of protonation falls from about 99% to about 0.1%. Thus, the weak acid accumulates in the blood compartment with the higher pH. By analogous reasoning, the absorption rates of weak bases such as aniline, with pKa = 5.0, are low because they exist as the less lipophilic conjugate acids in the low pH environment of the stomach.

An additional means by which substances are absorbed by passive diffusion involves the differential distribution of binding sites for the substance. If there are more binding sites on one side of the membrane compared to the other, the side of the membrane with the greater level of binding sites will have a greater amount of substance, regardless of whether both sides have equal concentrations of the free substance. For example, a substance that is introduced in the blood, which is rich in protein binding sites, will not accumulate in another fluid of another compartment such as the central nervous system where the level of protein binding sites is low. This differential distribution of binding sites for xenobiotics contributes to the barrier of accumulation of the substances in the central nervous system.

Although passive membrane transport processes primarily govern the access of most toxins to critical targets in the cell, **active transport** is critical in some important instances. In contrast to passive transport processes, active transport requires ATP for energy, works against substrate concentration gradients, and is saturable. Not surprisingly, the major active processes for transport of substances into cells are selective for nutrients and important endogenous substances. Included in this list are active transporters for sugars, neutral amino acids, basic amino acids, fatty acids, vitamin C, vitamin B_{12}, bile salts, and several metallic ions. As with the glucose facilitated diffusion transporter, toxic agents with structures similar to a nutrient can compete with the nutrient for active transport into the cell.

TOXIN ABSORPTION IN THE ALIMENTARY TRACT

In addition to molecular size and the lipophilic characteristics of the xenobiotic, the extent of absorption of xenobiotics in the alimentary tract depends on the physiological and anatomical features of each section (Figure 1.10), the residence time within each section, and the presence of food.

There is a reasonable correlation between logP value of certain well-known alkaloid drugs and their absorption following sublingual (oral) dosing. For cocaine, for example, a substance with relatively high lipid solubility, the ratio of sublingual effective dose to subcutaneous dose is approximately 2:1 (i.e., twice as much is required orally compared to direct injection) to produce the same response. In contrast, morphine, a substance with relatively poor lipid solubility, requires a sublingual dose roughly 10 times greater than the subcutaneous dose to give similar effects. These results are consistent with a primary role of the lipid-diffusion process in the absorption of these alkaloids.

The rate of absorption of xenobiotics in the stomach depends to a considerable extent on the acid/base properties of the substance. Weak acids are more lipid-soluble in their nonionized form in the highly acidic environment of the stomach, which, in humans, is normally in the range of pH 1 to 2 and rises to pH 4 during digestion. Since the pH of the blood is near neutral, the weak acids are deprotonated to the more polar conjugate base leading to the accumulation of the substance in the blood. Conversely, weak bases exist as protonated, conjugate acids in the low pH gastric juice and are not well absorbed.

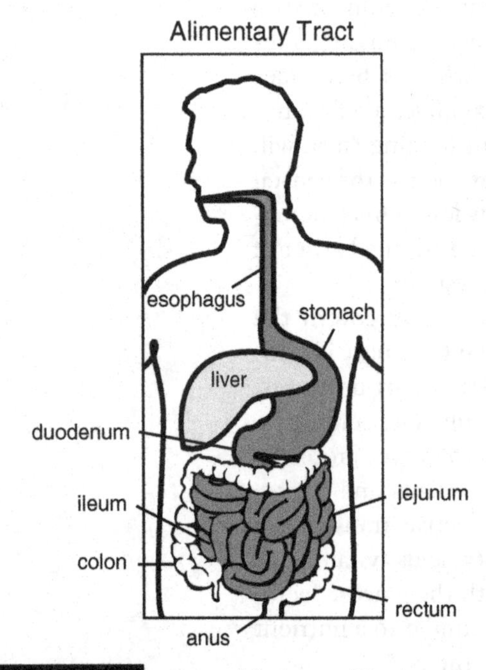

FIGURE 1.10 *General features of the human alimentary tract.*

The small intestine is the principal site of absorption of dietary xenobiotics. Since the pH of the contents of the small intestine ranges from pH 6 to 7 in the proximal small intestine to pH 7 to 8 in the distal ileum, acids and bases will be charged and relatively polar. According to the Henderson-Hasselbalch equation, however, unless the substances are strong acids or bases, a significant proportion of the xenobiotic will exist in the unionized, less polar form. Thus, since the small intestine is very long and is lined with highly absorptive enterocytes, absorption of weak acids and weak bases in this tissue can be nearly complete even though the charge equilibrium lies heavily in favor of the charged species.

The colon is the final major absorption section of the alimentary tract. Although the pH of the colonic contents is similar to that of the distal small intestine (i.e., pH 7–8), and colonic epithelial cells express a wide range of transport proteins, xenobiotic absorption in the large intestine is relatively minor compared to the small intestine because of the considerably reduced length and surface area of the colon. However, the large intestine can serve as a site of absorption for certain substances, especially those produced by bacterial action within the lower gut, as will be discussed later in this chapter.

In addition to physiological and anatomical characteristics of the alimentary tract that influence xenobiotic absorption, the effect of food can also be significant. The majority of important food-drug interactions are caused by food-induced changes in the bioavailability of the drug. Indeed, the most common effect of food is to reduce the bioavailability of orally administered drugs, sometimes to the point of blocking the activity of the drug. Such interactions frequently are caused by chelation with components in food that occur with common antibiotics such as penicillamine and tetracycline. In addition, the physiological response to food intake, in particular gastric acid secretion, may reduce the bioavailability of certain drugs through destruction of the drug or the capsule in which it is delivered. In some cases, however, administration of the drug with food may result in an increase in drug bioavailability either because of a food-induced increase in drug solubility or because of the secretion of gastric acid or bile in response to food intake. Such increases in drug bioavailability may result in serious toxicity.

INTESTINAL MICROFLORA

An additional unique feature of the human digestive tract is that it normally harbors a large and diverse community of mostly anaerobic microorganisms. The conditions for bacterial growth in the various sections of the gastrointestinal tract differ considerably, leading to large differences in the concentrations of bacteria. Thus, there are normally low concentrations of bacteria in the stomach and duodenum, increasing concentrations in jejunum and ileum, and the highest concentration in the colon (10^9–10^{12} /mL). The composition of the intestinal microbiota is relatively simple in infants but becomes more complex with increasing age, reaching a high degree of complexity in adults. Diet strongly influences the type of bacteria that are resident in the colon of an individual and of different human populations. Intestinal bacteria are important for the maturation and the maintenance of the immune system, and can influence colonic cell proliferation and contribute to the salvage of energy. In addition, the intestinal bacteria have a large metabolic capacity that can result in the conversion of dietary macromolecules to metabolites with beneficial or adverse health effects. For example, fermentation of dietary fiber can result in the formation of short chain fatty acids such as butyrate, and the fermentation of protein can result in the production of ammonia. Butyrate is thought to lower risk of colonic cancer, whereas high levels of ammonia can promote tumor development.

Intestinal bacteria also can mediate the metabolism of drugs and phytochemicals in the diet with consequences that may increase or decrease their biological activities. The gut microflora possess a diverse range of metabolic activities, including reductions, hydrolyses, and degradations. In many cases, these reactions can both complement and antagonize reactions of the liver, which are mainly oxidative and synthetic. For example, certain isoflavonoids, found for instance in soy, beer, or clover, have estrogenic properties and, depending on the developmental stage of growth at which they are administered, can either promote or inhibit mammary tumorigenesis in laboratory animals. In the human gastrointestinal tract, one of these isoflavones, daidzein, may undergo bacterial transformation to equol, which has a higher biological activity than daidzein. Differences in the gut microflora might account for the finding that approximately one of three individuals produces equol. Alternatively, daidzein may also be degraded by gut bacteria to the inactive metabolite, O-demethylangolensin. Thus, bacterial metabolism in the gut can strongly influence the activities of ingested xenobiotics and can account for a major proportion of individual variations in sensitivities to many orally administered substances.

THE BLOOD–BRAIN BARRIER

The blood–brain barrier (BBB) ensures an optimally controlled homeostasis of the brain's internal environment. The anatomical structure of the BBB is at the endothelial cell of arterioles, capillaries, veins, and at the epithelial cell surface of the choroid plexus, the area of the brain where cerebrospinal fluid (CSF) is produced. The endothelial cells allow a very selective transport of substances from blood to brain and brain to blood. In organs other than the brain, the extracellular concentrations of hormones, amino acids, and ions such as potassium undergo frequent small fluctuations, particularly after meals or bouts of exercise. If the brain were exposed to such fluctuations, the result might be uncontrolled neurological activity. The BBB and the choroid plexus are required to protect the brain from these fluctuations.

Several anatomical and physiological features are the basis for the BBB. These features are (1) very small junctions between vascular endothelial cells, (2) the envelopment of the endothelial cells with astrocyte glial cells, (3) the low concentration of binding proteins in the CSF compared to the blood, and (4) the expression of several active transport systems that control the influx and efflux of many endogenous and exogenous substances. The net result of these features is to highly regulate the influx of hydrophilic chemicals into the CSF. These characteristics, however, do little to limit the influx of many lipophilic drugs and toxic substances into the CSF. For example, nicotine, heroin, and chloramphenicol have considerable lipophilicity and readily pass through the BBB to produce toxic effects.

XENOBIOTIC ABSORPTION INTO LYMPH

Absorption of xenobiotics in the intestinal lymph system can strongly influence their potency and organ selectivity. The lymphatic system is an extensive drainage network distributed throughout all areas of the body. The lymph system functions mainly to return fluid in the interstitial space back to the blood. The intestinal lymphatics are essential in the absorption of products from lipid digestion, such as long chain fatty acids and lipid soluble vitamins. Each of the gut villi is drained by a central lacteal, which conducts fluid via the lymphatic capillaries to the mesenteric lymph duct. In contrast to blood capillaries, the intercellular junctions between endothelial cells in lymphatic capillaries are more open. Chylomicrons, which are lipoproteins produced within the enterocyte from absorbed dietary lipids produced from digestion, have an average diameter of 200–800 nm. Following secretion by the enterocyte, these colloidal particles are too large to be absorbed by the blood capillaries and are consequently selectively taken up into the lymphatic capillaries. Since the volume of blood flow versus lymph flow is estimated to be 500:1, xenobiotics that are not highly lipophilic tend to be diverted preferentially toward the blood system. However, the great tendency for lipophilic xenobiotics to associate with dietary lipids and intestinal lipoproteins results in their selective absorption into the lymph system. Furthermore, the conditions of exposures that increase the association of xenobiotics with lipids can strongly influence their absorption into the intestinal lymph system. For example, in a study with dogs, food was shown to markedly enhance lymphatic absorption in which the absorption of an orally administered lipophilic antimalaria drug, Hf, was increased from only about 1% in the fasted state to over 50% following a meal.

TRANSLOCATION

Translocation is the interorgan movement of substances in the body fluids and is responsible for the action of a toxin in a tissue that is not the site of exposure. The initial rate of translocation to distant organs and tissues is determined primarily by the volume of blood flow to that organ and the rate of diffusion of the chemical into the specific organ or tissue. As discussed previously, lymph can be an important vehicle for translocation of certain ingested lipophilic xenobiotics. Blood flow varies widely to the different organs and tissues in the body. Total blood flow (perfusion) is greatest in the liver, kidney, muscle, brain, and skin, and is much less in the fat and bone. Hence, the more highly perfused tissues receive the greater initial doses to xenobiotic.

The mammalian circulatory system has several features that can strongly influence the effects of the route of exposure on toxic action. A schematic representation of the major features of the circulatory system is presented in Figure 1.11.

Several features of the circulatory system that are important in toxin action are presented in the figure:

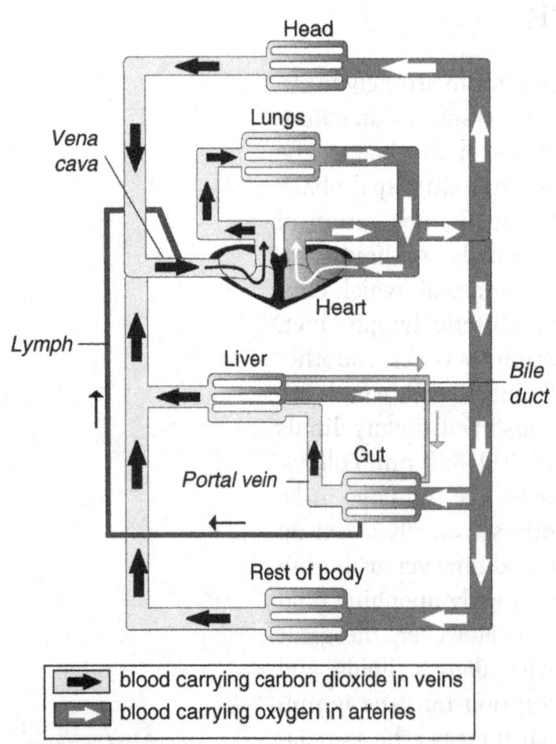

FIGURE 1.11 *General features of the mammalian circulatory system.*

1. Venous blood is pumped by the heart to the lung for oxygenation. By this process, the first capillary system encountered by substances that enter the venous blood is in the lungs. This is true for substances that are absorbed in the mouth, or into the abdominal lymph system. This also occurs for substances that are administered by intravenous injection.
2. Arterial blood is pumped directly to the capillary systems of each of the organs, from which the blood is returned to the

heart through the venous blood. Thus, xenobiotics that are administered intra-arterially can go directly to all the organs.

3. Blood returning from the gut does not go directly to the heart but primarily enters the hepatic portal vein and flows to the liver. Thus, substances absorbed into the blood through the small intestine encounter the capillary system and metabolic capabilities of the liver before entering the general circulation. This is also true for substances administered by intraperitoneal injection, since the hepatic portal vein drains this cavity as well. Because of its high metabolic capability, the liver can convert a wide range of xenobiotics into products that are more easily excreted. This process of metabolic conversion of xenobiotics before they reach the general circulation is called **first-pass metabolism**.

4. Abdominal lymph flows into the thoracic duct, from which the lymph fluid is eventually reunited with the blood at the junction of the left jugular and the left subclavian veins immediately before it enters the heart. Thus, xenobiotics absorbed into the lymph from the small intestines will avoid first-pass metabolism in the liver and encounter the less efficient xenobiotic metabolic apparatus of the lung.

5. Bile produced in the liver is transported to the upper small intestine (duodenum) where bile acids assist in digestion of lipids in the diet. Xenobiotic metabolites that are secreted into the bile will return to the small intestine, which can expose them to the metabolic capabilities of enzymes in enterocytes in the upper intestine and to bacteria in the lower intestine. Conversion of a xenobiotic metabolite produced in the liver into a product that is then absorbed in the intestine can result in a recycling process called **enterohepatic circulation**, which can significantly increase the residence time of the xenobiotic in the body.

DISTRIBUTION

Although translocation is the interorgan movement of xenobiotics, distribution is the resulting apportioning of the xenobiotic in the body tissues and compartments. In the adult human, approximately 38 liters of body water is apportioned into three distinct compartments: interstitial fluid (11 liters), plasma fluid (3 liters), and intracellular fluid (24 liters). The concentration of a xenobiotic agent in blood following exposure depends largely on its apparent volume of distribution. The volume of distribution (V_d) is defined as the ratio of the amount of compound in the body to the concentration of compound in plasma, or:

$$V_d = Dose_{iv}/C_p$$

where:

$Dose_{iv}$ is the amount of xenobiotic administered by intravenous injection
C_p is the concentration of xenobiotic in plasma

If the xenobiotic is distributed only in the plasma, a high concentration will be achieved within the vascular tissue and V_d will be small. In contrast, the concentration will be markedly lower if the same quantity of xenobiotic were distributed in a larger pool including the interstitial water and/or intracellular water. In this case the V_d will be large. For substances that distribute to both plasma and tissues, V_d values will be intermediate in magnitude.

Although for some substances the volume of distribution equates to a real physiologic volume, this is not always the case. For example, Evans Blue is a large, polar dye that does not pass through the capillary bed. As such, it is not able to pass out of the vascular system. The volume of distribution of Evans Blue, therefore, can be used as a measure of vascular volume. Bromide, moreover, is not able to cross cell membranes and distributes into extracellular space. Thus, bromide can be used to estimate the volume of extracellular water. In contrast, antipyrine and tritium labeled water are freely able to cross cell membranes and do not bind to cellular components. The apparent volume of distribution of these agents will equate to total body water. For other xenobiotics that extensively bind to intracellular components, however, the resulting apparent volume of distribution considerably exceeds the volume of total body water. The heart drug, digoxin, and the pesticide, DDT, and many other xenobiotics fall into this category. The volume of distribution of a drug that is much larger than total body water indicates that there is extensive tissue binding of the drug and that only a small fraction of the dose is in the vascular space.

The extent of binding to plasma proteins strongly influences the volume of distribution of many xenobiotic agents and their biological effects. Relatively few xenobiotics have sufficient solubility in blood for simple dissolution to regulate tissue distribution. Indeed, the distribution of most xenobiotics occurs in association with plasma proteins. Many organic and inorganic compounds of low molecular mass bind to lipoproteins, albumins, and other plasma proteins such as the iron binding protein transferrin. Since the protein-xenobiotic association is reversible, this binding can be an efficient means for transport of xenobiotics to various tissues. However, since the xenobiotic-bound plasma protein cannot cross capillary walls due to its high molecular mass, the availability of these protein binding sites usually decreases the toxic effect of the xenobiotic. In many cases, the xenobiotic can compete with another drug or an endogenous compound for protein binding. In these cases, the xenobiotic can displace the other bound substance from the binding site and increase its activity. An example of this kind of interaction is the displacement of bilirubin from plasma albumin by sulfa drugs, which can produce neurotoxicity in infants as a result of the unbound bilirubin crossing the immature blood–brain barrier.

STORAGE

Many xenobiotics can accumulate in tissues at higher concentrations than those in the extracellular fluids and blood. Such accumulation may be a result of active transport or, more commonly, nonspecific binding. Tissue binding of xenobiotics usually occurs with cellular constituents such as proteins, phospholipids, or nuclear proteins and generally is reversible. The tendency for storage of xenobiotics in the body depends to a large extent on their polarity. The more polar organic substances tend to bind to proteins in the blood and soft tissues. Inorganic substances can bind to selective metal binding proteins in liver and kidney or to nonselective sites in bone. Lipophilic substances are stored primarily in lipid tissues. If a large fraction of xenobiotic in the body is bound in this fashion its biological activity may be strongly affected.

Organ Storage

Both the liver and the kidneys exhibit transport and storage capabilities that can strongly influence the activities of certain xenobiotics. The liver plays a critical role in regulating iron homeostasis by maintaining a reservoir of the essential metal bound to the storage protein, ferritin. Accumulation of iron, exceeding the binding capacity of ferritin, can result in liver toxicity. Metallothionein is a small molecular weight, cysteine-rich protein important in maintaining zinc homeostasis and controlling the toxicity of metals such as cadmium. Cadmium can accumulate as a nontoxic complex with metallothionein in liver and kidney. Indeed, as much as 50% of the cadmium in the body is found in complex with metallothionein in the kidney. Cadmium toxicity is produced in the kidney when the storage capacity of metallothionein is exceeded.

Lipid Storage

Many lipid-soluble drugs are stored by physical solution in the neutral fat. In obese persons, the fat content of the body may be as high as 50%, and in lean individuals, fat may constitute as low as 10% of body weight. Hence, fat may serve as a reservoir for lipid-soluble xenobiotics. For example, as much as 70% of highly lipid-soluble drugs such as sodium thiopental, or environmental contaminants such as DDT, may be found in body fat soon after exposure. Fat is a rather stable reservoir because it has a relatively low blood flow, which results in a relatively long residence time in the body for the absorbed xenobiotic. Rapid loss of fatty tissue through disease or dietary restriction can produce a rapid release of stored xenobiotic, possibly with toxic consequences.

Bone Storage

Although the rate of perfusion of bone by blood is relatively low compared to other organs, the binding affinity of bone tissue for certain xenobiotics is great. Thus, bone is an important storage site for certain substances. Divalent ions, including lead and strontium, can replace calcium in the bone matrix, whereas fluoride can displace hydroxyl ion. The tetracycline antibiotics and other calcium chelating xenobiotics can accumulate in bone by adsorption onto the bone crystal surface. Bone can also become a reservoir for the slow release of toxic agents into the blood such as lead or radium. However, the residence time for lead in the bone can be very long (decades) with no apparent toxic effect. In contrast, fluoride accumulation in bone can result in increased density and other abnormalities of bone development (skeletal fluorosis), and accumulation of radioactive strontium can lead to bone cancer.

The many possible effects of tissue storage on toxicity may be summarized as follows:

1. Storage at a site that is not the site of toxic action will reduce toxic potency.
2. Storage at the site of toxic action increases toxic potency.
3. Slow release from a safe storage site can result in chronic toxicity that may be different from the acute toxicity of the xenobiotic.
4. Displacement from storage of a substance with greater potency can result in toxicity from the previously administered substance.
5. Prior treatment with a substance that binds with greater affinity to a storage site may block the binding of the second substance and result in increased toxic potency of the second substance.

EXCRETION

Toxicants or their metabolites can be eliminated from the body by several routes. The main routes of excretion are via urine, feces, and exhaled air. Thus, the primary organ systems involved in excretion of nongaseous xenobiotics are the kidneys and the gastrointestinal tract. Other avenues for elimination include the expired air, saliva, perspiration, and milk, which is important in exceptional circumstances, such as the exposure of nursing infants to certain pesticides and lipophilic drugs encountered by the mother.

KIDNEY

Elimination of substances by the kidneys into the urine is the primary route of excretion of toxicants in terms of both the number of substances excreted by this pathway and the amount of each substance. The functional unit of the kidney responsible for excretion is the nephron, which consists of three primary regions, including the glomerulus, proximal convoluted tubule, and the distal convoluted tubule (Figure 1.12).

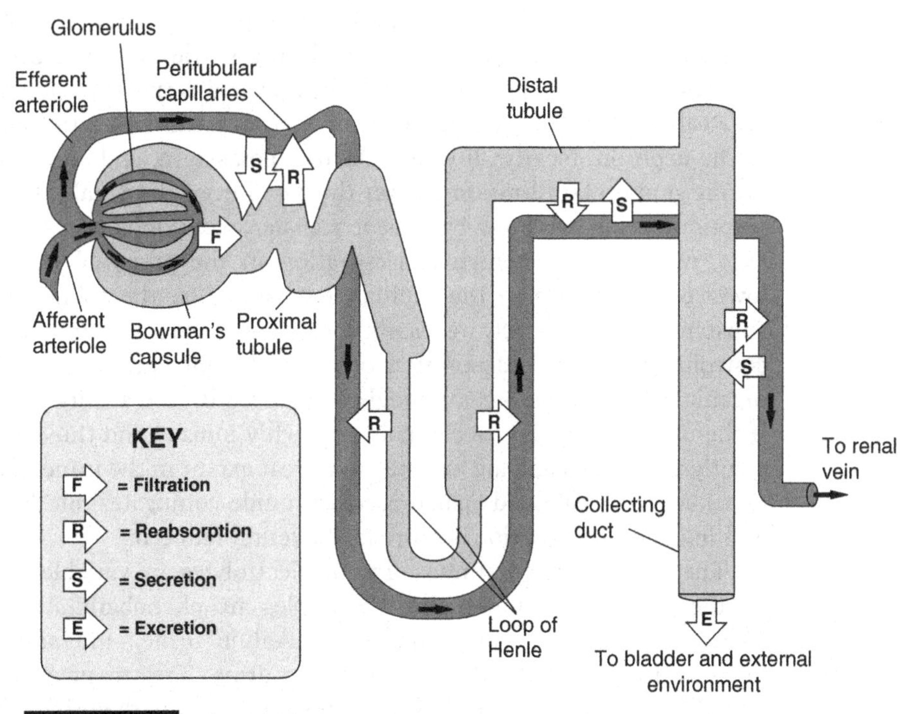

FIGURE 1.12 *General features of the nephron.*

Three physiological processes are involved in urinary excretion: filtration, secretion, and reabsorption. **Filtration** takes place in the glomerulus. A combination of characteristics lead to the production of approximately 45 gallons of filtrate each day for an adult person: a very large blood flow, a considerable hydrostatic pressure of the blood, and the large size pores in glomerular capillaries. The filtrate contains most of the lipid and water soluble substances in the blood, up to a molecular weight of approximately 60,000 Da. Thus, xenobiotics that are bound to albumen and other large proteins are not readily filtered from the blood, whereas substances that bind to small proteins such as metallothionein are filtered at the glomerulus to a large extent.

Secretion is an active process of transport of certain types of metabolites from the blood to the proximal tubule. Secreted substances include xenobiotics or their metabolites that are weak acids, including penicillin, uric acid, and many xenobiotic glucuronide conjugates, and weak bases, including histamine and many alkaloid drugs. The rate of active secretion of a xenobiotic can be reduced by treatment with a similar substance. For example, glucuronide conjugates secreted by organic acid active transporters can compete for excretion with uric acid and produce uric acid accumulation and toxicity in the form of gout. In another example, an organic acid of low toxicity called probenecid was developed during World War II to decrease the rate of excretion of penicillin by the organic acid active transporter.

Reabsorption takes place mainly in the proximal convoluted tubule of the nephron. Nearly all water, glucose, potassium, and amino acids lost during glomerular filtration reenter the blood from the renal tubules. Reabsorption occurs primarily by passive transfer based on concentration gradients, moving from a high concentration in the proximal tubule to the lower concentration in the capillaries surrounding the tubule. Lipophilic substances are passively reabsorbed to a large extent from the proximal convoluted tubule. Reabsorption of ionizable substances, especially weak organic acids and bases, is strongly influenced by the acidity of the urine. If the urine is alkaline, weak acids are highly ionized and thus are not efficiently reabsorbed and are excreted to a great extent in the urine. If the urine is acidic, the weak acids (such as glucuronide conjugates) are less ionized and undergo reabsorption with renal excretion reduced.

The urinary excretion rates of weak electrolytes are variable and depend on the pH of the tubular urine. Examples are phenobarbital and aspirin (acidic drugs), which are ionized in alkaline urine, and amphetamine (a basic drug), which is ionized in acidic urine. Treatment of barbiturate and aspirin poisoning may include changing the pH of the urine to facilitate excretion. The pH of the urine can be modified by several means. Acidosis (decreased pH of urine and other body fluids) can result from conditions that lead to decreased loss of carbon dioxide in respiration, such as lung damage or obstruction; depression of the central nervous system (CNS) by drugs or damage; or conditions that result in altered metabolic status, such as diarrhea and diabetes or a high protein diet. Alkalosis (increased pH of body fluids) can result from increased loss of carbon dioxide through respiration as occurs, for example, with CNS stimulation or in high altitude; or altered metabolic conditions induced by alkaline drugs (e.g., bicarbonate), excessive vomiting, and high sodium or low potassium diets.

EFFECTS OF MATURATION ON KIDNEY EXCRETION

Renal organic anion (OAT) and cation (OCT) transporters protect against endogenous and exogenous toxins by secreting these ionic substances into the urine as discussed previously in this chapter. However, these transporters are not fully developed for several weeks following birth and can account for the differential toxicities of many xenobiotics in adults and infants. For example, the nephrotoxicity of cephalosporin antibiotics, which are weak organic acids, has been ascribed to effects of immature OATs. Cephalosporins produce proximal tubular necrosis, apparently by induction of oxidative stress, only after their transport into proximal tubule cells. A relative decrease in nephrotoxicity of cephalosporins is observed in the young that has been ascribed to decreased tubular transport. However, this OAT deficiency leads to a different manifestation of cephalosporin toxicity due to competition for transport of essential nutrients by cephalosporin, which can result in carnitine deficiency. This lack of full development of renal transporters also results in the reduced excretion in neonates of both endogenous organic acids, such as benzoic acid, and exogenous acids such as p-aminohippuric acid and penicillin. In addition to increasing the adverse or beneficial effects of these substances, the elevated levels of xenobiotics are thought to accelerate the maturation of the associated transporters.

Fecal Excretion of Xenobiotics

Although renal excretion is the primary route of elimination of most toxicants, the fecal route also is significant for many substances. Fecal elimination of absorbed xenobiotics can occur by two processes: excretion in bile and direct excretion into the lumen of the gastrointestinal tract. The biliary route is an especially important mechanism for fecal excretion of xenobiotics and their metabolites. This route generally involves distinct active transport systems in liver for organic bases, organic acids, neutral substances, and metals. Whether a metabolite will be excreted into the urine or into the bile depends primarily on molecular size. Excretion via bile is the major excretory route for metabolites with molecular weights greater than approximately 350 Da. Similar metabolites with molecular weights smaller than about 350 Da are excreted preferentially in the urine. Examples of xenobiotics actively excreted in the bile include metal compounds such as dimethylmercury, lead, and arsenic, as well as metabolites of environmental toxins such as TCDD and drugs such as the estrogenic carcinogen diethylstilbestrol (DES).

DES provides an example of the major roles of biliary excretion and intestinal microflora in the activities of some xenobiotics. Research on experimental animals demonstrated that DES is excreted almost totally by the bile, and blockage of this pathway by bile duct cannulation resulted in a pronounced increase in the residence time of DES in the rodent body and an increase in the toxicity of DES by 130-fold! DES also is involved in the process of enterohepatic circulation. This process is initiated by the transport of the absorbed lipophilic xenobiotic to the liver via the portal vein where it undergoes a conjugation reaction to form a hydrophilic glucuronide or sulfate metabolite (Figure 1.13). If the conjugate is sufficiently large, it is then secreted into the bile. Compounds excreted in bile pass into the intestines where they may undergo deconjugation by the resident microorganisms. The deconjugated metabolite may be reabsorbed by the enterocytes and pass from the portal blood back to the liver. The processes of conjugation, biliary excretion, microbial deconjugation, and enterocyte absorption are repeated, thus comprising the enterohepatic cycle.

The efficiency of xenobiotic excretion in the bile and the effect of enterohepatic cycling can be influenced by several factors. The flow of bile in the liver usually is decreased with liver disease, whereas certain drugs such as phenobarbital can increase the rate of bile flow. Administration of phenobarbital has been shown to enhance the excretion of methylmercury by this mechanism, for example. The efficiency of enterohepatic circulation can be modified by conditions that reduce the intestinal microflora, as in antibiotic treatments, or decrease the reabsorption of xenobiotic metabolite such as by oral administration of binding agents.

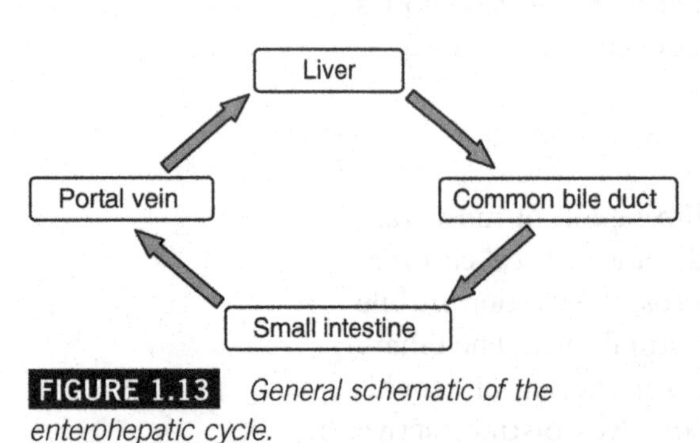

FIGURE 1.13 *General schematic of the enterohepatic cycle.*

A second manner by which xenobiotics can be eliminated via the feces is by direct intestinal excretion. Although this is not a major route of elimination, a large number of substances can be excreted into the intestinal tract and eliminated via feces. Some substances, especially those poorly ionized in plasma (such as weak bases), may be eliminated in feces by passively diffusing through the walls of the capillaries and enterocytes and into the intestinal lumen. Other substances such as cholesterol, plant sterols, and other lipophilic substances, as well as certain conjugated metabolites, are actively transported from the apical portion of the enterocyte into the intestinal lumen by transporters for neutral substances and for organic anions. Moreover, increasing the lipid content of the intestinal tract can enhance intestinal excretion of some lipophilic substances. Indeed, active efflux in the small intestine is thought to contribute to the poor oral availability of many drugs. Intestinal excretion is a relatively slow process and is therefore an important elimination route only for those xenobiotics that have slow rates of metabolism or slow rates of excretion by other means.

ADDITIONAL READINGS

Blaut, M., Clavel, T. (2007). Metabolic diversity of the intestinal microbiota: Implications for health and disease. *J. Nutr.* 137:751S-755S.

Katsura, T., Inui, K. (2003). Intestinal absorption of drugs mediated by drug transporters: Mechanisms and regulation. *Drug Metab. Pharmacokinet.* 18:1-15.

Klaasen, C.D. (2008). Casarett and Doull's Toxicology; The Basic Science of Poisons, 7th ed. McGraw-Hill, New York.

Moody, D.M. (2006). The blood-brain barrier and blood-cerebral spinal fluid barrier. *Semin. Cardiothorac. Vasc. Anesth.* 10:128-131.

O'Driscoll, C.M. (2002). Lipid-based formulations for intestinal lymphatic delivery. *Eur. J. Pharm. Sci.* 15:405-415.

Smart, R.C., Hodgson, E. (Eds.). (2008). Molecular and Biochemical Toxicology, 4th ed. John D. Wiley and Sons.

CHAPTER 2

Determination of Toxicants in Foods

CHAPTER CONTENTS

Analysis of food toxicants is somewhat different from that of straight chemistry. The chemicals of interest in food toxicology are the ones that give adverse effect toward animals and humans. In the case of food poisoning, it is important to detect the presence of a toxicant in foods. Therefore, two major processes are essential to the determination of toxicants in particular foods. One is to detect the presence of toxicant(s) in a food and the other is to qualitate and quantitate the toxicant(s). In the former case, a bioassay or animal test generally is conducted. If the toxicant is known, this step may be skipped.

Once the identity of the test toxicant has been carefully defined and the levels of exposure have been estimated, priorities for biological testing are established. Purified toxicants or complex mixtures with high rates of exposure will generally have the highest priority for further testing. Significant chemical structures may also be involved in the priorities assessment. Substances chemically related to a known toxicant are also likely to receive high priority for subsequent testing.

Food Science and Technology

Food safety assessment depends upon the determination of toxicants in foods. An important step in the initial phase of safety assessment is to estimate the levels of the toxicant to which the population is exposed. One method for making such an estimate is based on a dietary survey in which individual consumers are interviewed to obtain information on the types of foods they consume. Another method is a market basket analysis in which food is purchased from retail outlets, prepared by typical methods, and then analyzed for the components in question. Per capita disappearance of a particular food component is computed by dividing the annual domestic production plus import quantities by the number of people in the country.

It is important to develop accurate analytical methods to interpret the data correctly. The major tasks of chemical analysis in food toxicology involve separating a toxicant from other chemicals and then determining the amount. Almost by definition, toxicants are present in very low levels because substances with any significant level of toxicant are rejected as foods. This is illustrated by the fact that distaste for a particular food is developed after it is associated with an episode of illness.

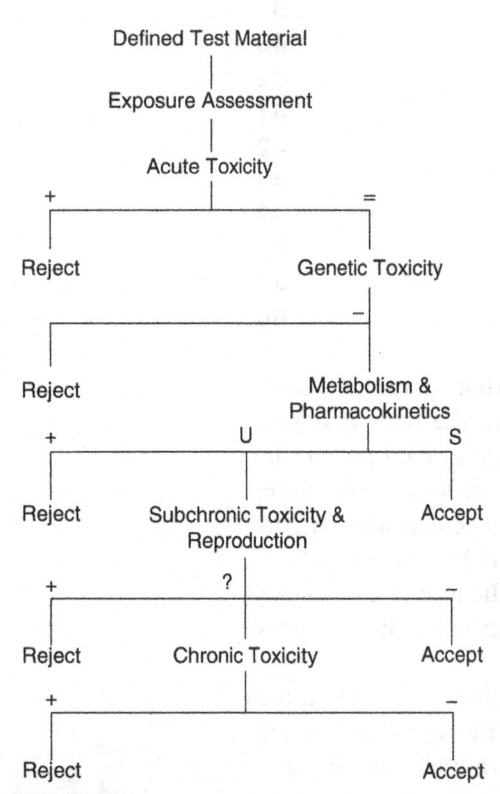

FIGURE 2.1 *Summary of decision-tree protocol proposed by US Food Safety Council. +, presents a socially unacceptable risk; –, does not present a socially unacceptable risk; S, metabolites known and safe; U, metabolites unknown or of doubtful safety; ?, decision requires more evidence.*

To provide a system for toxicity testing, the Scientific Committee of the Food Safety Council in the United States has proposed time requirements, or so-called decision-tree approaches. This approach provides state-of-the-art toxicity assessments and minimizes cost as well as the numbers of animals used. Although details of the protocol are likely to be modified, the overall scheme has received strong support from the scientific community. A summary of the decision-tree protocol proposed by the Food Safety Council is presented in Figure 2.1.

The initial phase of this protocol is the proper identification of the substance. In the case of pure substances, this is a relatively simple matter since procedures for chemical identification and criteria for purity are well established. However, determination of the safety of complex mixtures is more complicated. In these cases, it is ultimately desirable to establish the composition of the mixture and to determine which components of the mixture are responsible for its biological activity. In lieu of detailed information on the composition of a mixture, the process by which it is obtained must be described in as much detail as possible, so that the test material can be reproduced in other laboratories.

SAMPLING

Although a toxicant found in any food is significant, one of the purposes of using chemical analysis is to determine the amount of a toxicant and the probability of overexposure. To accomplish this, samples used for analyses must be taken according to a design. Well-developed statistical methods relevant to the types of conclusions are usually employed. It is helpful, for example, to take replicates at different points within the population to discover the variation. However, this variation should not be confused with the error that occurs within the analytical method.

Samples might be collected to screen for a certain class of compounds, but specific sample treatments required by different compounds may be contradictory and make a "total screen" impossible. For some chemicals, the sample might need to be treated with a base to prevent acid breakdown, whereas other chemicals might require an acid treatment to prevent base breakdown. In some cases, a general screen test may be desired in order to probe for a family of chemicals, and from that point, to test more specifically for chemicals discovered in the screen.

QUALITATIVE AND QUANTITATIVE ANALYSES OF TOXICANTS IN FOODS

The qualitative and quantitative analyses of toxicants in foods are the principal tasks of food toxicology. When toxicity is discovered in foods, the analyst's first job is to identify the toxic material(s) in the food. The analysis of toxicants requires both an assay for detecting the poison and a method for separating it from the rest of the chemicals in the food. The food first is separated into its components and each component is tested for toxicity. The active fraction is further separated and tested—and this continues until the pure toxicant can be completely isolated. At this point the structure of the compound can be identified by chemical analysis. In approximate decreasing order of importance, separation methods are:

1. Gas chromatography (GC)
2. High performance liquid chromatography (HPLC)

3. Column and thin layer chromatography (CC, TLC)
4. Distillation
5. Extraction

The method of elucidating the structure of unknown chemicals has improved dramatically since the development of analytical instruments such as ultraviolet spectroscopy (UV), infrared spectroscopy (IR), nuclear magnetic resonance (NMR), and mass spectroscopy (MS). Recently, MS interfaced to GC or HPLC and has become a powerful method for the isolation and identification of toxicants in food. In particular, the invention of the electron spray HPLC/MS promoted the use of the HPLC dramatically. Therefore, GC and HPLC may become equal in importance as separation methods today.

Once a toxic chemical has been identified, the quantitative analysis can be accomplished with a chemical analysis designed specifically for that chemical. In order to allow legally binding conclusions about toxicant levels in foods, the US government monitors a set of approved methods that have a certain set of criteria for quality. For example, unless circumstances warrant, the recovery achieved by the method in question must be at least 80%. Additionally, certain physical processes are required when preparing the testing samples.

Sample Preparation for Analysis of Toxicants

Distillation

When performing sample preparations for any analysis, distillation is the method used most often. Distillation is a method of separating chemical substances based on differences in their volatilities in a boiling liquid mixture. Generally, there are a several standard distillation methods:

1. Simple distillation: All the hot vapors produced are immediately channeled into a condenser, which cools and condenses the vapors. Usually this is used only to separate liquids whose boiling points differ greatly or to separate liquids from nonvolatile solids or oils. Most commonly, a volatile organic solvent used for extraction is removed using a simple distillation.
2. Fractional distillation: The separation of a mixture into its component parts, or fractions, such as in separating chemical compounds by their boiling point by heating them to a temperature at which several fractions of the compound will evaporate. Generally the component parts boil at less than 25 °C difference from each

other under a pressure of one atmosphere. This method is used to separate the components well by repeated vaporization-condensation cycles within a fractionating column. Therefore, many fractionating columns for increasing separation efficiency have been developed. In order to increase efficiency, the vapor-contacting surface in a column should be increased to increase the vaporization-condensation cycles. With each condensation–vaporization cycle, the vapors are enriched for a certain component. A larger surface area allows more cycles, improving separation.

3. Steam distillation: This method works on the principle that immiscible objects when mixed together can lower the boiling point of each other. A mixture of two practically immiscible liquids is heated while being agitated to expose the surfaces of both the liquids to the vapor phase. Each constituent independently exerts its own vapor pressure as a function of temperature as if the other constituent were not present. Consequently, the vapor pressure of the whole system increases. Boiling begins when the sum of the partial pressures of the two immiscible liquids just exceeds the atmospheric pressure. In this way, many organic compounds soluble in water can be purified at a temperature well below the point at which decomposition occurs. For example, the boiling point of bromobenzene is 150 °C and the boiling point of water is 100 °C, but a mixture of the two boils at 95 °C. Thus, bromobenzene can easily be distilled at a temperature 61 °C below its normal boiling point.

4. Vacuum distillation: To boil some compounds with high boiling points, it is often better to lower the pressure at which such compounds are boiled instead of increasing the temperature. This technique is very useful for compounds that boil beyond their decomposition temperature at atmospheric pressure. It is also useful in cases in which compounds change their structure under heat treatment or react readily with other components at higher temperatures.

Extraction

Once a representative sample of suitable size has been selected, the next step in analysis is usually to separate the analyte from the matrix. As with distillation, extraction may commonly be involved in many analytical processes. In particular, samples for gas chromatography must be in an organic solvent form. The ability of a specific analytical technique to detect a particular analyte is as dependent on the percent of chemical recovered by the

Table 2.1 Water Solubility of Typical Toxicants

Water Soluble Toxicants	Water Insoluble Toxicants
Glucosinolates	Bile acids
Cyanogenic glycosides	Vitamin A
Sodium cyanide	Tetrodotoxin
Cycasin	Saxitoxin
Nitrosamines	Estrogens
Sodium cyclamate	Polycyclic aromatic hydrocarbons
Sodium saccharine	Dioxins
ODPA	MAM
Saponin	Amino acid pyrolysates
Gossypol	Polychlorinated biphenyl
Ionized metals	Aflatoxins

method as it is on the sensitivity of the detector at the end of the process. Since all parts of the food sample need to be equally exposed to extraction, it is often necessary to blend or chop the sample so that it is homogeneous. The food can then be dissolved and the fibers and coarse insoluble materials filtered away.

Organic chemicals have varying degrees of water solubility, from organic acids like vinegar that are polar to the organic oils that are nonpolar and float in a separate layer on top of water. Therefore, it is extremely important to know the solubility of chemicals of interest in water. Table 2.1 shows the water-soluble and insoluble toxicants discussed in this book.

When a nonpolar organic phase is mixed with an aqueous phase, the two will separate into distinct layers. So many organic solvents are nonpolar that the term "organic phase" is used as equivalent to the nonpolar or "oil layer." Molecules that may be insoluble in water and thereby dissolve in the organic phase, as well as those that may be slightly soluble in water but have a greater affinity to the organic phase, will migrate into the organic layer. This step can yield a great deal of separation from the matrix if a nonpolar analyte is extracted from a polar matrix such as fruit, or in the opposite case, if nonpolar material such as fatty tissue is to be cleaned away from a polar analyte.

Many different types of organic phases exist, as well as interactions with organics (such as acetone) that pass between the phases and affect the relative solubility of the molecule of interest in each phase. Hence, solvent choice is a crucial factor in extraction. The important factors for solvent choice are as follows:

- Good solubility for the target chemicals
- High purity (no additional contamination)

- Low boiling point (easy to remove)
- Low cost (large amount of solvent is often required)
- Low toxicity

Table 2.2 shows the most commonly and widely used solvents for food analysis.

If the analyte has an acidic or basic group that might be charged at one pH and neutral at some other, it is the neutral molecule and therefore the pH at that point that will facilitate the movement of most of the chemicals to the organic phase. The addition of salt can also increase the polarity of the water phase and drive solvents like acetone and chemicals that were associating with them into the water from the water phase. These differences can allow fine-tuning of extraction methods so that molecules that are less polar than the analyte can be cleaned off in one step. The analyte can then be brought out by changing conditions to drive it into the organic phase.

Cleanup by Solid Phase Extraction (SPE)

After an extraction, any further separation of the analyte from the matrix done before placing the sample on the final analytical device is called **cleanup**. The term comes from the need to minimize the amount of

Table 2.2 Commonly Used Solvents in Order of Increasing Polarity and Their Physical/Toxicological Natures

Solvent	Solubility in Water (%)	B.P. (°C)	Toxicity
Hexane	Insoluble	69.0	Lethal conc. in air for mice: 40,000 ppm
Heptane	Insoluble	98.4	Lethal conc. in air for mice: 14,000 ppm
Cyclohexane	Insoluble	80.7	Lethal conc. in air for mice: 20,000 ppm
Carbon tetrachloride	0.0005	76.8	Lethal conc. in air for mice: 10,000 ppm
Carbon disulfide	0.005	46.3	LD_{50} rabbits: 300 mg/kg (oral)
Benzene	0.075	79.6	LD_{50} rats: 5.7 g/kg (oral)
Dichloromethane	2.00	40.5	LD_{50} rats: 1.6 g/kg (oral)
Diethyl ether	6.50	34.5	LD_{50} rats: 1.2 g/kg (oral)
Isopropyl ether	0.20	68.5	Lethal conc. in air for mice: 16,000 ppm
Ethyl acetate	10.0	77.0	LD_{50} rats: 5.6 g/kg (oral)
Pyridine	Soluble	115.5	Lethal conc. in air for mice: 4000 ppm
Isopropanol	Soluble	80.4	LD_{50} rats: 5.8 g/kg (oral)
Acetone	Soluble	56.5	LD_{50} rabbits: 5.3 g/kg (oral)
Tetrahydrofuran	Soluble	66.0	Irritating to skin, eyes, mucous membranes
n-Propanol	Soluble	97.5	LD_{50} rats: 1.87 g/kg (oral)
Ethanol	Soluble	78.5	LD_{50} rats: 13.7 g/kg (oral)
Methanol	Soluble	64.7	LD_{50} rats: 6.2 ~ 13 g/kg (oral)
Acetonitrile	Soluble	81.6	LD_{50} rats: 3.8 g/kg (oral)

extraneous chemicals placed on the sensitive analytical devices to keep the injection ports and columns clean for as long as possible. Cleanup is also a preparative step in an analytical method. A preparative separation is designed to yield some chemical sample for further use, whereas an analytical separation is designed to quantitate the target analyte. Most cleanup methods are chromatographic separations optimized for complete recovery of the analyte with less resolution of chemicals in the mixture.

Recently, solid phase extraction (SPE), a separation process that is used to extract compounds (analyte) from a mixture, has become a more and more important technique for sample preparation. SPE is used to concentrate and purify samples for analysis. SPE can be used to isolate analytes of interest from a wide variety of matrices, such as food samples.

The separation ability of SPE is based on the theory of chromatography. It consists of mobile phase (eluting solvent) and stationary phase (packing materials in a cartridge) through which the sample is passed. Impurities in the sample either are washed away while the analyte of interest is retained on the stationary phase, or vice versa. Analytes that are retained on the stationary phase can then be eluted from the solid phase extraction cartridge with the appropriate solvent.

SPE cartridges are commercially available with a variety of different stationary phases, each of which can separate analytes by a different chemical mechanism. Most stationary phases are based on a bonded silica material that is derivatized with a specific functional group. Some of these functional groups include hydrocarbon chains of variable length (for reversed phase SPE), quaternary ammonium or amino groups (for anion exchange), and sulfonic acid or carboxyl groups (for cation exchange).

Isolation and Identification by Chromatography

Chromatography is an elegant method of chemical separation that, using only a few simple principles, has given chemists tools for separating and purifying practically all chemicals. Because of its simplicity, efficiency, and wide range of applications, chromatography has had a great impact on chemical toxicology. Since the invention of gas chromatography/mass spectrometry in 1952, tremendous numbers of unknown chemicals, including toxicants, have been identified. However, these chemicals were limited to ones with relatively low boiling points (less than 500 °C) due to the nature of gas chromatography. Recently, the development of liquid chromatography/mass spectrometry has advanced the identification of relatively high boiling point chemicals such as proteins and carbohydrates.

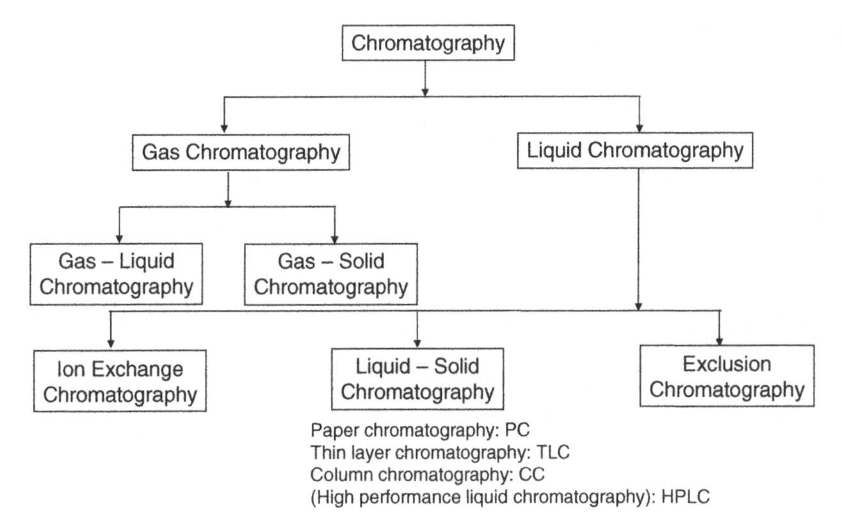

FIGURE 2.2

Systematic summary of chromatography.

This separation principle in chromatography is based on a mobile phase and a stationary phase. The mobile phase contains a mixture of chemicals, one of which is the target analyte. When the mobile phase moves through the stationary phase, the chemicals will have a tendency to move more slowly than the mobile phase because of their affinity for the stationary phase. The different affinities for the stationary phase will cause them to slow at different rates, and this will separate them in the mobile phase.

The great diversity and power of the method derives from the many types of mobile phases, including properties of the solvents, gaseous mobile phases and gas temperature; and the great variety of stationary phases, such as silica, paper, and gel as well as nonpolar, oil-like stationary phases. Figure 2.2 shows a systematic summary of chromatography.

BIOLOGICAL DETERMINATION OF TOXICANTS

Acute Toxicity

The assay for detecting a poison is usually by observing the toxic effect itself. Since it is rarely desirable to use humans, a "model system," usually rats or mice, must be selected to use in the identification process. The first toxicity test is generally an acute toxicity test using experimental animals most commonly in a single dose. The toxic effect that occurs within 24 hours of exposure is recorded. The primary purpose of an acute toxicity test is to determine the level of the substance that induces mortality in experimental animals. This is the point at which the median lethal dose (LD_{50}) is determined. Information obtained from these acute toxicity tests generally is used as the basis for establishing dose and route of exposure for subsequent prolonged toxicity tests. Except in the rare instance, if the substance is found to be too acutely toxic for food use, it will be tested for genetic toxicity, metabolism, and pharmacokinetics.

Genetic Toxicity

The primary objective of genetic toxicity testing is to determine the tendency of the substance to induce mutations in the test organism. A mutation is an inheritable change in the genetic information of a cell. Approximately 10% of all human diseases may have a genetic component and thus may arise from a mutation of one form or another. It is well known, for example, that Down's syndrome, Kleinfelter's syndrome, sickle-cell anemia, and cystic fibrosis arise from specific genetic changes. Most if not all cancers are thought to have their origin in one or more mutations. With the exception of hormones, most substances, approximately 85 to 90%, are carcinogenic in animal species and have been shown to be mutagenic by one assay or another. Although much more information is required before a converse correlation can be established, the fact that a substance is shown to be mutagenic by appropriate tests places it under considerable suspicion as a possible carcinogen.

The decision-tree approach proposes a battery of genetic tests early in the testing scheme. It is suggested that there is a high degree of correlation between mutagenicity and carcinogenicity. Therefore, a substance can be banned from use in food because of its carcinogenic probability based on the results of mutagen tests alone. If the substance does not show high carcinogenic probability based on mutagenicity tests, it must be tested further, including long-term carcinogenicity tests.

Although the details of an appropriate array of mutagenicity tests are the subject of continuing controversy, the general outline seems to be fairly well established. Assays for which there seem to be general support include analyses of point mutations (localized changes in DNA) in micro-organisms and in mammalian cells, investigation of chromosomal changes (major recombination of genetic material) in cultured mammalian cells and in whole animals, and investigation of cell transformation (tumors produced by implantation in animals) using cultured human or other mammalian cells.

Bioassay

Bioassay is a commonly used shorthand term for biological assays. Bioassays typically are conducted to measure the effects of a substance on living organisms, including microorganisms and experimental animals. Generally, bioassay is a simple and convenient method to detect toxicants but it may not be useful for quantitative analysis for low levels of substances compared with advanced instrumental methods.

Bioassay has been used for various purposes including measurement of the pharmacological activity of new or chemically undefined substances; investigation of the function of endogenous mediator; determination of the side-effect profile, including the degree of drug toxicity; assessing the amount of pollutants being released by a particular source, such as wastewater or urban runoff; and assessing the mutagenicity of chemicals found particularly in foods.

Bacterial Reverse Mutation Assay

Among many bioassays used in various fields, the bacterial reverse mutation assay is one of the most widely and commonly used methods in food toxicology research. The purpose of the bacterial reverse mutation assay is to evaluate a chemical's genotoxicity by measuring its ability to induce reverse mutations at selected loci in several bacterial strains. This assay, commonly called the Ames assay (test), was developed originally by Dr. Bruce Ames (Professor at University of California, Berkeley) in the early 1970s. It is sensitive to a wide range of mutagenic chemicals.

The Ames assay measures genetic damage at the single base level in DNA by using tester strains of bacteria, such as *Salmonella typhimurium* strains. These strains have a unique mutation that "turn off" histidine biosynthesis in *Salmonella*. Because of these original mutations, the bacteria require exogenous histidine to survive and will starve to death if grown without these essential nutrients (auxotrophy). The key to the assay is that the bacteria can undergo a reverse mutation, turning the essential gene back on and permitting the cell to grow in the absence of histidine. Each bacterial strain was created by a specific type of mutation—either a base-pair substitution or frame-shift mutation. Because a reverse, compensating mutation usually must occur by the same mutagenic mechanism, mechanistic toxicological data is also available from the Ames assay results based on the pattern of strain(s) reverted. The bacteria used to trace histidine will undergo several cell divisions, but will stop growing once they have run out, leaving a characteristic "background lawn" that decreases in density with increasing toxicity. After 48 hours, only those cells that have undergone a reverse mutation, or turn the essential gene back "on," have survived producing mutant colonies. Mutation results are reported as revertants per plate. Figure 2.3 shows colonies with strain TA 100 (base-pair mutant).

After the development of the Ames assay, many food additives were tested for mutagenicity using this method. Interim, it was also discovered that most carcinogens exhibited positive reactions toward the Ames assay. Figure 2.4 shows the chronological changes in overlap of known carcinogens and mutagens assessed by the Ames assay. As the figure shows, this assay is a simple and convenient method to screen carcinogenic compounds. For example, some nitrofuran compounds were used widely as a safe preservative for food products, in particular for fish pastes such as *Kamaboko* in the 1960s. However, nitrofuran compounds such as 2-(2-furyl)-3-(5-nitro-2-furyl)acrylamide (AF-2) (Figure 2.4), exhibited mutagenic activity by the Ames assay. Consequently, a long-term animal study proved that AF-2 was a carcinogen, and therefore the use of nitrofurans in food products was prohibited in 1976.

In the 1970s, it was thought that a short-term Ames assay was sufficient to detect unknown carcinogens. However, it was found that a mutagen is not always a carcinogen. Therefore, conventional methods, along with prolonged testing with experimental animals, have remained an important method to assess carcinogenicity of chemicals. However, with the discovery of numerous mutagens, as shown in Figure 2.5, short-term bioassays still remain a convenient method to screen for possible carcinogens.

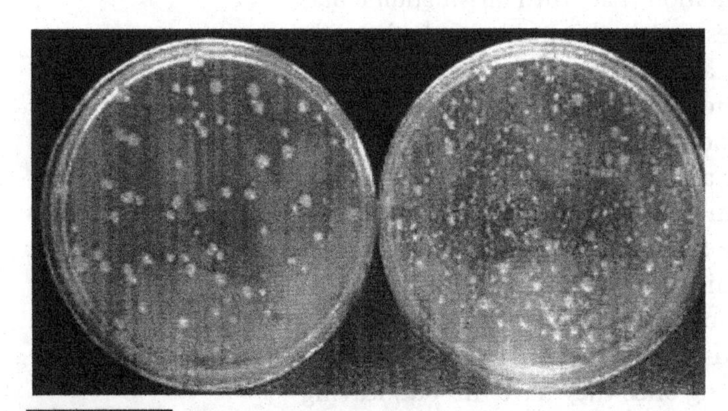

FIGURE 2.3 *Colonies with strain TA 100 (base-pair mutant) appeared on agar plate in the Ames assay.*

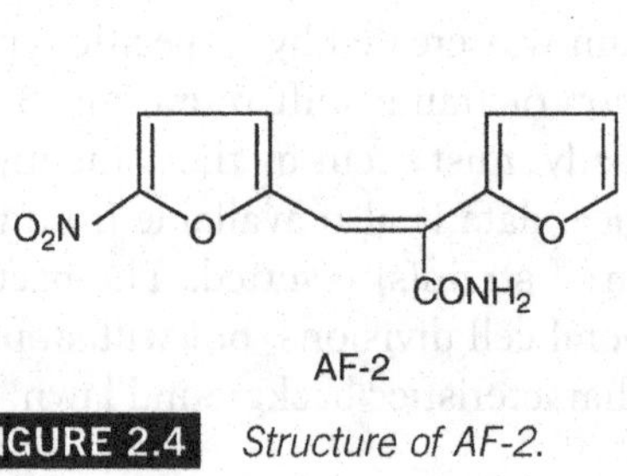

O_2N O O $CONH_2$

AF-2

FIGURE 2.4 *Structure of AF-2.*

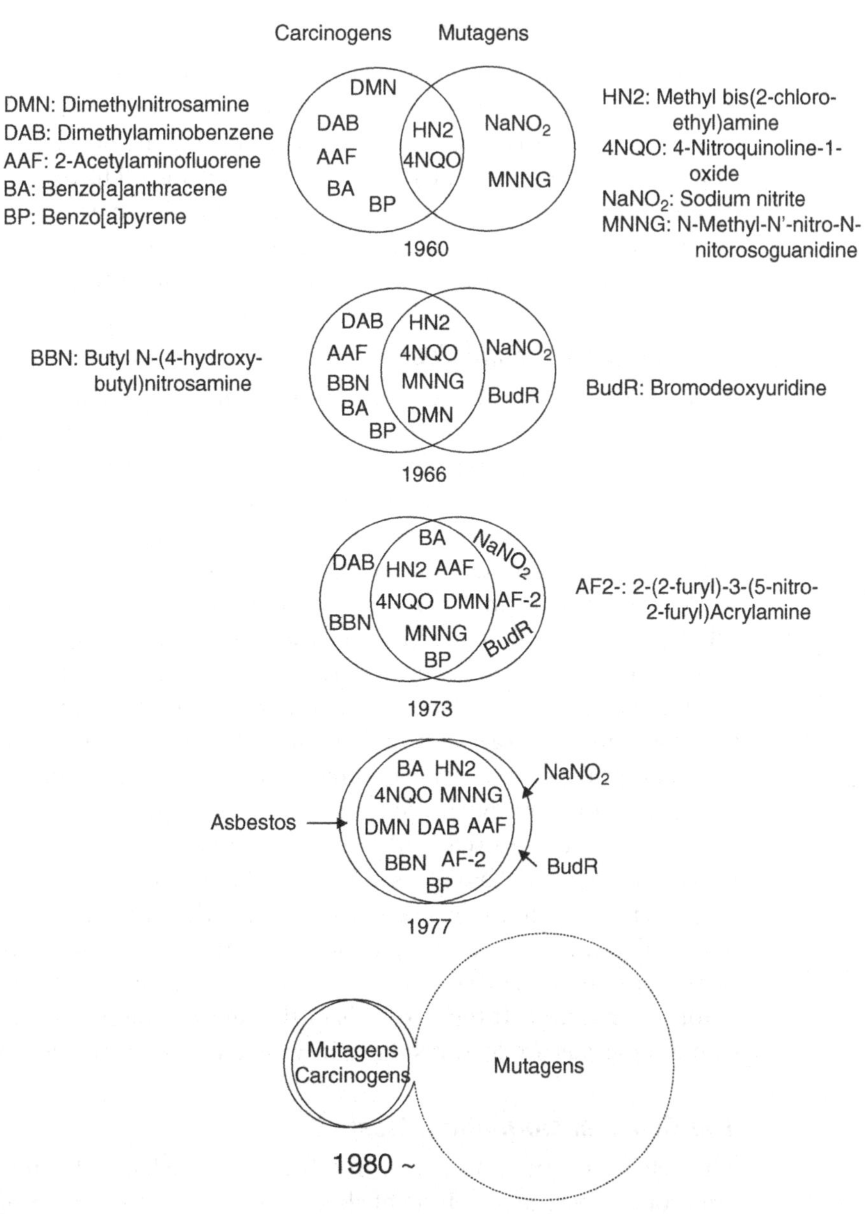

FIGURE 2.5 *The chronological changes in overlap of known carcinogens and mutagens assessed by the Ames assay.*

The Host-Mediated Assay

Another test that uses microbial organisms to determine the mutagenic potential of a substance is known as the host-mediated assay. In this test, a bacterial organism is injected into the peritoneal cavity of a mammal, usually a rat, and the animal is treated with the test substance. The test substance and its metabolites enter the circulation of the animal, including the peritoneal cavity. After an appropriate period, the test organism is removed from the peritoneal cavity and examined for induction of mutations.

The Dominant-Lethal Test

A third mutagenesis assay, known as the dominant-lethal test, determines genetic changes in mammals. In this test males are treated with the test substance and mated with untreated females. The dominant-lethal mutation will arise in the sperm and may kill the zygote at any time during development. Females are dissected near the end of gestation and the number of fetal deaths and various other reproductive abnormalities are noted.

Assays for point mutations also have been developed using mammalian cell lines. One cell line that has been used extensively is the Chinese hamster ovary cell for which resistances to various substances such as 8-azaguanine is used as markers. In contrast to the *Salmonella* system, the Chinese hamster ovary cell lines mainly detect forward mutations. However, one problem often encountered with such assays is the variability in metabolic capability of the cell line. Thus, in some cases, tissue homogenates such as those used in the Ames Assay or test cells are incorporated in the host-mediated assay, as is most commonly done with bacterial cells.

Mutation of the more general type (those that are not point mutations) may be determined by scoring induced chromatid and chromosomal aberrations. Structural changes in chromosomes may be caused by breaks in the chromosomal unit. If the two ends of the break remain separated, chromosomal materials are lost, resulting in visible breaks in the chromosome.

The Cell Transformation Assay

The cell transformation assay, in which mammalian cells are used, is an important aspect of any array of short-term genetic toxicity tests. Many cell lines have been developed for the measurement of malignant transformation following exposure to a test substance. A commonly used cell line is embryo fibroblasts from rats, hamsters, and mice. After a period of normal growth, cells are suspended in an appropriate buffer, treated with a test substance, and then portions of cells are tested for survival rates. The remaining material is plated out on an appropriate medium, and the transformed cells are observed at the colony stage. Malignancy of cells can be confirmed by the production of tumors following transplantation of transformed cells into the appropriate host.

If the genetic toxicology studies lead to the finding of mutagenesis, with the implication of possible carcinogenicity, a risk assessment is applied. If the substance is mutagenic in several assays that are correlated with human carcinogenesis, and the intended use of the substance results in appreciably high human exposures, then with no further testing, the substance may be banned from further use. If the substance is determined to be of low mutagenic risk because, for example, it is mutagenic in several assays but only at very high doses, or the mutagenic activity is observed in only one of the assays, then further studies must be conducted.

Metabolism

Generally, following mutagenicity tests, metabolic studies would be conducted. The objective of this phase of testing is to gain both a general and quantitative understanding of the absorption, biotransformation, disposition (storage), and elimination characteristics of an ingested substance after single and repeated doses. If the biological effects of metabolites are known, the decision to accept or reject the substance can be made on this basis. For example, if all the metabolites can be accounted for and they are all known to be innocuous substances, then the test substance is considered safe. However, if certain metabolites are toxic or if most of the parent substance is retained within certain tissues, then further testing may be indicated. Further evidence of the potential hazard of a substance may also be derived from the knowledge that if the substance has appreciable toxicity in the metabolism of a test species, it may have a similar effect in human metabolism. Thus, knowledge of the metabolism and pharmacokinetics of a substance is essential for establishing the relevance of results from animal testing to projecting likely hazards in humans.

Subchronic Toxicity

Based on the results of these initial investigations, subchronic toxicity studies may be designed. The objective of the subchronic studies is to determine possible cumulative effects on tissues or metabolic systems. Generally, subchronic tests are performed for several months' duration and may extend to one year. Conventional subchronic studies designed to evaluate the safety of food components usually are limited to dietary exposure for 90 days in two laboratory species, one of which is a rodent. Subchronic tests include daily inspection of physical appearance and behavior of the test animal. Weekly records of body weight, food consumption, and characteristics of excreta are maintained. Periodic hematological and eye examinations are performed in addition to biochemical tests of blood and urine. Under certain circumstances, tests are run for hepatic, renal, and gastrointestinal functions along with measurements of blood pressure and body temperature. All animals are autopsied at the termination of the experiment and examined for gross pathologic changes, including changes in the weights of the major organs and glands.

Teratogenesis

Teratogenesis testing is an important aspect of subchronic testing. Teratogenesis may be defined as the initiation of developmental abnormalities at any time between zygote formation and postnatal maturation. Relatively little is known about the mechanisms of teratogenesis since it may be caused by radiation, a wide range of chemicals, dietary changes, infection, temperature extremes, or physical trauma. Furthermore, we cannot predict whether a specific substance will be teratogenic based on chemical structures. Since our knowledge of mechanisms of teratogenesis is relatively primitive, teratogenesis assays rely primarily on prolonged testing periods in animals. Administration of substances to bird embryos has been used with some success. However, since the embryos develop with no metabolic interchange with the outside environment, in contrast to the placenta-mediated interchange for mammalian embryos, teratogenesis testing in mammals is much preferred.

The phase of embryonic development most susceptible to adverse influences is organogenesis. As illustrated in Figure 2.6, the human fetus is most susceptible to anatomical defects at around 30 days of gestation. That is, exposure to a teratogenic influence around this period is most likely to produce anatomical defects in the developing fetus. One of the major problems in teratogenesis testing is that organisms may be susceptible to teratogenesis for only a few days during the growth of the fetus. If the test substances are not administered precisely at this time, the teratogenic effect will go undetected. Exposure to a teratogen prior to organogenesis may produce no effect or may lead to fetal death and no teratogenic response will be seen. Exposure to a teratogen following the period of organogenesis may lead to functional problems that may be relatively difficult to observe and may not be detected as teratogenic effects.

Factors that determine the effective dose of the substance to which the fetus is exposed are (1) the efficiency of the maternal homeostatic processes and (2) the rate of passage of a teratogen across the placenta. The maternal

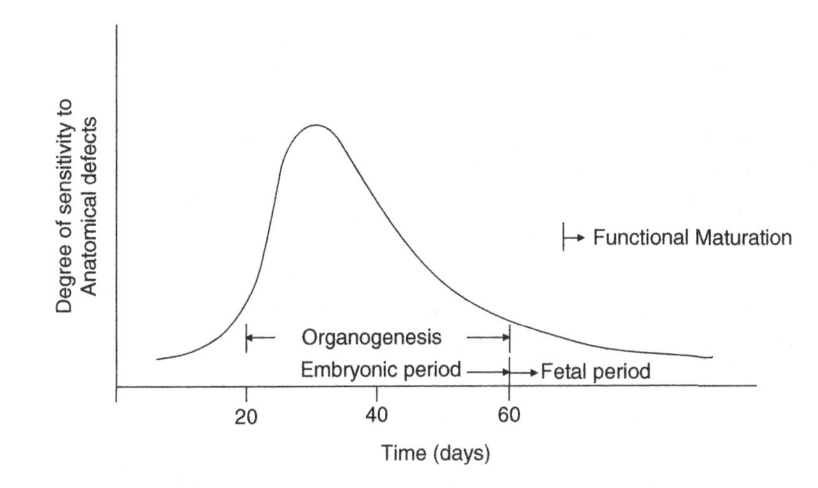

FIGURE 2.6

Degree of sensitivity of human fetuses to anatomical defects at various times during gestation.

homeostatic processes depend on several factors, including the efficiency of liver metabolism and possible excretion of the substance into the bile, possible metabolism and urinary excretion by the kidney, and tissue storage and protein binding. These processes work together in the maternal system to reduce the overall concentration of the substance to which the developing fetus is exposed. The placenta can also serve as an effective barrier in the passage of certain water-soluble substances of large molecular weight into the fetal circulatory system. However, in the case of certain more lipid-soluble compounds (e.g., methyl mercury) the placenta does little to retard passage into the fetal system.

Teratogenesis testing protocols should include both short-term (1–2 days) treatments of pregnant females during organogenesis and continuous treatments during gestation. Teratogenesis tests that include short-term dosing avoid effects of maternal adaptive systems such as induction of metabolic pathways in the liver. This testing protocol also avoids preimplantation damage, increases the likelihood that the embryos will survive to the period of organogenesis, and ensures that critical periods of organ development are covered. Furthermore, this continuous dosing protocol monitors cumulative effects both in the maternal and fetal systems. For example, changes in concentrations and composition of metabolites to which the fetus is exposed during gestation vis-à-vis the diminished metabolic activity of the maternal liver are closely monitored, and the level of saturation of maternal storage sites in relation to a rise in the concentration of the test substance in the fetal system may be screened.

Since adverse effects on the reproductive system may arise from many causes, tests of reproductive toxicity may include treatment of males prior to mating, short-term dosage of females starting prior to mating and continuing on to lactation, short- and long-term dosing of females during the period of organogenesis and in other periods, and pre- and postnatal evaluation of the offspring. These tests can involve large numbers of animals and periods of time comparable to what would be required for carcinogenesis tests. As a result, measurement of reproductive toxicity can be a very time consuming and expensive procedure. Both mechanistic understanding and testing efficiency are sorely needed in this important area of toxicology.

Toxic effects observed in this battery of acute and subchronic tests are evaluated to determine if the tests are relevant to actual conditions of exposure. Many substances at this point in the testing procedure can be rejected from use if their toxicity is sufficiently high. On the other hand, a final decision to accept a relatively nontoxic substance cannot be made if the substance did not satisfy various additional requirements including not being consumed at a substantial level, not possessing a chemical structure leading to suspicion of carcinogenicity, no effects on subchronic toxicity testing that would suggest the possibility that long-term exposure would lead to increased toxicity, or no positive results in tests of genetic toxicity.

Chronic Toxicity

The objective of chronic toxicity testing is to assess toxicity resulting from long-term, relatively low-level exposure, which would not be evident in subchronic testing. Testing protocols require administration of the test substance by an appropriate route and in appropriate dosages for the major portion of the test animal's life.

Chronic toxicity tests are designed so that each treated and control group will include sufficient numbers of animals of both sexes of the chosen species and strain to have an adequate number of survivors at the end of the study for histopathological evaluation of tissues and for statistical treatment of the data. Selecting the proper size of the test group is a major problem in chronic toxicity tests. Table 2.3 indicates the required group sizes as determined by statistical theory. Large numbers of animals must be used if low percentage effects are to be detected. To reduce the numbers of animals required in theory to detect small percentage effects, protocols involving large doses generally are used. However, this practice is coming under increasing scrutiny since the test organism is likely to respond quite differently to high doses of the test material than to low doses. For example, the

Table 2.3 Theoretical Sizes of Test Groups Required to Determine Toxicity at Indicated Frequencies and Level of Significance

True Frequency of Toxic Effect	1 in 20		1 in 100	
Level of significance	0.05	0.001	0.05	0.001
Least number of animals for each dose	58	134	295	670

rates of enzymatic processes such as absorption, excretion, metabolism, and DNA repair are highly sensitive to substrate concentration and are saturable. Thus, high doses of a substance may produce toxic effects by overwhelming a system that readily disposes of low doses.

Despite prudent food-additive laws in the United States, it is still viewed that there is no safe dose of a carcinogen. However, research continues into the existence of a threshold dose below which exposure to a carcinogen may be safe.

In most cancer tests, 50 animals of each sex are used for each dose level. Body weights are recorded periodically throughout the testing period, and the level of food consumption is monitored. Animals are examined for obvious tumors, and at the end of the experiment the animals are autopsied and subjected to detailed pathological examination.

Rats and mice are widely used in chronic testing because of their relatively low cost and the large volume of knowledge available concerning these animals. The strain of animals used for the test depends on the site of toxicity of the test substance and the general susceptibility of the strain to various toxic agents. Generally, strains with some known sensitivity to a range of carcinogens are used. It is likely that a carcinogenic effect will be shown in these animals if the substance is indeed carcinogenic.

Variations in diet can also considerably complicate interpretation of the results from chronic toxicity testing. Administration of semisynthetic diets can result in increased tumor yield with several types of carcinogens compared to experiments using unrefined diets. Diets that provide insufficient calories result in decreased tumor incidence whereas protein deficiency retards tumor growth. For example, dimethylaminoazobenzene-induced carcinogenesis is enhanced with riboflavin deficiency in rats. The influences of various dietary components on carcinogenesis are often complex and the mechanism of action is often specific to the carcinogen in question. Many dietary components, such as certain indoles, flavonoids, and certain pesticides that induce xenobiotic metabolizing systems in the liver and other tissues, will decrease the carcinogenic potency of many substances.

Even when the various aspects of chronic toxicity tests mentioned in the previous discussion are considered, several other more or less incidental factors can influence the outcome. For example, the temperature and humidity of the room in which the animals are housed must be carefully controlled, as must the type of bedding used in the cages. Cedar wood used as bedding has influenced the outcome of cancer testing, perhaps due to induction of xenobiotic metabolizing enzymes by volatiles from the cedar. Furthermore, cancer tests that are said to differ only with respect to the time of year during which they were performed have produced different results. Thus, it is necessary for even the most well-designed chronic toxicity tests that reproducibility of experimental results be determined.

The chronic toxicity test provides the final piece of biological information on whether to accept or reject a substance suggested for food use. If no carcinogenic effects are found, this information, along with all previous data and the estimations of exposure, will be used in the overall risk assessment of a substance. If a substance is determined to be a carcinogen, then in most instances current US law prohibits its use as a food additive. Further testing is needed only if some tests are considered faulty or if unexpected findings make the test design retrospectively inadequate to answer the questions raised.

SUGGESTIONS FOR FURTHER READING

Kister, H. Z. (1992). "Distillation Design," 1st Ed. McGraw-Hill.

Maron, D. M., Ames, B. N. (1983). Revised Methods for the Salmonella Mutagenicity Test. *Mutation Res.* 113:173-215.

McCann, J., Ames, B. N. (1976). Detection of carcinogens as mutagens in the Salmonella/microsome test: Assay of 300 chemicals: Discussion. *Proc. Natl. Acad. Sci. USA* 73:950-954.

Robinson, J. W., Skelly Frame, E. M., Grame, G. M., Frame II, G. M. (2004). Undergraduate Instrumental Analysis, 6th Ed. CRC Press, Boca Raton, FL.

Seader, J. D., Green, D. W. (1998). "Separation Process Principles." Wiley, New York.

CHAPTER 3

Biotransformation

CHAPTER CONTENTS

The composition of the cellular membrane in living organisms confers a higher degree of selectivity in absorption of most water-soluble or highly polar substances. This selectivity of absorption allows for specific routes of uptake of certain water-soluble nutrients and provides a significant level of resistance to the toxicity of most water-soluble xenobiotics. In contrast, however, the same properties that allow for selectivity in absorption of water-soluble substances allow an almost unimpeded uptake of many lipophilic substances. Thus, although most living organisms must expend energy to actively absorb water-soluble nutrients, their ability to prevent absorption of most lipophilic toxins is very limited.

Food Science and Technology

These membrane features result in pronounced differences in the potential effects of lipid-soluble substances, compared to water-soluble substances released into the environment. Following the release of a water-soluble substance, it often will be distributed at low concentrations throughout the aqueous components of the environment. In contrast, a lipid soluble substance, even if released into a very large aqueous system such as a large lake or an ocean, is likely to be concentrated in living tissue by nonselective passive absorption into membranes and lipid tissues.

A major component of the body's defense system against nonselective and high capacity absorption of lipophilic xenobiotics, lipid storage, and multidrug resistance or MDR transporters is a highly efficient metabolic apparatus that converts lipophilic xenobiotics to hydrophilic metabolites that are excreted into the urine or the bile. This apparatus consists of phase I and phase II metabolic reactions.

PHASE I REACTIONS

Phase I reactions are oxidative, reductive, or hydrolytic enzymatic reactions that introduce a reactive moiety capable of further reactions by phase II processes. Phase I reactions usually introduce a reactive oxygen or nitrogen atom into the xenobiotic that can serve as the site of a subsequent phase II conjugation reaction. Examples of phase I reactions are provided in Figure 3.1. A special note to consider is in these reactions, with one exception, lipophilic and relatively unreactive precursors are converted to somewhat less lipophilic and now nucleophilic products. The major reactive cellular macromolecules, which include proteins, DNA, and RNA, are characteristically electron-rich, and thus are also nucleophilic. Thus, conversion of xenobiotics to nucleophilic metabolites decreases their ability to react with cellular macromolecules and is considered a deactivation or detoxification reaction. The exception is the production of an epoxide from an unsaturated precursor. Epoxides are generally electrophilic, and, depending on the presence of other substituents on the molecule, can be highly reactive with cellular macromolecules. Thus, the conversion of a xenobiotic to an epoxide often increases its ability to react with cellular macromolecules and is considered to be an activation or a toxification reaction. In many cases, it is an epoxide metabolite that is the ultimate mediator of the toxic effects of the parent xenobiotic.

The metabolic reactions presented in Figure 3.1 provide examples of phase I reactions that can be used to predict likely phase I metabolites of other xenobiotics. However, many characteristics of the substrate in

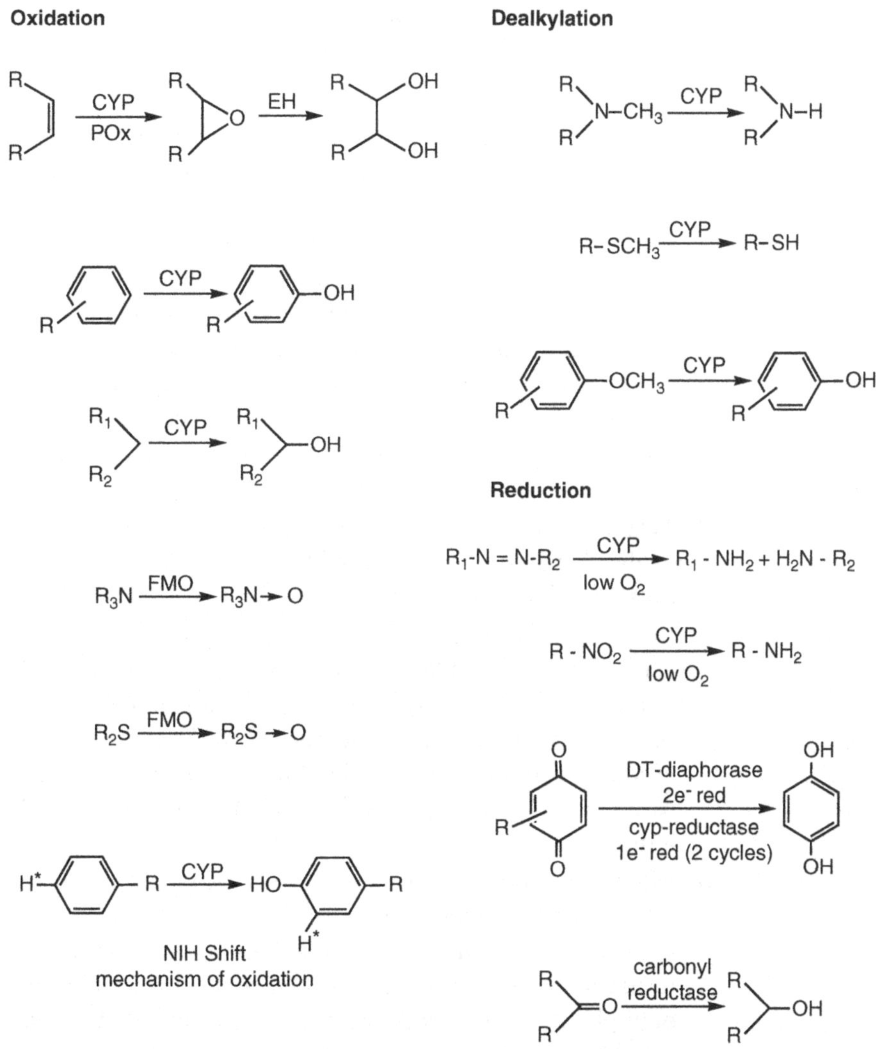

Oxidation

R–CH=CH–R $\xrightarrow[POx]{CYP}$ epoxide $\xrightarrow{EH}$ R–CH(OH)–CH(OH)–R

R–benzene $\xrightarrow{CYP}$ R–phenol (–OH)

$R_1R_2CH_2$ $\xrightarrow{CYP}$ R_1R_2CH–OH

$R_3N \xrightarrow{FMO} R_3N \rightarrow O$

$R_2S \xrightarrow{FMO} R_2S \rightarrow O$

H^*–C₆H₄–R $\xrightarrow{CYP}$ HO–C₆H₃(H^*)–R

NIH Shift
mechanism of oxidation

Dealkylation

$R_2N–CH_3 \xrightarrow{CYP} R_2N–H$

$R–SCH_3 \xrightarrow{CYP} R–SH$

R–C₆H₄–OCH_3 $\xrightarrow{CYP}$ R–C₆H₄–OH

Reduction

$R_1\text{-}N = N\text{-}R_2 \xrightarrow[\text{low } O_2]{CYP} R_1 - NH_2 + H_2N - R_2$

$R - NO_2 \xrightarrow[\text{low } O_2]{CYP} R - NH_2$

R-quinone (O, O) → hydroquinone (OH, OH)
DT-diaphorase
$2e^-$ red
cyp-reductase
$1e^-$ red (2 cycles)

$R_2C{=}O$ $\xrightarrow{\text{carbonyl reductase}}$ R_2CH–OH

FIGURE 3.1 *Xenobiotic oxidations and reductions.*

addition to the presence of reactive moieties, such as lipophilicity, hydrogen bonding, surface charge distribution, molecular size, and molecular conformation, combine to determine the rate of its enzymatic reaction. Although advances in computer modeling rapidly are improving the validity of projections of products of xenobiotics from a given enzyme, the reliable prediction of the metabolic fate of xenobiotics in an organism remains an important future goal.

(S)-Nicotine (1)

FIGURE 3.2 *Nicotine oxidation.*

Nevertheless, some general rules are helpful in predicting the qualitative nature of likely phase I metabolites of xenobiotics.

1. The most reactive moieties in a xenobiotic are usually the heteroatoms, S, N, and P, which upon oxidation yield more polar sulfoxides, nitroxides, hydroxylamines, and phosphine oxides. An example of a nitrogen oxide formation is the conversion of nicotine to the inactive N-oxide (Figure 3.2).

   In addition, oxidative desulfurization can occur with the conversion of thiophosphonates to phosphonates, and thioamides to amides. An important example of this type of reaction is the conversion of the inactive and environmentally more stable pesticide, parathion, to the active product, paraoxon.
2. Terminal carbon-carbon double bonds are usually more readily oxidized than are internal double bonds or aromatic rings, and the initial products are epoxides.

   In most cases the epoxide intermediate produced from an aromatic xenobiotic is unstable and rapidly rearranges to a phenol with retention of the ring hydrogen atom by a reaction known as the NIH rearrangement. In some cases, however, the oxidation of aromatic carbons can occur by direct insertion of oxygen between carbon and hydrogen, as is the case for oxidation of aliphatic carbon atoms.
3. Aliphatic carbons are usually less reactive than alkene and arene carbons and the oxidations result in direct insertion of an activated oxygen atom between carbon and hydrogen with the formation of a hydroxyl group. The stereochemistry of hydroxylation of aliphatic carbons depends on the selectivity with which the xenobiotic substrate fits into the active site of the enzyme. Examples of this kind of stereospecific enzymatic oxidation of aliphatic carbon atoms are found in the CYP-mediated conversion of cholesterol to bile acids. This example also illustrates the general principle that activated carbons such as those in allylic positions (i.e., located adjacent to a double bond) are generally more electron rich and thus more readily oxidized than nonallylic carbons.

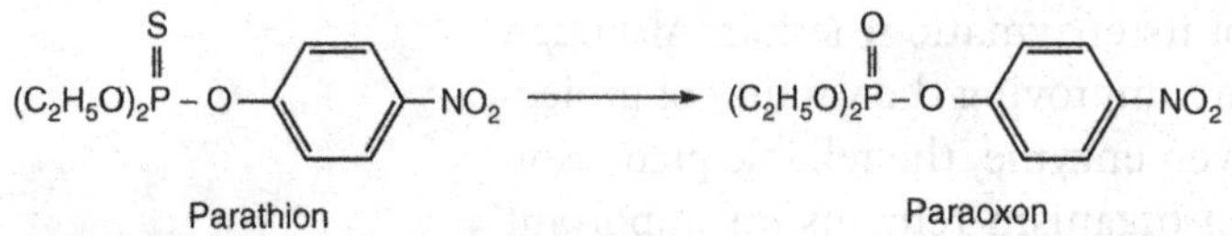

Parathion

Paraoxon

FIGURE 3.3 *Parathion desulfurization.*

4. Dealkylation reactions mediate the conversions of alkyl-substituted heteroatoms (N, S, or O) by processes that involve initial oxidations of the first carbon atom of the alkyl group and its loss as an aldehyde or ketone. An important example of oxidative dealkylation is the conversion of the street drug, ecstasy, to inactive products.

5. Reductive phase I reactions of greatest importance mediate the conversions of aldehydes and ketones to alcohols and of quinones to hydroquinones. In addition, under certain conditions, xenobiotics containing nitro and azo groups can be reduced to primary amines. A large number of compounds containing carbonyl groups are produced endogenously, are present in the diet, and can produce toxic effects by binding to and inactivating certain enzymes by cross-linking DNA and by inducing oxidative stress. Although nitro and azo substances are not as widely prevalent as the carbonyl compounds, nitro compounds are present in automobile exhaust and azo compounds are used as coloring food additives. In contrast to the protective effects of reduction of the carbonyl compounds, reduction of these compounds can increase their toxicities by facilitating their conversion to electrophilic products. An example of carbonyl reduction is the conversion of the essential oil component, carvone, to carveol. Examples of nitro and azo reductions are the conversions of chloramphenicol and prontosil to the corresponding aromatic amines.

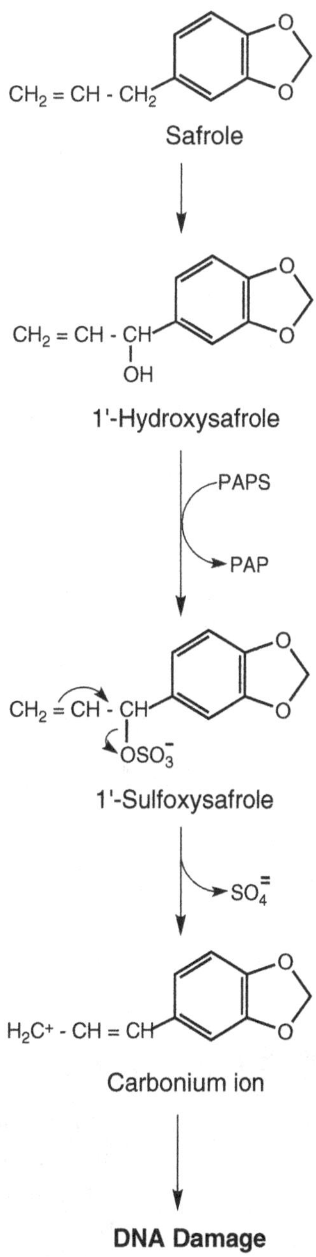

$CH_2 = CH - CH_2$

Safrole

$CH_2 = CH - CH(OH)$

1'-Hydroxysafrole

PAPS

PAP

$CH_2 = CH - CH(OSO_3^-)$

1'-Sulfoxysafrole

$SO_4^=$

$H_2C^+ - CH = CH$

Carbonium ion

DNA Damage

FIGURE 3.4 *Safrole metabolic activation.*

PHASE II REACTIONS

Phase II reactions involve the enzymatic addition of an endogenous compound to the xenobiotic or the phase I product that usually increases the hydrophilicity and rate of excretion of the xenobiotic. The most important phase II reactions, in terms of the quantities and number of xenobiotics metabolized, are the additions of glucuronic acid, sulfate, or glutathione to the xenobiotic or phase I product.

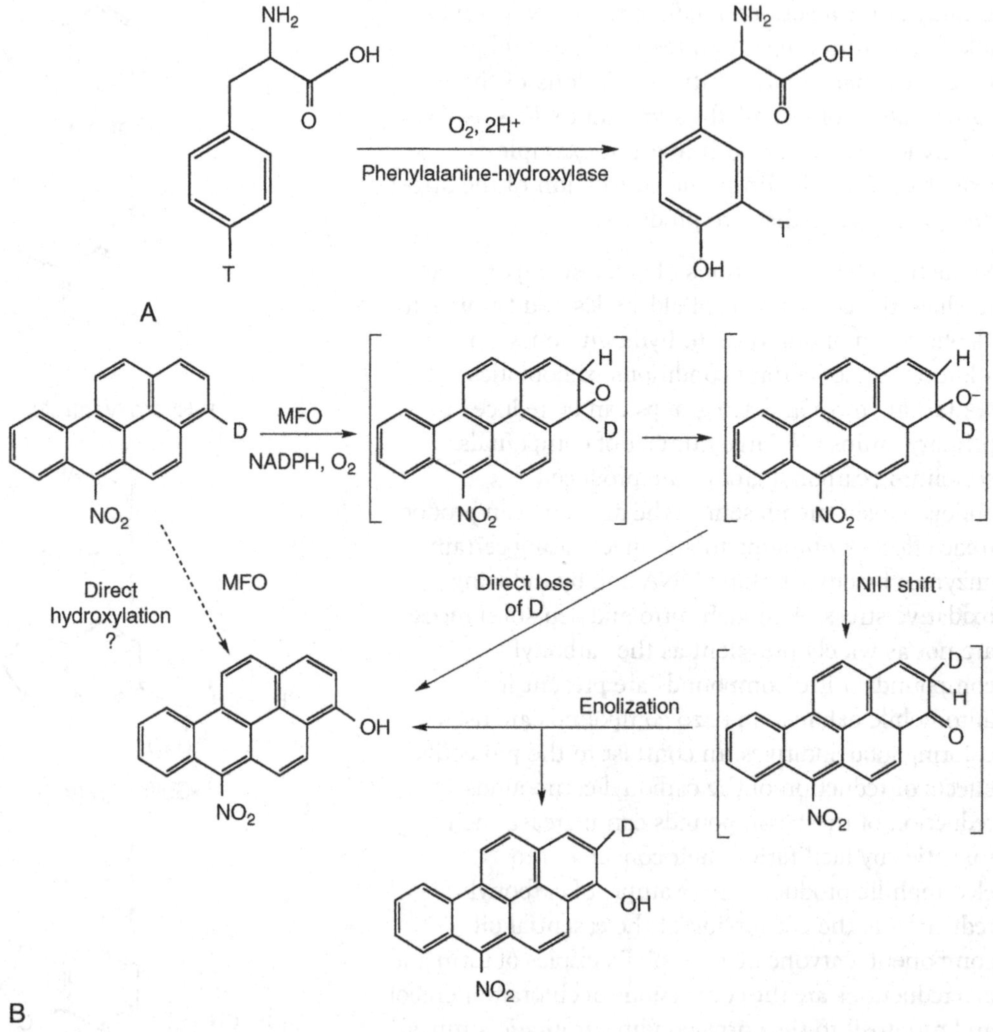

FIGURE 3.5 *A) Phenylalanine oxidation and NIH shift; B) Nitro-benzo[a]pyrene oxidation and NIH shift.*

The xenobiotic substrates for the glucuronidation and sulfation reactions are nucleophiles, whereas the substrates for the glutathiolation reactions are electrophilic. Less common but in some cases important phase II reactions involve the additions of amino acids, usually glycine, glutamine, or taurine, to carboxyl moieties of xenobiotics. Additional phase II reactions are the additions of methyl or acetyl groups. The products of the latter two reactions are unusual for phase II reactions because, with the exception of the methylation of amines with the formation of charged quaternary products, they often produce more lipophilic products.

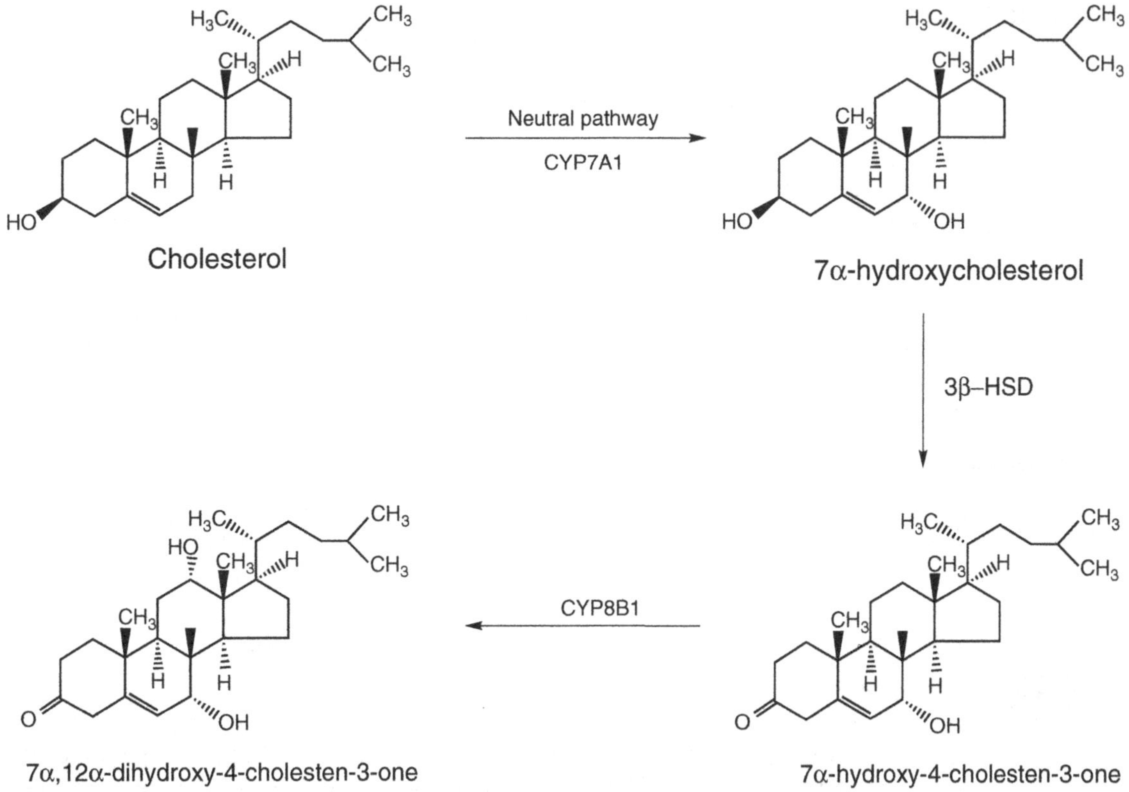

Cholesterol
Neutral pathway
CYP7A1
7α-hydroxycholesterol
3β–HSD
CYP8B1
7α,12α-dihydroxy-4-cholesten-3-one
7α-hydroxy-4-cholesten-3-one

FIGURE 3.6 *Cholesterol oxidation.*

Demethylation
Ecstasy
N-Ethyl-3,4-dihydroxyamphetamine
N-Dealkylation

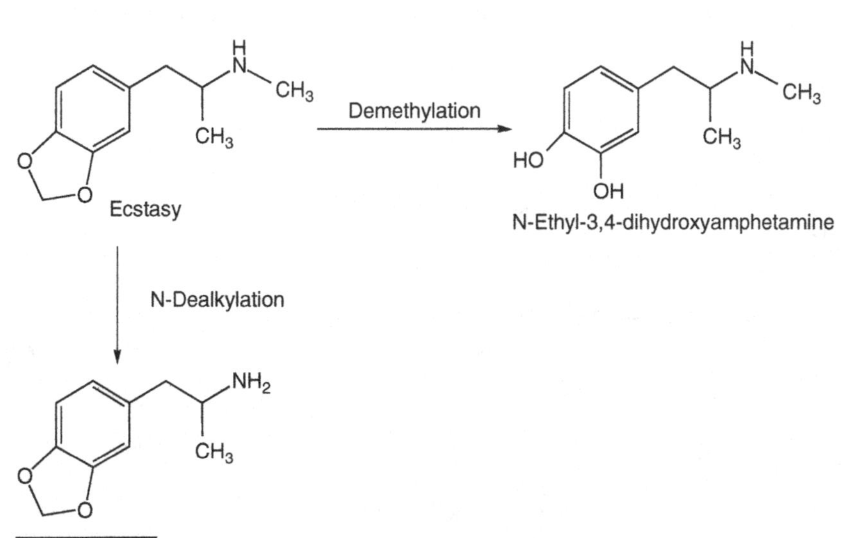

FIGURE 3.7 *Oxidative dealkylation.*

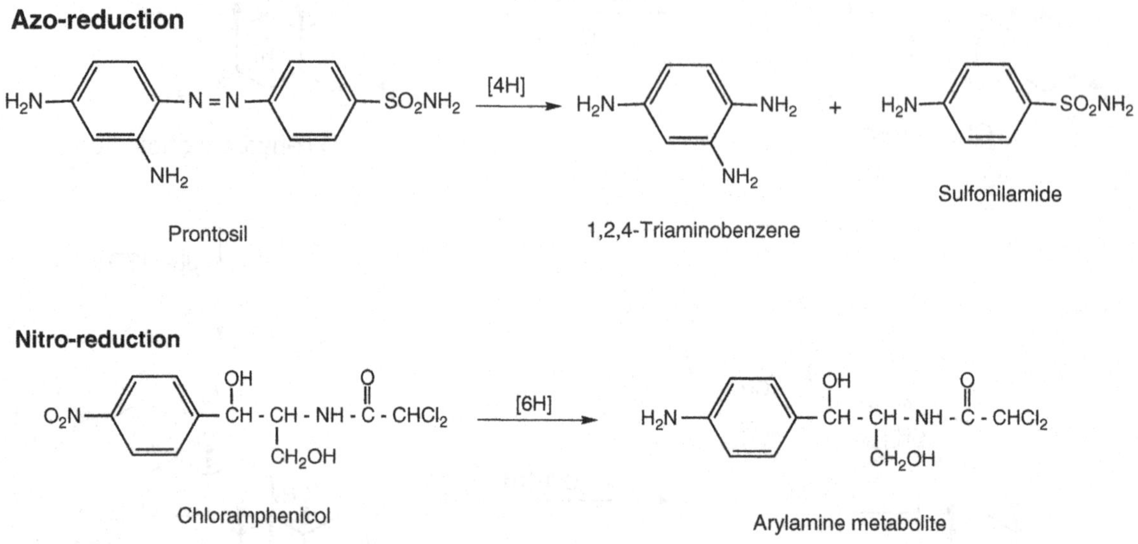

Azo-reduction

H_2N–C6H3(NH_2)–N=N–C6H4–SO_2NH_2 $\xrightarrow{[4H]}$ H_2N–C6H3(NH_2)–NH_2 + H_2N–C6H4–SO_2NH_2

Prontosil — 1,2,4-Triaminobenzene — Sulfonilamide

Nitro-reduction

O_2N–C6H4–CH(OH)–CH(CH_2OH)–NH–C(=O)–$CHCl_2$ $\xrightarrow{[6H]}$ H_2N–C6H4–CH(OH)–CH(CH_2OH)–NH–C(=O)–$CHCl_2$

Chloramphenicol — Arylamine metabolite

FIGURE 3.8 *Azo and nitro group reductions.*

Examples of the more common phase II reactions of phase I products are provided in Figure 3.9. Common substrates for the formal sulfate and glucuronide addition reactions are types of alcohols, hydroxylamines, phenols, amines, and thiols, all of which are nucleophiles.

Common substrates for the displacement reactions with glutathione are epoxides, halides, sulfates, nitro groups, sulfonates, and thiocyanates. A further important class of phase II reactions involves the reductive addition of

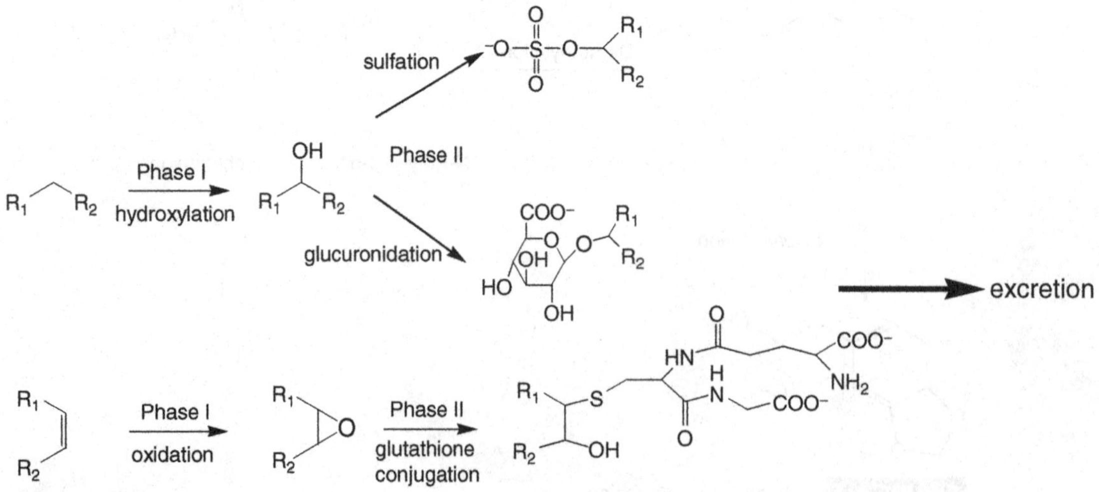

Phase I hydroxylation; Phase II: sulfation, glucuronidation; Phase I oxidation; Phase II glutathione conjugation; excretion

FIGURE 3.9 *Phase I and phase II metabolism.*

glutathione to α,β-unsaturated systems such as quinones, maleates, and certain azo compounds. All these substrates are electrophilic and thus, potentially reactive with cellular molecules.

PHASE I ENZYMES

Cytochrome P450

The phase I and II reactions are mediated by large families of enzymes that are present primarily in the smooth endoplasmic reticulum of cells situated in the portals of entry to the body, including the small intestine, liver, and lung.

Several groups of enzymes mediate phase I reactions of endogenous and exogenous substances, and the cytochrome P450 enzymes mediate by far the greatest proportion of these reactions. These enzymes, called CYPs, are heme-containing iron complexes that absorb light maximally at 450 nm when exposed to carbon monoxide. The CYP super family consists of over 400 enzymes in 36 families that are differentially expressed in apparently all species from bacteria and fungi to mammals. Sequencing of the human genome revealed the existence of 59 human CYPs, three-quarters of which mediate the metabolism of endogenous substances such as sterols, vitamins, and other lipids, and show few interindividual differences in expression across the human population. As much as one-quarter of the human CYPs apparently are involved primarily in the metabolism of xenobiotics, and the levels of expression of enzymes are often highly variable among individuals, and from time to time, in one individual. The enzymes are expressed selectively in most mammalian tissues with the highest level and greatest diversity occurring in the liver.

The differences in the CYP repertoires of different species can be quite large. For example, the plants *Arabidopsis* and rice normally express 239 and 458 different CYPs, respectively. Moreover, rodents express nearly twice as many CYPs (108 active genes) as the human. The large number and diversity of the plant CYPs seem to account for the large diversity of secondary products (phytochemicals) produced by the plants, many of which function in plant defense against other organisms. The much smaller groups of CYPs expressed by the mammals appear to have evolved largely to promote the detoxification and excretion of the plethora of plant secondary products. Consistent with this notion is the larger number of CYPs produced by rodents, perhaps to deal with the very large diversity of potentially toxic substances in their diets compared to the more selective human diet.

CYP nomenclature is based on homologies of related enzymes. CYPs with less than 40% amino acid identity are assigned to different families, such as 1, 2, 3, and so on. CYPs with 40 to 55% sequence identity are assigned to the different subfamilies, like 2A, 2B, 2C, and so on. Enzymes with greater than 55% identity are assigned to the same subfamily—2A1, 2A2, 2A3, and so on.

Of the over 50 CYP enzymes expressed in humans, the members of families 1 to 4, including only six enzymes, mediate the metabolism of as much as of 90% of clinically useful drugs, and likely a large proportion other xenobiotics as well. The most prevalent enzymes in this group expressed in the human liver are, in decreasing order of expression level, CYP3A4, CYP2C9, CYP2C8, and CYP2A6. The enzymes responsible for the metabolism of most of the useful drugs are, in order of decreasing importance, CYP3A4, CYP2D6, CYP2C9, and CYP2C19. It is interesting to note that two of the enzymes that are most important in the metabolism of drugs, CYP3A4 and CYP2C9, are also highly expressed in human liver. However, the two other enzymes that are most important in drug metabolism, CYP2D6 and CYP2C19, are expressed at comparatively low levels. On the assumption that the levels of enzyme expression should correspond to the level of substrate, the roles, perhaps in phytochemical metabolism, of the more prevalent CYPs such as CYP2C8 and CYP2A6 are yet to be determined.

The cytochrome P450 mediated reactions occur in a membrane bound protein complex consisting of a CYP enzyme in close association with NADPH-cytochrome P450 reductase (CYP reductase). Under normal fully oxygenated conditions, the reaction catalyzed by CYP is a monooxygenation, in which one atom from molecular oxygen is incorporated into a substrate and the other is reduced to H_2O with reducing equivalent derived from NADPH. As indicated in Figure 3.10, CYP must be in the reduced ferrous (Fe^{+2}) state to bind substrate and O_2. This occurs in a two-step process that involves first the binding of the substrate to the ferric form of the enzyme, followed by a one-electron transfer to yield the reduced CYP bound to substrate. Electrons are relayed from NADPH to CYP via the flavoprotein CYP reductase, which transfers electrons from NADPH to CYP through redox reactions with flavin adenine dinucleotide (FAD) and flavin mononucleotide (FMN). Following a series of electron transfers, water, the Fe^{+3} oxidized form of the CYP, and the oxidized form of the xenobiotic are produced in the lumen of the smooth endoplasmic reticulum. The oxidized

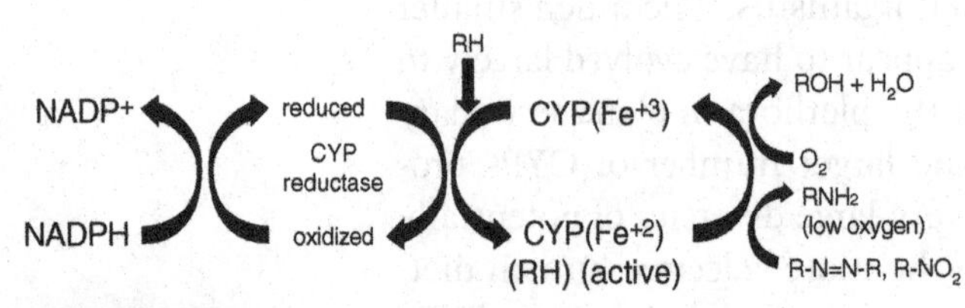

FIGURE 3.10 *NADPH-CYP oxidative pathway.*

xenobiotic may then pass into the cytosol by passive diffusion or undergo additional phase I and phase II reactions in the endoplasmic reticulum.

Although the CYP enzyme complex is most active in the oxidation of xenobiotics, this system also contributes to the reduction of certain substances, especially those containing oxidized nitrogen moieties. Thus, under conditions of low oxygen tension as occurs normally in the lower gastrointestinal tract and in some diseased tissues such as tumors, the CYP reductase and CYP can mediate the transfer of electrons from NADPH, not to oxygen but to azo, nitro, nitroso, hydroxylamine, and nitrogen oxide groups in many xenobiotics, ultimately leading to the production of the fully reduced amine (Figure 3.10). This reaction sequence mediates the biological activity of the parent nitro compound in several ways, as described next.

The initial one electron addition to the nitro group produces a radical intermediate that can react with molecular oxygen to produce superoxide and other reactive oxygen species (ROS). This induced oxidative stress is thought to account for both the adverse effects of certain nitro compounds in normal tissues, and the beneficial effects of antitumor and antibacterial drugs. Indeed, a class of therapeutic agents, called bioreductive pro-drugs, exploits the reducing environments of tumors to target their activities. An example of a promising bioreductive N-oxide-containing pro-drug is AQ4N, which is reduced in hypoxic tumor tissues to the more active amino derivative, AQ4 (Figure 3.11). This reaction is mediated directly by CYP3A4, CYP1A1, and CYP1B1.

Also, the fully reduced amino compound produced in the gut may be absorbed and transported, for example, to the liver where it is metabolically converted to a strong electrophile that can react with cellular macromolecules to produce mutations or other toxic effects. The potent hepatocarcinogen, 2,6-dinitrotoluene, is activated by this enterohepatic mechanism.

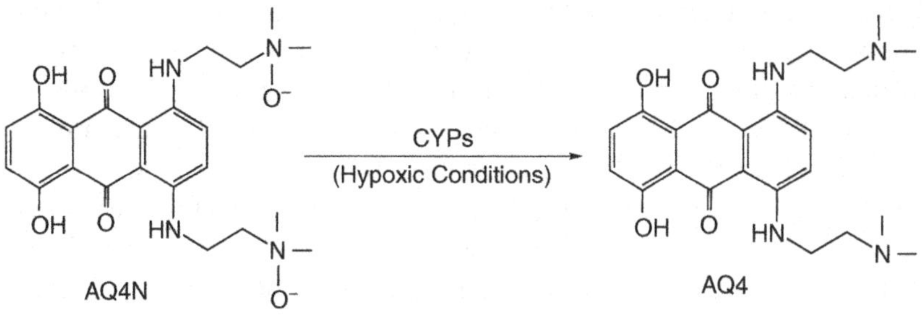

FIGURE 3.11

Amine oxide reduction.

CYP3A4

CYP3A4 is the most important of the phase I enzymes in humans because it is the CYP that is expressed in the highest levels in humans and mediates the metabolism of over 50% of the therapeutic agents of many kinds and an unknown proportion of environmental chemicals such as flavonoids, mycotoxins such as aflatoxin B_1, pesticides, and a number of food additives, and possibly also endogenous peptides. Although there is a high level (40- to 50-fold) of interindividual variability in CYP3A4 expression, this enzyme accounts for as much as 25% of total CYP protein in the liver and up to 70% of total CYP expression in the intestine. Although the intestinal content of CYP3A4 is only about 1% of liver content, the intestine contributes equally to the metabolic first-pass effect for CYP3A4 substrates. In addition, CYP3A4 appears to cooperate with the neutral xenobiotic transporters (MDR) with similar substrate specificities and inducibilities, leading to an efficient first-pass drug extraction in the intestine. Protein expression of CYP3A4 increases slightly from the duodenum to the jejunum and then decreases by approximately 75% toward the ileum and colon.

As is the case for most of the members of the CYP1–4 families, the cellular activity of CYP3A4 is subject to modulation by several mechanisms, including effects on enzymatic activity and regulation of expression of the enzyme. CYP3A4 enzymatic activity is inhibited by several xenobiotics, including ketoconazole, troleandomycin, and the furanocoumarin, 6,7-dihydroxybergamottin, found in grapefruit juice, by classical mechanisms that involve the binding of the inhibitor to the active site and displacement of the substrate. In addition, an unusual property of CYP3A4 is that it is highly sensitive to other, so-called effector substances. This type of regulation results from the fact that the enzyme has a very large active site that can accommodate two molecules (of either the same or different compounds) with the result that the second molecule influences the enzyme activity on the first substance leading to increased or decreased metabolism. Examples of the results of this complex of interactions are that the rate of metabolism of a single substrate can increase in a nonlinear manner with the substrate concentration. In addition, effectors such as α-naphthoflavone and certain other flavonoids can accelerate the oxidation of aflatoxins, polycyclic aromatic hydrocarbons, and sex hormones. Conversely, erythromycin can inhibit CYP3A4-mediated metabolism of testosterone by a mechanism that does not reduce the binding of the substrate with the enzyme.

In addition to these effects on CYP3A4 activity, the level of expression of this enzyme is inducible by several different classes of compounds, both therapeutic agents and dietary chemicals. Chemical inducers of CYP3A4 include a diverse group of substances that include rifamycin, carbamazepine, glucocorticoids, and the phytochemical, hyperforin, from St. John's wort (Figure 3.12). Induction of CYP3A4 is mediated by the pregnane X receptor (PXR), the constitutive androstane receptor (CAR), and the bile acid receptor (FXR). On binding of an inducer ligand, the inducer-receptor dimerizes with the retinoic acid X receptor (RXR) and the dimer binds to a cognate 5-upstream response element of the CYP3A4 gene, resulting in transcriptional activation. All three receptor heterodimers, PXR/RXR, CAR/RXR, and FXR/RXR, can bind to the same promoter site in DNA. The upstream region of the CYP3A4 gene has binding sites for a number of transcription factors including AP-3, hepatocyte nuclear factors (HNF)-4 and -5, and a glucocorticoid response element (GRE). HNF-4 plays a role in the inducible and constitutive expression of CYP3A4, the former through modulation of PXR-mediated gene expression (Figure 3.13). Only a few compounds, including the antiepileptic drug, phenytoin, have been shown to induce CYP3A4 preferentially through CAR instead of PXR.

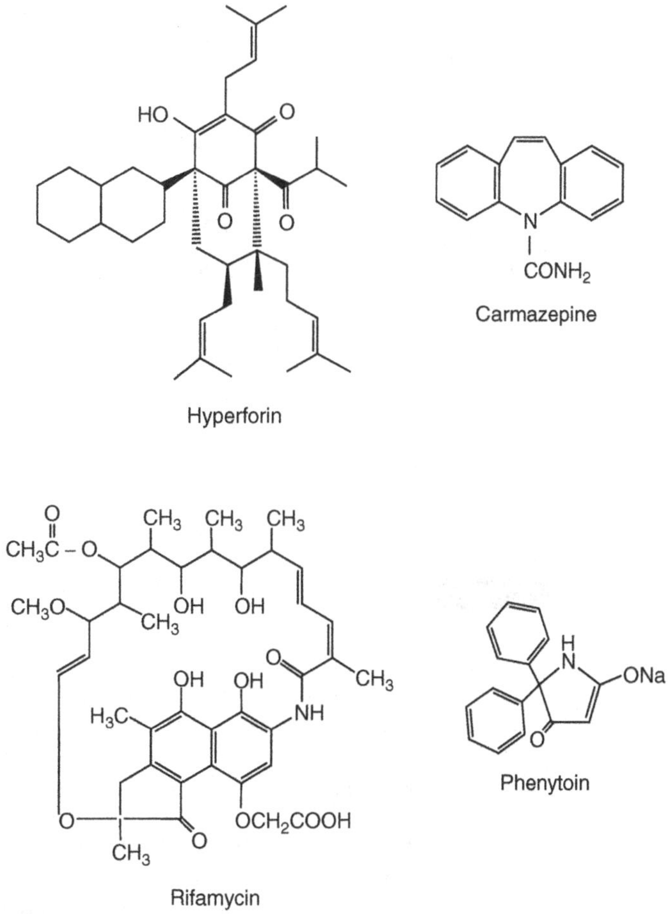

FIGURE 3.12 *CYP3A4-inducing agents.*

CYP3A4 is critical to the metabolic clearance to vitamin D and provides an example of an important drug-nutrient interaction. It is well recognized that long-term therapy with some antiepileptic drugs, including phenobarbital, phenytoin, and carbamazepine, and the antimicrobial agent rifampicin, can cause a metabolic bone disease, osteomalacia, which is characterized as a painful syndrome of bone softening in adults. The effects are very similar to the effects of vitamin D deficiency. Studies have established that these drugs activate CYP3A4 expression via the PXR and accelerate the catabolism of the active form of vitamin D, $1{,}25(OH)_2D_3$, in human liver and intestine. Indeed, contrary to prior suppositions, the steroid and xenobiotic

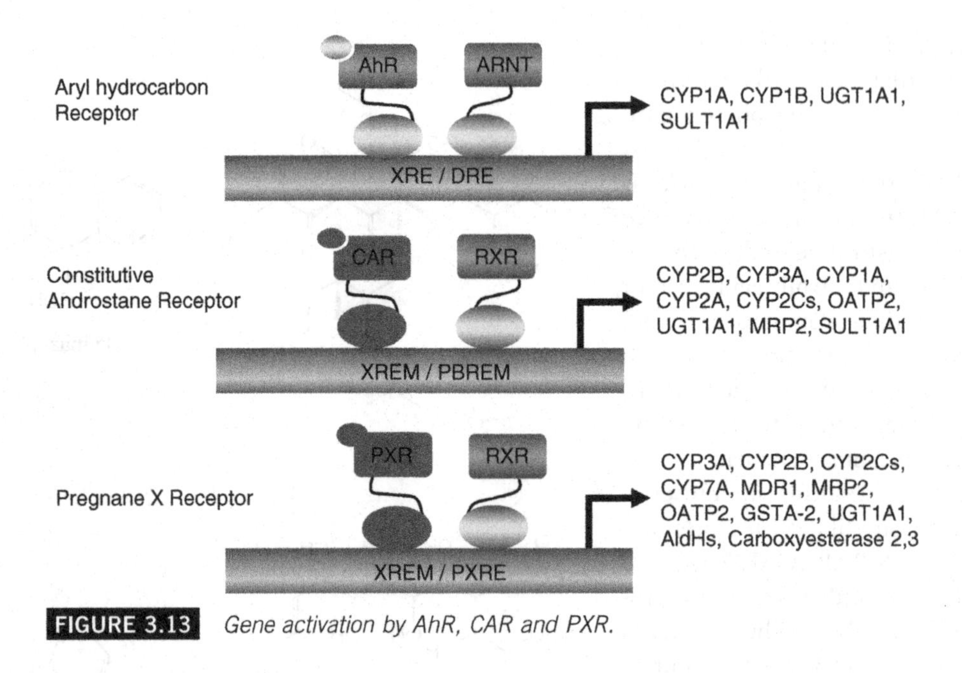

FIGURE 3.13 *Gene activation by AhR, CAR and PXR.*

receptor (SXR) represses CYP24 expression in the liver and intestine and thereby blocks vitamin D catabolism in these tissues.

CYP1B1

Human CYP1B1 has been shown to be an important enzyme in the activation of diverse procarcinogens such as nitroarenes, polycyclic aromatic hydrocarbons, and arylamines to reactive metabolites that cause DNA damage. CYP1B1 is constitutively expressed in steroidogenic tissues such as uterus, breast, ovary, testis, prostate, and the adrenal gland. The enzyme also is present in many other extra hepatic tissues including kidney, thymus, spleen, brain, heart, lung, colon, and intestine, and is expressed at a higher level in a wide range of human cancers including cancers of the skin, brain, testis, and breast, as compared to the nontransformed tissue. Although the significance of this latter observation for tumor development is not clear, it suggests that CYP1B1 may be involved in the metabolism of an endogenous substrate that is critical to the survival or behavior of the transformed cells.

Substrates for CYP1B1 include endogenous and exogenous substances. This enzyme is active in the metabolism of estradiol, as well as of carcinogenic hydrocarbons such as benzo[a]pyrene and 7,12-dimethylbenz[a]anthracene. It mediates the oxidation of caffeine and theophylline, and the O-dealkylation of ethoxycoumarin, ethoxy-trifluoromethyl-coumarin, and ethoxyresorufin. It is less active in the metabolism of ethoxyresorufin than CYP1A1, which is not expressed in human liver. However, CYP1B1 is inactive in the oxidation of many of the other drugs that are specific substrates for other CYPs.

Because of the possible importance of CYP1B1 activity in tumorigenesis, there is considerable interest in the identification of selective inhibitors of this enzyme. Naturally occurring substances that exhibit this inhibitory activity include xanthotoxin, certain linear furanocoumarins, and the flavonoid homoeriodictyon. Synthetic selective inhibitors of CYP1B1 include certain polycyclic aromatic hydrocarbon derivatives such as pyrene, 2-ethynylpyrene, and 3,3′,4,4′,5′-pentachlorobiphenyl, and certain methoxylated stilbene derivatives related to resveratrol. There is a high level of constitutive expression of the CYP1B1 gene in liver, which apparently is dependent on endogenous or exogenous ligands for the aryl hydrocarbon receptor (AhR).

As is true for other members of the CYP1 family, transcriptional activation of the CYP1B1 gene is stimulated primarily by polycyclic hydrocarbons by activation of the AhR (Figure 3.13). The promoter region of the CYP1B1 gene contains five xenobiotic responsive elements (XRE)/dioxin responsive elements (DRE), which are the binding sites of the heterodimer, composed of the activated AhR and Aryl Hydrocarbon Nuclear Translocator (Arnt). Arnt is a nuclear protein that is important in the cell's response to xenobiotics and hypoxia. As indicated in Figure 3.14, AhR exists in a dormant state in the cytoplasm in association with a complex of HSP90, XAP2, and p23 chaperone proteins. Upon ligand binding, AhR in the complex is activated by a conformation change that induces the dissociation of HSP90 and exposes a nuclear localization signal(s). The ligand-activated AhR then translocates into the nucleus and forms a heterodimer with the closely related Arnt protein already present in the nucleus. This complex (xenobiotic-AhR/Arnt) binds the XRE/DRE in the promoter regions of responsive genes.

An additional important component of the regulation of expression of AhR responsive genes is the AhR repressor (AhRR). This protein localizes in the nucleus as a heterodimer with Arnt. The AhRR/Arnt heterodimer also recognizes the XRE, but functions as a transcriptional repressor. Therefore, AhRR functions as a negative regulator of AhR by competing with AhR for formation of a heterodimer with Arnt and binding to the XRE sequence. Since the AhRR gene in rodents contains several XRE sequences, AhRR is inducible in an AhR-dependent manner. Thus, AhRR can negatively regulate the expression of AhR-responsive genes as part of a regulatory feedback loop.

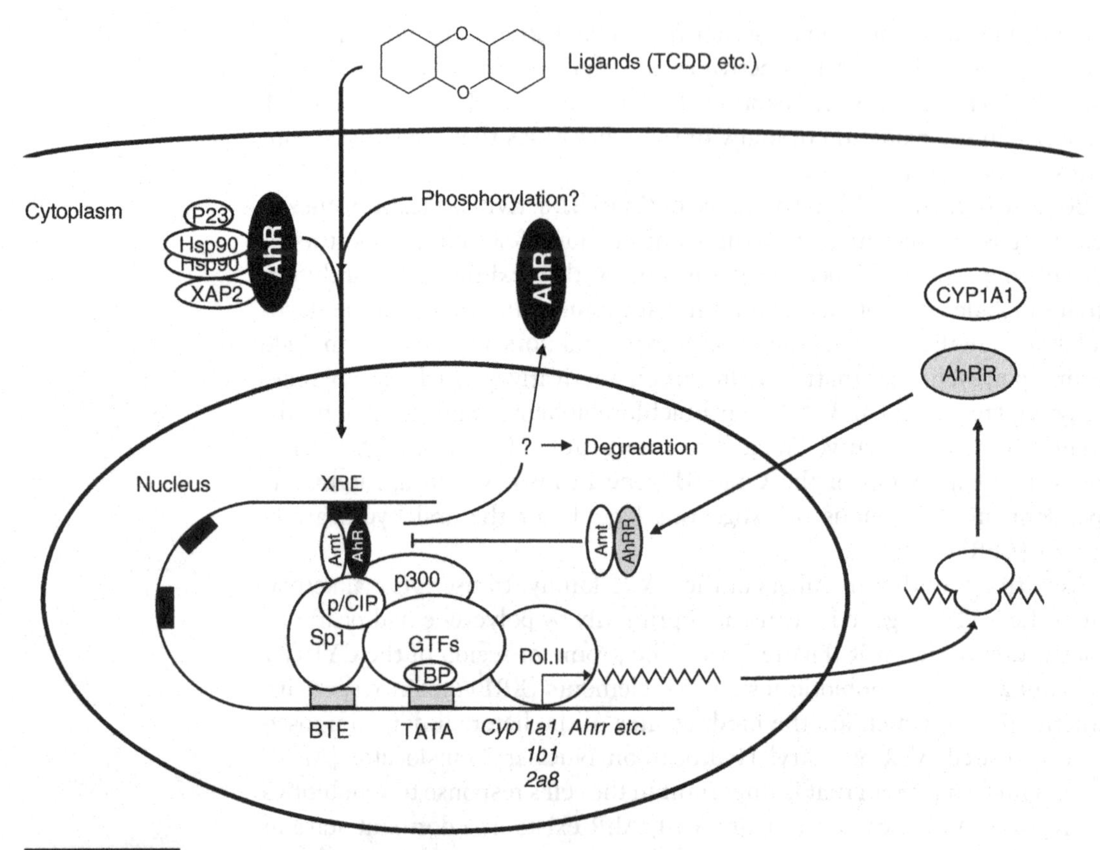

FIGURE 3.14 *Regulation of AhR activity.*

CYP2E1

CYP2E1 is primarily a mammalian enzyme that is of great importance because it metabolizes several significant low molecular weight endogenous substances and a large number of xenobiotics, in some cases by processes that result in the activation of carcinogens. CYP2E1 is expressed constitutively in the liver and is inducible by its endogenous substrates, ethanol, acetone, and other endogenous ketones. Although the enzyme is expressed constitutively at low or undetectable levels in extra hepatic tissues, following induction the protein is clearly present in many tissues including the mucosa of nose and mouth, esophagus, kidney, lung, brain, bone marrow, lymphocytes, and proximal colon.

One of the unusual features of CYP2E1 is its selectivity for many low molecular weight substrates. Its physiological substrates appear to be the gluconeogenic precursors acetone and acetol, as well as fatty acids.

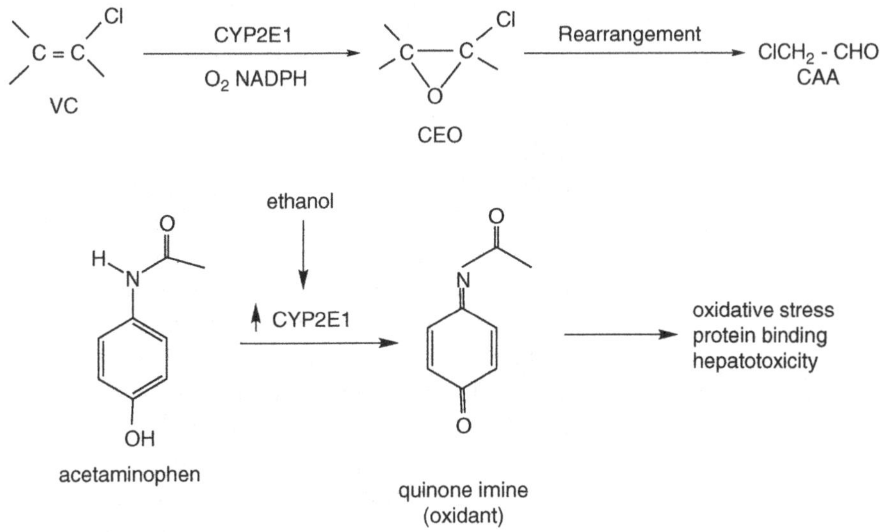

FIGURE 3.15 *Reactions catalyzed by CYP2E1.*

Examples of the over 80 xenobiotic substrates include: ethanol, acetaldehyde, acetaminophen, acrylamide, aniline, benzene, carbon tetrachloride, N-nitrosodimethylamine, and vinyl chloride. Some examples of the types of reactions that are mediated by this versatile enzyme are provided in Figure 3.15, including the dehydrogenation of acetaminophen and the epoxidation of vinyl chloride.

There is considerable interest in means to reduce CYP2E1 activity because of the important role of this enzyme in the activation of several important carcinogens. Standard pharmaceutical inhibitors of CYP2E1 are relatively small aromatic molecules related in structure to acetaminophen and include chlorzoxazone (competitive inhibition), isoniazid (noncompetitive inhibition), 4-methylpyrazole (suicide inhibitor), and 1-phenylimidazole (competitive) (Figure 3.16). The grape phytochemical resveratrol, the soy isoflavones genestein and equol, and the garlic product diallyldisulfide are noncompetitive inhibitors of CYP2E1.

The mechanism of regulation of the expression of CYP2E1 is complex and is modulated at several molecular levels by endogenous and exogenous influences. Evidence that the level of CYP2E1 protein is stringently regulated is the observation that although the level of CYP2E1 protein is only a few percent of the levels of CYP3A4 in liver, the level of CYP2E1 mRNA is as much as a 1000-fold greater than the mRNA level for CYP3A4. Studies in rats have demonstrated increases of CYP2E1 protein arising from either transcriptional or posttranscriptional effects following chemically induced

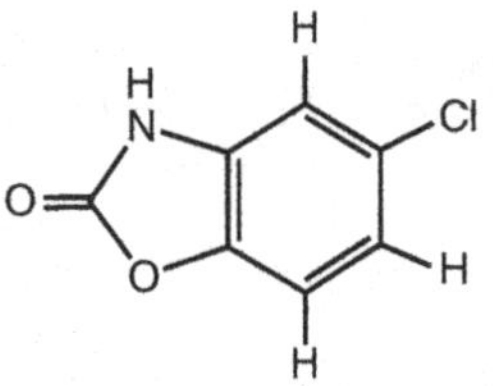

Chlorzoxazone (CLZ)

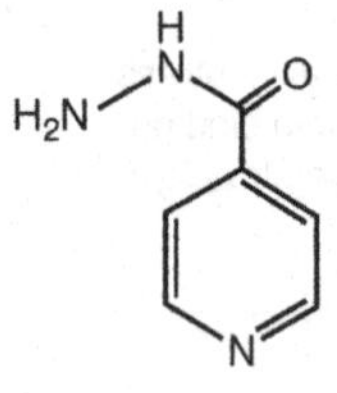

Isoniazid

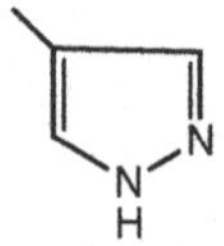

4-Methylpyrazole

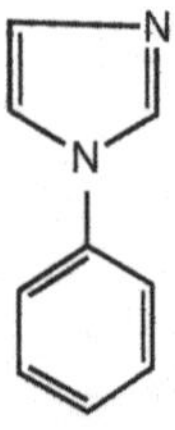

1-Phenylimidazole

FIGURE 3.16 *CYP2E1 inhibitors.*

diabetes and starvation. Because diabetes is associated with high serum levels of ketone bodies, which are substrates for CYP2E1, the increase in protein may be due in large part to protein stabilization. Insulin treatment causes a decrease of the CYP2E1 protein in chemically induced diabetes by increasing the rate of mRNA turnover. Induction of CYP2E1 protein is greater in both liver and fat of obese rats and in normal rats on a high fat diet, compared to normal rats on normal lower fat diets. The expression of CYP2E1 in human hepatocyte culture is reduced by several cytokines, including IL-1, IL-6, and TNF-α, by a mechanism that involves decreased transcription and enhanced mRNA stability. The major mechanisms by which the levels of CYP2E1 protein are increased involve posttranslational stabilization to proteosomal degradation as a result of protein binding to substrates. The exact activators for this highly efficient means of regulation have yet to be precisely characterized.

PEROXIDASES

The peroxidases are a relatively small but important group of heme-containing proteins that mediate the oxidation of endogenous and exogenous substances by a mechanism that involves peroxide as the oxidizing agent (Figure 3.17). The family of human peroxidases includes myeloperoxidase (MPO), eosinophil peroxidase (EPO), uterine peroxidase, lactoperoxidase (LPO) present in mammary gland, salivary peroxidase, thyroid peroxidase, and prostaglandin H1/2 synthases (PHS) present in inflammatory cells, brain, lung, kidney, GI tract, and urinary bladder. MPO, EPO, and LPO are water-soluble enzymes expressed primarily in the lysosomes, respectively, of neutrophils, eosinophils, and secretory cells of exocrine glands. Lysosomes are membrane bound cytoplasmic organelles of eukaryotic cells that contain many digestive enzymes

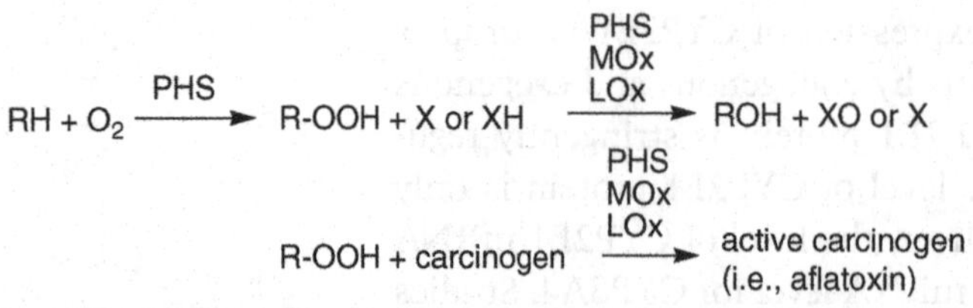

$$RH + O_2 \xrightarrow{PHS} R\text{-}OOH + X \text{ or } XH \xrightarrow{PHS,\ MOx,\ LOx} ROH + XO \text{ or } X$$

$$R\text{-}OOH + \text{carcinogen} \xrightarrow{PHS,\ MOx,\ LOx} \text{active carcinogen (i.e., aflatoxin)}$$

FIGURE 3.17 *Peroxidase reactions.*

maintained at a pH of about 5.0. Neutrophils and eosinophils are types of white blood cells important in the body's defense against microorganisms and viruses. MPO and EPO are released into the phagocytic vacuole and the plasma, whereas LPO is secreted into milk, saliva, and tears. These peroxidases function by oxidizing chloride, bromide, or thiocyanate to form strong oxidants (e.g., hypochlorous acid) that kill microorganisms as part of a defense function against bacteria and parasites. In contrast to the CYP enzymes, which require NADPH in the metabolism of xenobiotics, the peroxidases couple the reduction of hydrogen peroxide and lipid hydroperoxides to the oxidation of other substrates.

PHS has two catalytic activities: the first reaction is a **cyclooxygenation** that converts arachidonic acid to the cyclic endoperoxide-hydroperoxide (PGG_2). The second reaction is a **peroxidation** that converts the hydroperoxide to the corresponding alcohol (PGH_2), which is coupled to the oxidation of an electron donor (Figure 3.18).

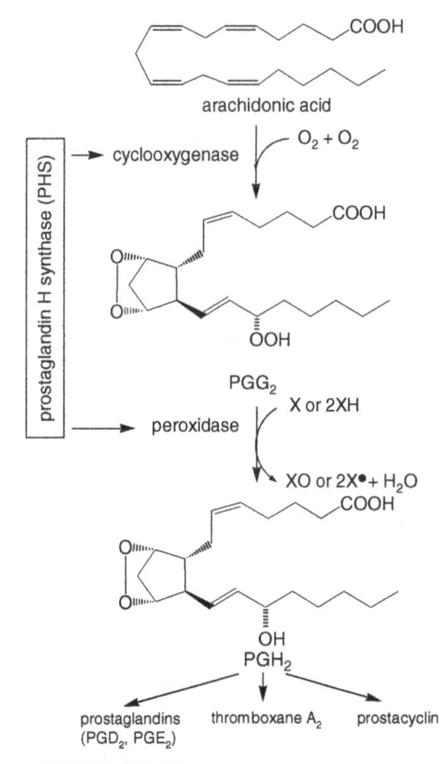

FIGURE 3.18 *Prostaglandin H synthase/arachidonic acid reaction.*

The peroxidases are important in the production of oxidative stress and the metabolism of many xenobiotics. Xenobiotics can act as electron donors and are cooxidized to form radicals that in turn cooxidize various physiological electron donors, for example, proteins, nucleic acids, and lipids. PHS can bioactivate carcinogens such as β-naphthylamine, a bladder carcinogen, and certain polycyclic aromatic hydrocarbons to electrophilic, radical intermediates that can react with cellular macromolecules (Figure 3.19). The production of the strong oxidant hypochlorite as part of the defense response results in the nonselective oxidation of certain molecules in both parasite and host. A common result of this induced oxidative stress is agranulocytosis, which is an acute blood disorder, often caused by radiation or drug therapy, and is characterized by severe reduction in the granulocyte-type of white blood cells.

Hypochlorite oxidizes glutathione, protein thiols and thioethers, and converts amino acids to haloamines. Hypochlorite forms oxidation products by reacting with the unsaturated carbon bonds of fatty acids, cholesterol, and some amino acids, which disrupts membranes and protein functions leading to cell lysis and death. In addition, the secondary acute myelogenous leukemias produced by chemotherapy with etoposide, a phenolic

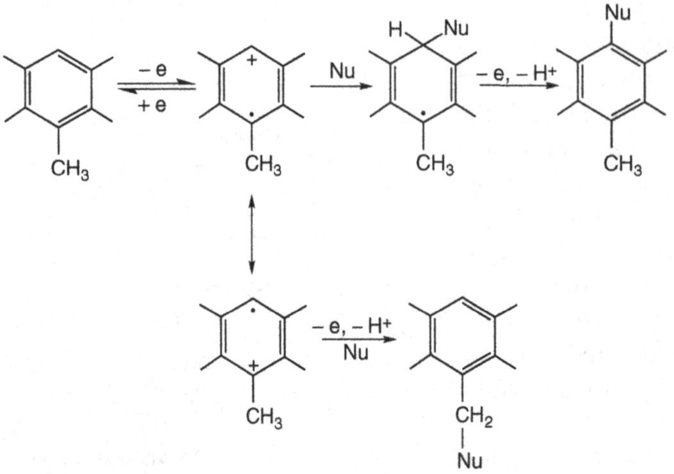

FIGURE 3.19 Prostaglandin H synthase activation of aromatic hydrocarbons.

topoisomerase inhibitor, or the leukemias induced by chronic exposure to benzene have been attributed to DNA pro-oxidant phenoxyl radicals formed by metabolism of etoposide or phenol, respectively, by MPO/H_2O_2. The triphenylethylene drug tamoxifen is used in the treatment of metastatic breast cancer or as an adjuvant therapy, and is one of the safest anticancer drugs available, but is a hepatocarcinogen in rats and can induce uterine cancer in women. Peroxidase catalyzes the activation of 4-hydroxytamoxifen, a major metabolite that forms DNA adducts and covalently binds to proteins.

The peroxidase-mediated conversion of a major metabolite (DMBI) of the anti-inflammatory agent indomethacin to an electrophilic product is illustrated in Figure 3.20. This reaction illustrates the net dehydrogenation

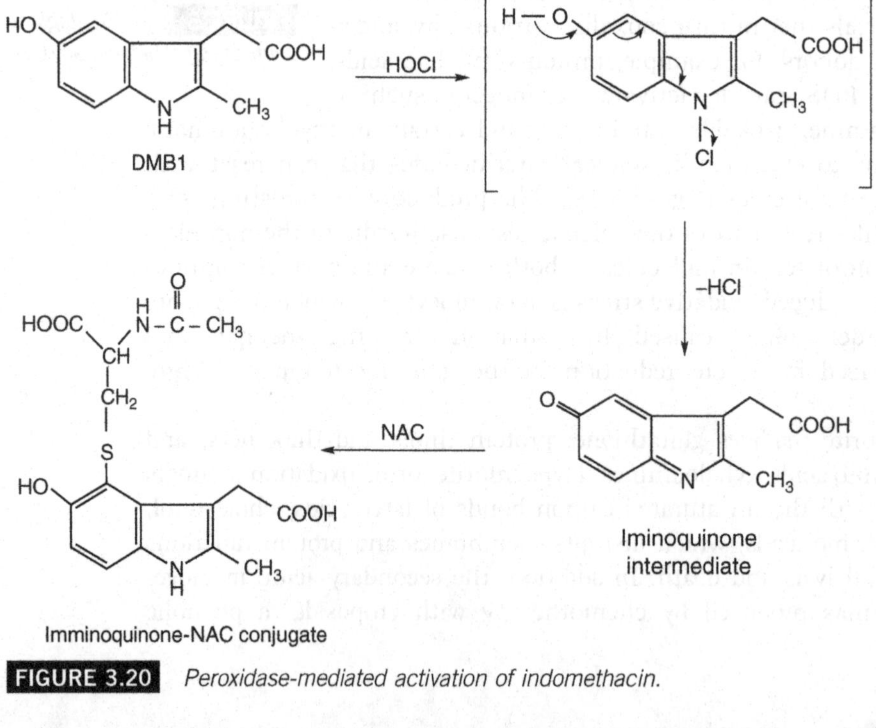

FIGURE 3.20 Peroxidase-mediated activation of indomethacin.

reaction with the formation of reactive quinones and imminoquinones of potentially many endogenous and exogenous substances.

FLAVIN-CONTAINING MONOOXYGENASE (FMO)

The human FMOs comprise a group of five membrane-bound enzymes that selectively mediate the oxidation of nucleophilic heteroatoms of lipophilic xenobiotics. Common substrates include trimethylamine, nicotine, and cimetidine. The products of FMO oxidations are generally more polar and more readily excreted. Although this process usually reduces the toxicity of most substrates, it also can reduce the therapeutic usefulness of substances such as sulindac, which is a pro-drug used in the chemoprevention of colorectal cancer (Figure 3.21A). Under normal conditions, the sulfoxide moiety of sulindac is reduced to the active sulfide. Reoxidation of the sulfide to the sulfoxide and then to the sulfone by FMO thus blocks the required activation step of the drug. A further potentially adverse effect of FMO is the initial oxidation of certain aromatic amines that leads to the production of reactive electrophilic intermediates that can bind to cellular macromolecules, as is illustrated for fluorenylamine in Figure 3.21B. Indole-3-carbinol and N,N-dimethylamino stilbene carboxylate are among the few known competitive inhibitors of FMO.

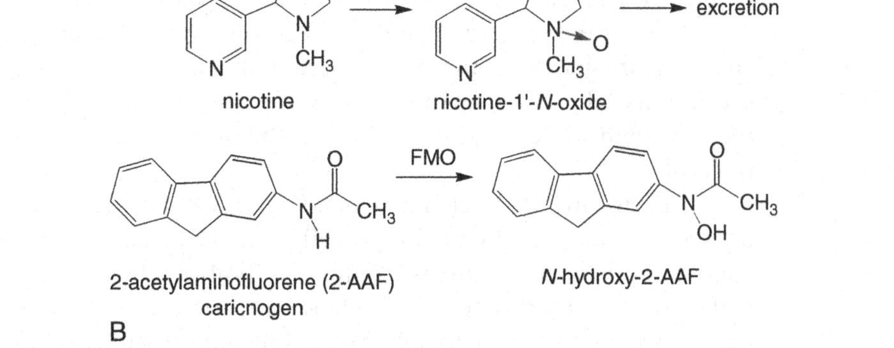

A

FMO

excretion

nicotine

nicotine-1'-*N*-oxide

FMO

2-acetylaminofluorene (2-AAF) caricnogen

N-hydroxy-2-AAF

B

FIGURE 3.21 *A) Sulfur oxidation of sulindac catalyzed by FMO; B) amine oxidations catalyzed by FMO.*

The FMO isozymes are differentially expressed with the highest levels of FMO1 and FMO2 occurring, respectively, in the adult kidney and lung, whereas the highest levels of expression of FMO3, FMO4, and FMO5 occur in the adult liver. Curiously, the expressions of FMO2–5 in fetal liver are only about 10% or less than the expression in the adult liver, whereas expression of FMO1 is approximately 10-fold greater in fetal than in adult liver.

FMO expression shows little response to xenobiotics and is more responsive to physiological state. Certain hormones, however, can influence FMO activity. For example, FMO expressions in rat lung and kidney are positively regulated by testosterone and repressed by estradiol. Cortisol regulates hepatic FMO activity in female mice. FMO2 mRNA and protein expression during gestation correlates with progesterone and corticosterone plasma concentrations. Up to a 20-fold variation of FMO activity has been observed during the estrous cycle in pigs. Diet, age, gender, diabetes, and ethnicity are reported to have relatively small effects on FMO activities.

EPOXIDE HYDROLASE (EH)

Epoxides (oxiranes) of many types are present in the environment or are produced endogenously by oxidative enzymes such as the CYPs and peroxidases. Although most of these substances are relatively stable in cells, several are strong electrophiles that can react with cellular macromolecules and damage cells. In some cases, epoxides of endogenous lipids are important signaling molecules for organismal responses. For example, a role of sEH in the development of adult respiratory distress syndrome and multiple organ failure has been established by the identification of leukotoxin diol, a sEH metabolite of the fatty acid epoxide leukotoxin, as the important trigger of those often fatal effects. The levels of these epoxides in cells are regulated to a large extent by the family of epoxide hydrolase enzymes. These enzymes are included in the discussion of phase I metabolic enzymes because, like the esterases that are generally thought of as phase I enzymes, the epoxide hydrolases mediate the *trans*-addition of water to the substrate molecule.

There are five distinct forms of EH enzymes in mammals, including microsomal epoxide hydrolase (mEH), soluble epoxide hydrolase (sEH), cholesterol epoxide hydrolase (chEH), LTA4 hydrolase, and hepoxilin hydrolase. mEH mediates the metabolism primarily of xenobiotic epoxides and exhibits a remarkably broad substrate specificity. This enzyme can mediate the inactivation of xenobiotic epoxides such as aflatoxin

B_1-9,10-epoxide, whereas it mediates an epoxide hydrolysis in the metabolic pathway that activates benzo[a]pyrene to the ultimate carcinogen, benzo[a] pyrene-7,8-dihydrodiol-9,10-epoxide. sEH complements the activity of mEH in the hydrolysis of a wide array of mutagenic, toxic, and carcinogenic xenobiotic epoxides. For example, sEH preferentially hydrolyzes antisubstituted epoxides, whereas mEH preferentially hydrolyzes syn-substituted epoxides. In addition, sEH mediates the hydrolysis of certain endogenous fatty acid epoxides such as arachidonic acid epoxides, also called epoxyeicosatrienoic acids or EETs, and linoleic acid epoxides, also called leukotoxins, involved in the regulation of blood pressure and inflammation. Typical substrates for mEH (class A) and sEH (class B) are presented in Figure 3.22. The latter three hydrolases, chEH, LTA4 hydrolase, and hepoxilin hydrolase, are specific for individual endogenous substrates as their names indicate. The activity of mEH shows a pH dependence in which the enzyme preferentially hydrolyzes the *cis*-isomer of stilbene oxide at pH 9.0, whereas the preference is for the *trans*-isomer at pH 7.4.

Interest in EH inhibitors has grown in recent years with the realization that selective inhibitors are useful in studies of the mechanism of enzyme action and that inhibition of the hydrolysis of certain endogenous epoxides can be of therapeutic benefit. Inhibitors that are selective for mEH or sEH are indicated in Figure 3.23.

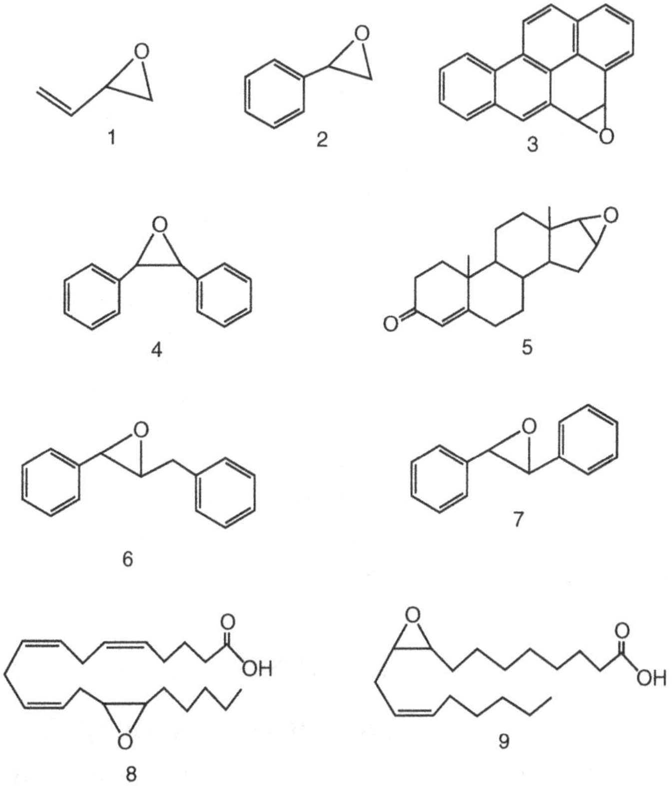

FIGURE 3.22 *Epoxide substrates for epoxide hydrolase.*

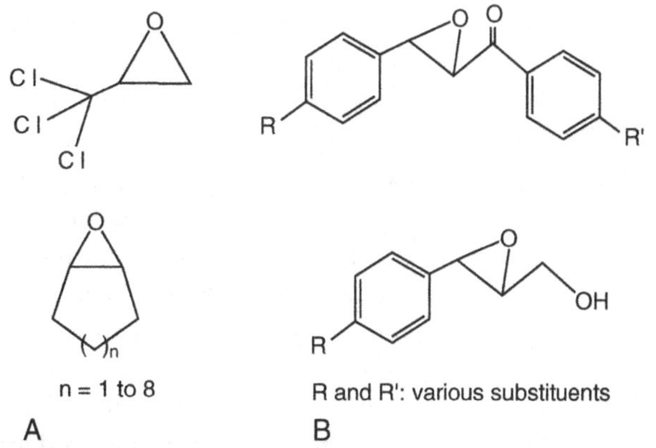

FIGURE 3.23 *Class A and class B inhibitors of epoxide hydrolase.*

Class A inhibitors are selective for mEH, whereas the class B inhibitors are selective for sEH.

The epoxide hydrolases are found in virtually all mammalian tissues, with the highest levels of mEH in the liver, kidney, and testis, and 0.1 to 0.01 times lower expression in lung and lymphocytes. mEH is expressed primarily in the endoplasmic reticulum in close association with the CYP enzymes, which facilitates the efficient hydrolysis of CYP products. Interindividual variation in the expression of EH has been reported to be as high as 40-fold, and individuals expressing low levels of EH are more susceptible to adverse effects of certain pharmaceuticals such as acetaminophen and phenytoin. The expression of mEH is induced 2- to 3-fold by CAR ligands such as phenobarbital and by AhR ligands such as 3-methylcholanthrene, and up to 10-fold by activators of the antioxidant response pathway such as BHA and BHT, and ionizing radiation. The expression of sEH is induced by substances known as peroxisome proliferators, including clofibrate, as well as by common environmental contaminants such as the phthalate esters plasticizers.

ESTERASES

Carboxylesterases (CES)

The hydrolysis of xenobiotic esters is mediated primarily by a group of carboxylesterases (CES), organophosphatases, and cholinesterases. The cholinesterases will be discussed further in Chapter 9. The mammalian CES comprise a multigene family, the gene products of which are localized in the endoplasmic reticulum of many tissues. These enzymes efficiently catalyze the hydrolysis of a variety of ester- and amide-containing chemicals, as well as drugs and pro-drugs, to the respective free acid and an alcohol. They are involved in detoxification or metabolic activation of various drugs, environmental toxicants, and carcinogens, such as methyl-2,4-D, permithrin, vinyl acetate, and cocaine (Figure 3.24).

Carboxylesterase-mediated hydrolysis of vinyl acetate is necessary to generate the active intracellular cross-linking agent, acetaldehyde, and the cytotoxic metabolite, acetic acid. Conversions of the pro-drugs CPT-11 and paclitaxel-2-ethylcarbonate to their active metabolites, SN-38 and paclitaxel, respectively, also are mediated by CES. CES also catalyze the hydrolysis of endogenous compounds, such as short- and long-chain acyl-glycerols, long-chain acyl-carnitine, and long-chain acyl-coenzyme A (CoA) esters.

The expression of CES is ubiquitous, with high levels in the liver, small intestine, kidney, and lung. The limited body of work on the regulation of expression of these enzymes indicates that rodent isozymes are induced by phenobarbital, aminopyrine, Aroclor, polycyclic aromatic hydrocarbons, glucocorticoids, pregnenolone-16, α-carbonitrile, and peroxisome proliferators, indicating the involvement of many receptor systems in the regulation of expression of CES.

There is interest in the development of nontoxic selective CES inhibitors that might have applications in either the prevention of drug toxicity, for example, by inhibiting morphine production from heroin, or increasing the half-lives of drugs that are inactivated by CES. Such drugs include flestolol, a beta-adrenergic blocking agent, which has a dramatically reduced half-life in vivo due to CE-mediated hydrolysis. By administration of a specific inhibitor in combination with the drug of choice, prolonged plasma concentrations and hence greater efficacy might be achieved.

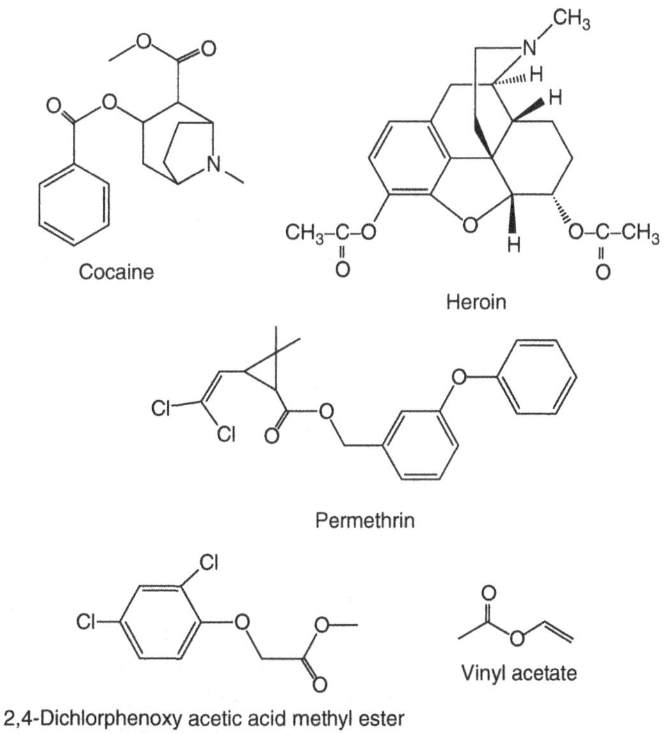

FIGURE 3.24 *Substrates for carboxylesterases.*

Paraoxonase

Paraoxonase-1 (PON1) is an enzyme belonging to a three-member gene family with PON2 and PON3, each member of which is highly conserved in mammalian evolution. Whereas there is consensus that the paraoxonase family members have a general protective influence, their precise biological role has remained elusive. A toxicological role, protecting from environmental poisoning by organophosphate derivatives, drove much of the earlier work on the enzymes. More recently, clinical interest has focused on a protective role in vascular disease via a hypothesized impact on lipoprotein lipid oxidation. Recent confirmation that the primary activity of the paraoxonases is that of a lactonase considerably expands the potential sources of biological substrates for the enzyme. Studies on such substrates may shed further light on different mechanisms by which paraoxonases beneficially influence atherosclerosis, as well as defining possible roles in limiting bacterial infection and in innate immunity.

The PON enzymes are differentially expressed. PON1 and PON3 are expressed in the liver and secreted in the blood where they are associated with high-density lipoprotein (HDL) particles. PON2 is not present in the blood, but is expressed in many tissues, including heart, lungs, liver, and brain.

Comparatively little is known about the regulation of PON expression compared to the other xenobiotic metabolizing enzymes. One study showed that fenofibric acid could increase PON1 transcription by 70% by a mechanism that apparently involved activation of the liver X receptor (LXR). In addition, dietary antioxidants, such as polyphenols, increase PON1 mRNA expression and activity, apparently by an aryl hydrocarbon receptor-dependent mechanism.

Studies of the enzymatic activities of purified PONs have shown them to be lactonases/lactonizing enzymes, with some overlapping substrates (e.g., aromatic lactones), but also to have distinctive substrate specificities. All three PONs metabolize very efficiently the hydrolysis of 5-hydroxy-eicosatetraenoic acid-1,5-lactone and 4-hydroxy-docosahexaenoic acid-1, 4 lactone, which are products of both enzymatic and nonenzymatic oxidation of arachidonic acid and docosahexaenoic acid, respectively, and may be the enzyme's endogenous substrates. Organophosphates are hydrolyzed almost exclusively by PON1, whereas bulky drug substrates such as lovastatin and spironolactone are hydrolyzed only by PON3. Of special interest is the ability of the human PONs, especially PON2, to hydrolyze and thereby inactivate N-acyl-homoserine lactones, which are quorum-sensing signals of pathogenic bacteria.

PHASE II XENOBIOTIC METABOLISM

The function of phase II metabolism is to convert the parent xenobiotic or a functionalized phase I metabolite to a water soluble product that can be excreted in urine or bile. In almost all cases, the phase II metabolites are less toxic than the parent compound or phase I metabolite. The great majority of phase II reactions involves the addition of a polar endogenous molecule such as glucuronic acid, sulfate, or glutathione to the phase I product (Figure 3.9) to produce larger molecular weight products. These products often are called conjugates to reflect the fact that they result from the combination of two precursor molecules. An important distinction between these reactions is that the substrates for the glucuronidation and sulfate reactions are nucleophiles, whereas the substrates for glutathione conjugation are electrophiles. Thus, the combined phase II metabolic capability of an organism can mediate the conversion to water-soluble products of phase I products with a diverse array of chemical substrates.

Glucuronide Conjugation

Glucuronidation is a major pathway of xenobiotic biotransformation in most mammalian species, including humans. Xenobiotic substrates for the reaction include a large number of alcohols, phenols, carboxylic acids, amines, and thiols. This enzymatic reaction is mediated by a group of uridine diphosphate-glucuronyl transferases (UGTs) that are located in the membrane of the endoplasmic reticulum (ER) of liver, intestine, skin, brain, spleen, and nasal mucosa. The cosubstrate for the reaction is uridine diphosphate–glucuronic acid (UDP–GA), which is synthesized in the cytosol from glucose and is actively transported into the ER.

As pointed out in Figure 3.25, the β-glucuronide product that is produced in the reaction is actively transported out of the liver by the organic anion transporter (OAT) into the blood. If the product has a molecular weight of less than about 350 Da, it will be subject to kidney clearance, which is assisted by the multidrug resistance transporter (MDR) or possibly inhibited by the peptide (PEP) and organic cation transporters (OCT). Similar processes can mediate the biliary excretion of xenobiotic phase II metabolites with molecular weight of greater than about 350 Da.

Approximately 15 UGTs have been characterized in humans. Results of recent molecular epidemiology studies suggest that UGT2B7, UGT1A9, and UGT1A7 are important in the metabolism and excretion of certain

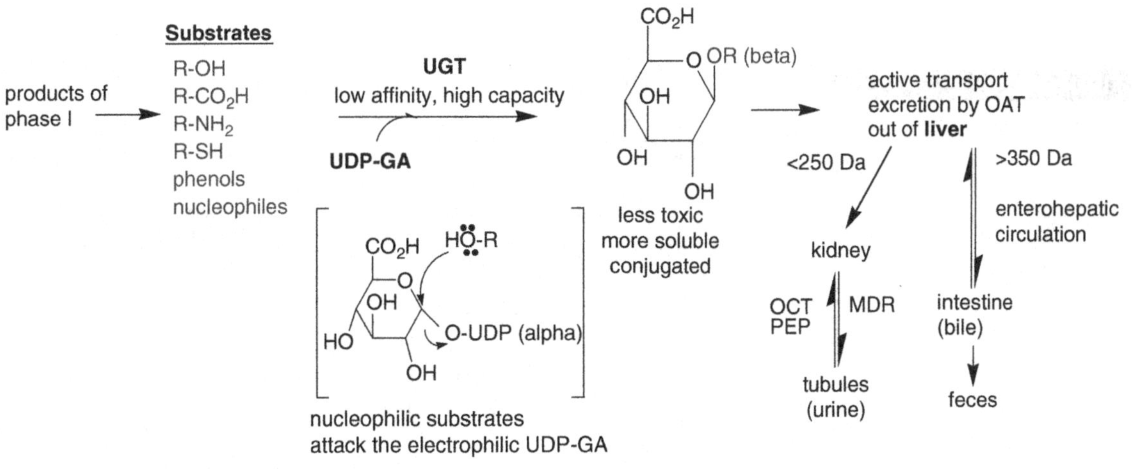

FIGURE 3.25 *Phase II glucuronidation pathway.*

tobacco carcinogens, including 4-(methylnitrosamino)-1-(3-pyridyl)-1-butanone (NNK) and benzo[a]pyrene (BaP), and that deficiencies in the expression or activities of these enzymes is related to increased susceptibility to lung cancer in smokers.

As mentioned previously in Chapter 1, glucuronidation often plays an important role in the process of enterohepatic circulation. Thus, deconjugation of a glucuronide product can be mediated by β-glucuronidase activity in gut bacteria, which frees the xenobiotic to be reabsorbed into the hepatic portal blood for transport to the liver where the glucuronide may be reformed. This process can greatly increase the residence time of the xenobiotic in the body.

Sulfate Conjugation

Many of the xenobiotics and endogenous substrates that undergo O-glucuronidation, including alcohols, amines, phenols, and hydroxylamines, also will undergo sulfate conjugation (Figure 3.26). In contrast to the glucuronidation reactions however, phase II sulfation reactions do not produce stable sulfate products of carboxylic acids. The endogenous cosubstrate for the reaction is the activated sulfate derivative, phospho-adenosine-phosphosulfate (PAPS). The sulfate conjugate products, being salts of strong acids, are less lipid soluble than the glucuronides, and thus are more readily excreted, usually by active OAT in the kidney.

As illustrated in Figure 3.27 for acetaminophen, the sulfotransferases, which are high affinity and low capacity enzymes compared to the UGTs, mediate the conjugation of low concentrations of xenobiotic to the

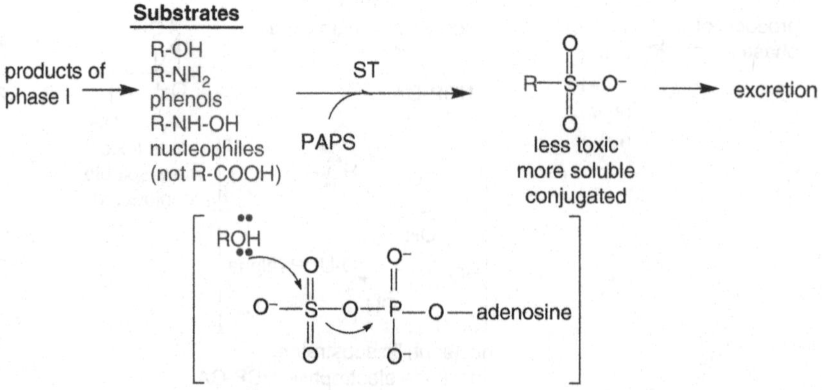

FIGURE 3.26
Phase II sulfation pathway.

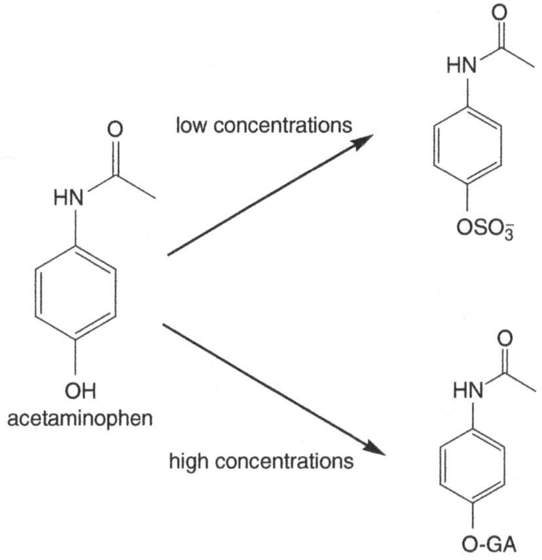

FIGURE 3.27 *Phase II reactions with acetaminophen.*

corresponding sulfate product, whereas the products from UGT-mediated reactions predominate at high substrate concentrations.

As is the case for glucuronide conjugates, many sulfate conjugates that are excreted into the bile may be subject to enterohepatic circulation following hydrolysis by arylsulfatases or arylsulphate sulphotransferases in gut microflora.

There are nine genes encoding cytosolic sulfotransferases in humans, and they belong to the SULT1 or SULT2 gene families. Recent studies have shown that a loss of function mutation for SULT1A1 is associated with a 3.5-fold increase in esophageal cancer risk in males who smoke tobacco and are heavy consumers of ethanol.

Glutathione Conjugation

Substrates for glutathione conjugation include an enormous array of electrophilic xenobiotics or xenobiotic metabolites. Since xenobiotic electrophiles are responsible for the toxic effects of many environmental toxicants, the glutathione conjugation reactions are highly important in the protection of organisms from these toxic agents. Substrates for the glutathione S-transferases (GSTs) share three common features: they (1) are hydrophobic, (2) are electrophilic, and (3) react nonenzymatically with glutathione at a slow but measurable rate. That these reactions are of considerable biological importance is

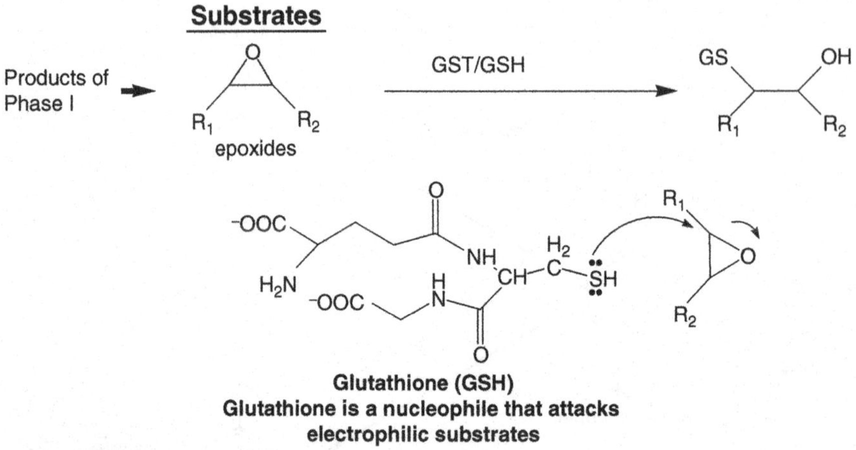

FIGURE 3.28 *Phase II reaction of glutathione with epoxides.*

indicated by the facts that the concentration of glutathione is very high in liver (10 mM), and GSTs make up about 10% of total cellular protein. The general GST-mediated reaction of an epoxide with glutathione is illustrated in Figure 3.28.

GSTs are homodimers composed of relatively small proteins (23–29 KDa) that primarily are expressed as soluble, cytosolic proteins (95%), with a small proportion expressed as membrane bound enzyme in microsomes (5%). GSTA1 is a highly abundant soluble protein that was originally called ligandin because of its considerable binding capacity for many xenobiotics.

Glutathione conjugates can be formed in the liver and can be excreted intact in bile or can be converted to mercapturic acids in the kidney and excreted in the urine as illustrated in Figure 3.29. This conversion to the mercapturic acid increases the efficiency of excretion in the kidney through the OAT active transporter.

Studies of the role of GST polymorphisms in disease indicate that a deficiency of another soluble GST, GSTM1, is associated with increased risks of cancers of the lung, bladder, stomach, colon, and skin, which suggests that this isozyme may mediate the detoxification of certain carcinogens. Studies with carcinogens from tobacco smoke have shown this to be true.

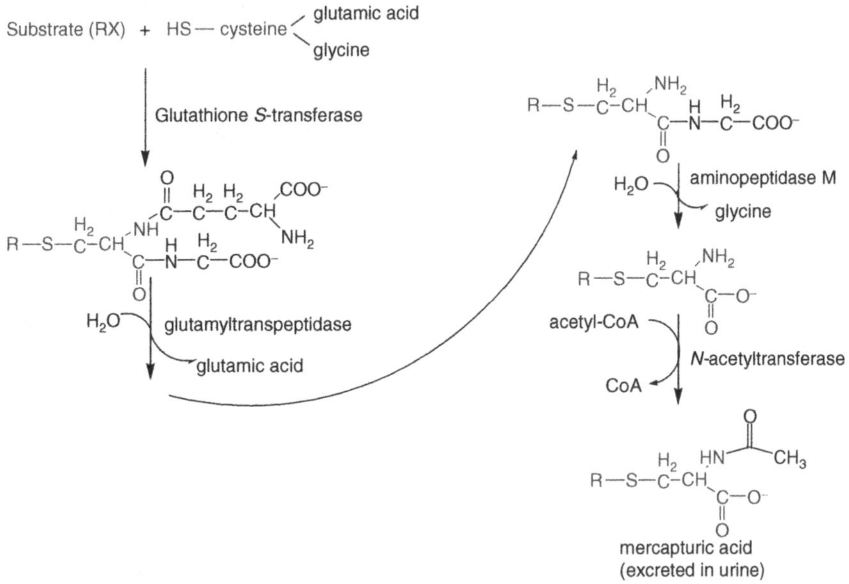

FIGURE 3.29 *Pathway of mercapturic acid formation.*

ADDITIONAL READING

Brown, C.M., Reisfeld, B., Mayeno, A.N. (2008). Cytochromes P450: A structure-based summary of biotransformations using representative substrates. *Drug Metab. Rev.* 40:1-100.

Copple, I.M., Goldring, C.E., Kitteringham, N.R., Park, B.K. (2008). The Nrf2-Keap1 defence pathway: Role in protection against drug-induced toxicity. *Toxicol.* 246:24-33.

Finel, M., Kurkela, M. (2008). The UDP-glucuronosyltransferases as oligomeric enzymes. *Curr. Drug Metab.* 9:70-76.

Furness, S.G., Lees, M.J., Whitelaw, M.L. (2007). The dioxin (aryl hydrocarbon) receptor as a model for adaptive responses of bHLH/PAS transcription factors. *FEBS Lett.* 581:3616-3625.

Jana, S., Paliwal, J. (2007). Molecular mechanisms of cytochrome P450 induction: Potential for drug-drug interactions. *Curr. Protein Pept. Sci.* 8:619-628.

Johnson, W.W. (2008). Cytochrome P450 inactivation by pharmaceuticals and phytochemicals: Therapeutic relevance. *Drug Metab. Rev.* 40:101-147.

Lindsay, J., Wang, L.L., Li, Y., Zhou, S.F. (2008). Structure, function and polymorphism of human cytosolic sulfotransferases. *Curr. Drug Metab.* 9:99-105.

Mitchell, S.C. (2008). Flavin mono-oxygenase (FMO)—The 'other' oxidase. *Curr. Drug Metab.* 9:280-284.

Moskaug, J.O., Carlsen, H., Myhrstad, M.C., Blomhoff, R. (2005). Polyphenols and glutathione synthesis regulation. *Am. J. Clin. Nutr.* 81:277S-283S.

Pool-Zobel, B., Veeriah, S., Böhmer, F.D. (2005). Modulation of xenobiotic metabolising enzymes by anticarcinogens—Focus on glutathione S-transferases and their role as targets of dietary chemoprevention in colorectal carcinogenesis. *Mut. Res.* 591:74-92.

Shimada, T. (2006). Xenobiotic-metabolizing enzymes involved in activation and detoxification of carcinogenic polycyclic aromatic hydrocarbons. *Drug Metab. Pharmacokinet.* 21:257-276.

Testa, B., Krämer, S.D. (2007). The biochemistry of drug metabolism—An introduction: Part 2. Redox reactions and their enzymes. *Chem. Biodivers.* 4:257-405.

CHAPTER 4

Chemical Carcinogenesis

CHAPTER CONTENTS

Cancer is a class of diseases in which cells have been altered to display the characteristics of uncontrolled growth, invasion into other tissues, and the development of new areas of uncontrolled cell growth at distant sites from the original site of cell growth. Most cancers form a tumor but some, like leukemia, do not. Current statistics indicate that cancer of one type or another claims the life of one in every four or five Americans. Cancer is second only to heart disease as a leading cause of death in the United States, and is responsible for close to 500,000 deaths per year. Cancer is recognized as a highly complex, multifactorial disease caused, in part, by endogenous metabolic or other imbalances associated with age or genetic makeup and, in part, by a wide variety of exogenous factors including diet, lifestyle, and exposure to ionizing radiation and chemicals of natural or man-made origin. Chemical carcinogenesis focuses on the role and mechanisms of action of exogenous chemicals in the cause of cancer.

DEFINITIONS

Because of the complex nature of cancer, it is helpful of provide definitions of important terms that are associated with this disease. Thus,

Food Science and Technology

- Cancer is a subset of neoplastic lesions.
- A neoplasm is defined as a heritably altered, relatively autonomous growth of tissue, which can either be benign or malignant.
- Cancers are malignant neoplasia.
- Tumors are space-filling neoplasia. A wart is an example of a benign tumor.
- A carcinogen produces malignant neoplasia in previously untreated animals.
- Carcinogenesis is the process of development of malignant neoplasia.

PHASES OF CARCINOGENESIS

Many decades of research have resulted in the characterization of cancer as a progression of specific stages. Although the number of defined stages is variable for different types of cancer, there is general agreement that there are at least three stages for all cancers: initiation, promotion, and progression. Each stage involves genetic or nongenetic alterations in cell functions and may be induced by different kinds of substances. The fact that a single treatment with a single carcinogen, such as aflatoxin B_1 or benzo[a]pyrene, can lead to metastatic cancer indicates that at a sufficiently high dose, these so-called complete carcinogens can mediate all three stages of carcinogenesis.

Initiation

The first step in carcinogenesis, initiation, generally is caused by a mutation in one or more genes that control key regulatory pathways involved in the proliferation of the cell. Initiation requires one or more rounds of cell division for the "fixation" of the lesion in DNA. The process is irreversible since the effect persists after removal of the causative substance even though the initial initiated cell may eventually die during the development of the neoplasm. Some of the better known cancer initiating agents are shown in Figure 4.1.

Mutational alterations in the function or expression of a large number of genes can lead to cancer initiation. Thus, the overexpression of genes that activate cell proliferation, the so-called proto-oncogenes, can transform normal cells to the carcinogenic state. Protein products in this category include growth factors and their receptors (e.g., IGF and IGFR), and certain kinases

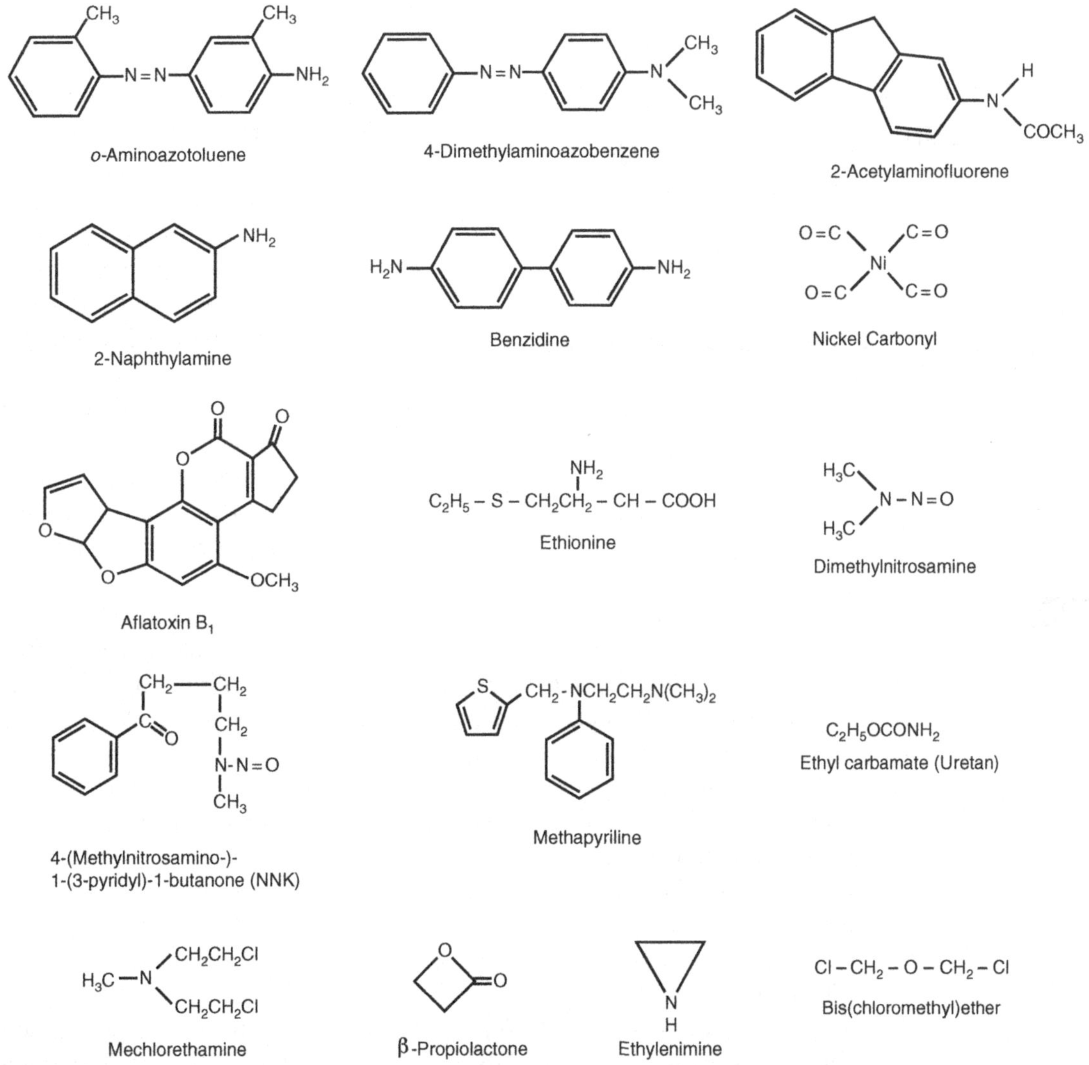

o-Aminoazotoluene

4-Dimethylaminoazobenzene

2-Acetylaminofluorene

2-Naphthylamine

Benzidine

Nickel Carbonyl

Aflatoxin B_1

Ethionine

Dimethylnitrosamine

4-(Methylnitrosamino-)-
1-(3-pyridyl)-1-butanone (NNK)

Methapyriline

$C_2H_5OCONH_2$
Ethyl carbamate (Uretan)

Mechlorethamine

β-Propiolactone

Ethylenimine

Cl – CH_2 – O – CH_2 – Cl
Bis(chloromethyl)ether

FIGURE 4.1 *Examples of cancer initiating agents.*

that regulate cell proliferation (e.g., Ras). Conversely, inactivation of genes, the protein products of which normally inhibit cell proliferation (e.g., p53), the so-called tumor suppressors, also can lead to abnormal cell proliferation. In addition, the inactivation of genes involved in cells' defenses against cell damage, including the genes that mediate DNA repair and the antioxidant defense, increase the cells' sensitivity to initiating agents.

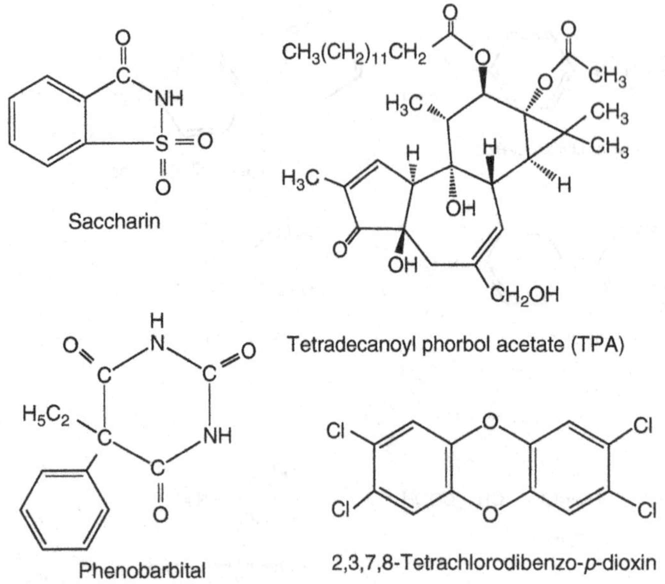

Saccharin

Tetradecanoyl phorbol acetate (TPA)

Phenobarbital

2,3,7,8-Tetrachlorodibenzo-*p*-dioxin

FIGURE 4.2 *Examples of cancer promoting agents.*

Promotion

Promotion, the next distinct step in carcinogenesis, is the induction of selective growth enhancement in the initiated cell and its progeny by the continuous exposure to a promoting agent. This is an epigenetic event since it does not involve further changes in DNA and is thus reversible because the removal of the promoting agent results in a decrease in cell proliferation. Nevertheless, the lag time between initiation and exposure to the promoting agent may be very long, up to years after exposure to the initiating agent, in the development of a cancer.

Cancer promoters represent a large group of mitogenic agents from many structural classes that can induce cell proliferation by many and sometimes as yet poorly understood mechanisms. Some of the major classes of such substances include high doses of redox active synthetic substances such as BHA and BHT, which can induce the formation of increased levels of cellular reactive oxygen species under certain conditions; biological irritants such as the synthetic sweetener, saccharin, and the plant derived phorbol esters; certain activators of the Ah receptor such as dioxin; peroxisome proliferators such as oxidized fats and certain drugs; and various endocrine active substances such as estradiol and diethylstilbestrol (Figure 4.2).

The hepatitis B virus (HBV) is a potent tumor promoter that functions through the expression of the HBx protein. This protein can alter the expression of several genes involved in the regulation of cell proliferation and apoptosis, including p53, AP-1, NF-κB, c-myc, as well as in DNA repair. The result of HBV infection is impaired apoptosis, genetic instability, and defective cell cycle regulation.

Progression

Progression is an irreversible phase of carcinogenesis that results in further degrees of independence, invasiveness, and metastasis, and results from the continuing mutagenic evolution of a basically unstable karyotype of the tumor cells. The developing tumor begins to recruit inflammatory immune cells and acquires "wound-healing" characteristics, such as the secretion of chemo-attractants for other cell types, angiogenesis factors and proteases that facilitate the spread to new tissue and independent growth of the tumor. Metastatic cells that enter the bloodstream must be coated by macrophages and fibroblasts to survive. Examples of progressor agents include asbestos fibers, benzene, benzoyl peroxide, other peroxides, oxidative stress, and inflammation, all of which are weak mutagens in normal cells but exhibit significant genotoxicity in the initiated and rapidly growing tumor cells. Although in principle, cancer initiation may result from the change in expression of only a single gene, recent studies have shown that fully metastatic tumor cells have up to 100 mutated genes.

Angiogenesis

Angiogenesis in normal tissues is a carefully controlled process of vascular development to meet the needs to the growing or healing tissues. Insufficient angiogenesis can lead to stroke, heart disease, infertility, ulcers, and a thickening of the skin (scleroderma). Excessive angiogenesis can result in blindness, psoriasis, rheumatoid arthritis, AIDS complications, and metastatic cancer. Indeed, tumor angiogenesis plays a central role in both primary tumor growth and in metastasis. The role of angiogenesis in tumor growth and metastasis is illustrated schematically in Figure 4.3.

Growth of a tumor beyond 2 to 3 mm^3 requires development of a microvessel network to facilitate delivery of nutrients and oxygen to the tumor. The density of the microvasculature often is used as an indicator of biological aggressiveness and metastatic potential in many primary tumors because neovascularization facilitates metastasis by allowing access of cancer cells to the circulation. The abilities of primary breast, prostate, and colorectal carcinomas to metastasize to the lymph nodes have been correlated directly to the degree of angiogenesis within the primary tumor. Drugs that inhibit angiogenesis have been shown to successfully inhibit tumor growth both in rodent models and in human patients. For example, systemic administration of angiostatin causes regression of primary murine tumors of the mammary gland, colon, and prostate. Recombinant endostatin inhibits angiogenesis and the growth of primary and metastatic tumors in mice. Interferon-α treatment causes regression of human hemangiomatous disease. A variety of phytochemicals, such as those found in soybeans (isoflavones and saponins), red wine and grapes (resveratrol), green tea (catechins), and *Brassica* vegetables (3,3′-diindolymethane, DIM) have been shown to suppress tumor growth by mechanisms that at least in part involve the inhibition of tumor angiogenesis.

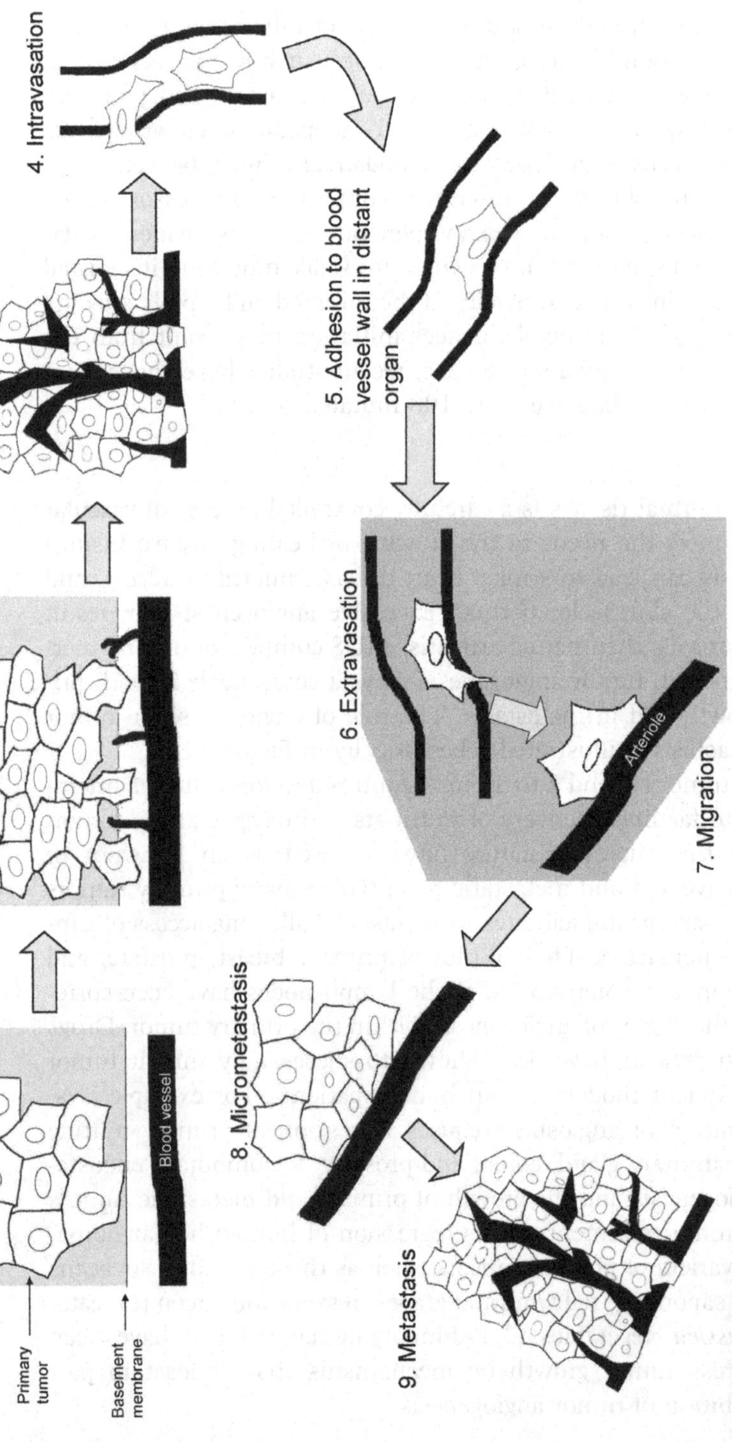

FIGURE 4.3 *The role of angiogenesis in tumor growth and metastasis.*

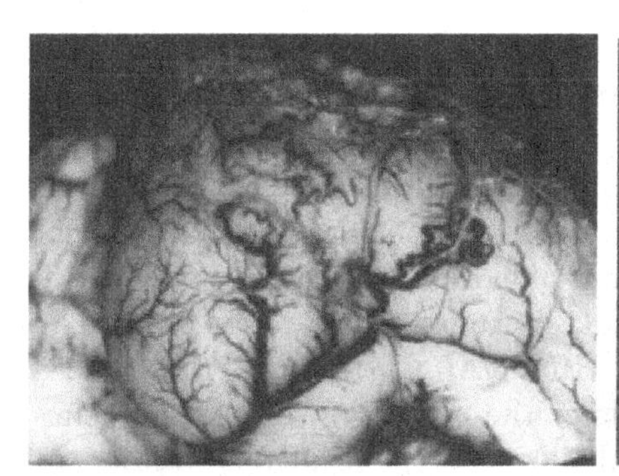
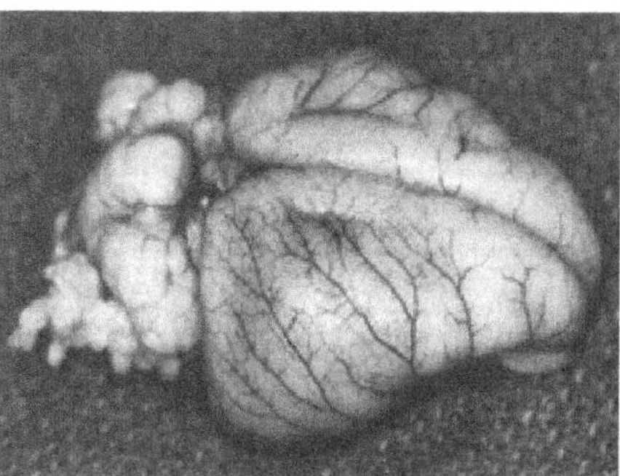

FIGURE 4.4 *Normal vasculature (on right) is very orderly, unbranched, nearly parallel vessels compared with tumor vasculature (on left).*

The structures of tumor vessels differ from those of normal tissues. Differences in cell signaling and composition, tissue integrity, and vascular permeability result in an often grotesquely shaped, fragile, and leaky circulatory network in the tumor (Figures 4.4 and 4.5). During tumor progression, the neovasculature displays a high degree of plasticity to meet the growing tumor's growth requirements. Endothelial and other stromal cells are actively recruited into tumors by factors that are secreted by the developing tumor.

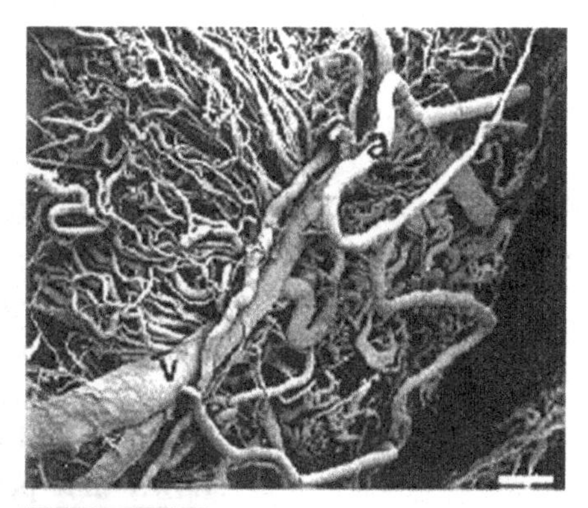

FIGURE 4.5 *Tumor vasculature. Note coiling, irregularity, and size heterogeneity of vessels.*

Angiogenesis is regulated by a large number of pro-angiogenic and anti-angiogenic factors. The net balance between these two classes of factors determines whether angiogenesis will occur. Increased production of positive angiogenic factors (activators) is necessary but not sufficient for induction of neovascularization. Angiogenesis also requires suppression of negative regulators (inhibitors). For example, transcripts of the established angiogenesis inhibitors, endostatin, and PF-4, as well as TSP2, were upregulated in the tumor specimens from mice treated orally with the *Brassica* phytochemical, DIM. TSP2 is a member of the thrombospondin family of secreted proteins, many of which are endogenous angiogenesis inhibitors that localize to the extracellular matrix and inhibit matrix remodeling proteases and growth factors. Through inhibition of proteases, particularly matrix metalloproteinase-2 (MMP-2), TSP2 prevents angiogenesis by regulating cellular adhesion and inhibiting endothelial cell growth and proliferation. Concurrent with the increase in TSP2 expression (expression of an important protease) plasminogen activator urokinase, which mediates breakdown of the extracellular matrix and facilitates angiogenesis, also was significantly downregulated by DIM. Expression of ephrin B2, a transmembrane ligand that activates Eph receptors on endothelial cells, also was reduced by DIM treatment.

The development of hypoxic conditions at the core of tumors reaching a critical size of a few millimeters in diameter is considered to be the initial stimulus that triggers tumor angiogenesis. The hypoxia-induced factor-1α (HIF-1α) accumulates rapidly in tumor cells exposed to hypoxic conditions and heterodimerizes with HIF-1β/ARNT to form HIF-1. HIF-1 is a transcription factor that regulates the expression of over 60 genes, including genes that encode several angiogenesis regulatory factors and enzymes involved in energy metabolism. Important angiogenic activators that are expressed in hypoxic tumor cells include vascular endothelial growth factor (VEGF), fibroblast growth factor (FGF), and placental growth factor (PlGF). Angiogenesis inhibitors that are downregulated in hypoxic tumor cells include pigment epithelium-derived factor (PEDF), transforming growth factor-β (TGF-β), endostatin, angiostatin, and thrombospondin-1.

HIF-1 has become an attractive target for the development of anticancer drugs. HIF-1α is extremely short-lived under normoxic conditions with a half-life of less than five minutes. Certain prolyl residues of HIF-1α are hydroxylated by specific HIF-prolyl hydroxylases (PHD1–3) to facilitate an interaction with von Hippel-Lindau tumor suppressor protein (pVHL), which leads to ubiquitination followed by proteosomal degradation. Under hypoxic conditions, the level of HIF-1α protein increases due to a decreased rate of ubiquitin-targeted degradation. Studies in many laboratories have shown that potential therapeutic agents can disrupt the HIF-1 signaling pathway through a variety of mechanisms, including the inhibition of HIF-1α protein synthesis, nuclear translocation, transactivation of target genes, or activation of HIF-1α protein degradation under hypoxic conditions. Destabilization of HIF-1α protein appears to be the major mechanism by which certain phytochemicals inhibit HIF-1 activity, since inhibition of prolyl hydroxylases or proteosomal degradation can reverse the inhibitory effects of these substances on HIF-1α accumulation.

CANCER EPIDEMIOLOGY

Epidemiological data on geographical and temporal variations in cancer incidence, as well as studies of migrant populations and their descendants who acquire the pattern of cancer risk of their new country, indicate that environmental exposures make a substantial contribution to human cancers. These studies, along with investigations of lifestyle factors and habits, led to the conclusion that as much as 80%, by some estimates, of cancer deaths in Western industrial countries can be attributed to factors such as tobacco, alcohol, diet, infections and occupational exposures, with diet (35%) and tobacco (30%) as the major contributors. Recent estimates suggest that about 75% of cancer deaths of smokers or 50% of cancer deaths of nonsmokers in the United States could be avoided by elimination of these risk factors. Recent international studies show a similar or higher involvement of environmental and lifestyle factors in the cause of major cancers.

Data presented in Table 4.1 lists gender differences in the death rates for the major cancers in the United States. Note that deaths from lung cancer, which is largely preventable by cessation of smoking, claim the highest proportion of deaths from all cancers for both men and women (33% and 24%). Deaths from prostate and breast cancers are second in importance, respectively, for men and women (14% and 18%), followed by colorectal cancers (10–11%). Cancers of the liver (3%), brain (2%), esophagus, and stomach (3–6%) are relatively minor for both genders, whereas the proportions of leukemias and lymphomas (7–8%) are similar to proportions of colorectal cancers.

International and generational data on incidence and mortality of breast and prostate cancer are presented in Tables 4.2 and 4.3. Note in Table 4.2 that the lowest rates of breast cancer occur in Asian women living in Asia

Table 4.1 Gender Differences (%) in the Death Rates for the Major Cancers in the United States

	Death Rate (%)	
Cancer Type	**Males**	**Females**
Brain	2	2
Esophagus and stomach	6	3
Lung	33	24
Liver	3	3
Leukemia and lymphomas	8	7
Pancreas	4	5
Colon and rectum	10	11
Urinary	4	1
Prostate	14	—
All others (e.g., oral, skin, and bladder)	16	17
Breast	—	18
Ovary	—	6
Uterus and cervix	—	4

Table 4.2 Breast Cancer Cases and Acculturation[a]

Group and Place	Cases	Rate[b]
Chinese		
Shanghai, China	6084	26.5
Singapore, China	2187	39.5
Hong Kong, Japan	5392	34.0
Los Angeles, USA	266	36.8
San Francisco, USA	459	55.2
Hawaii, USA	159	57.6
Japanese		
Osaka, Japan	7544	24.3
Miyagi, Japan	2440	31.1
Los Angeles, USA	319	63.0
San Francisco, USA	138	68.4
Hawaii, USA	903	72.9

[a] *Source: Parkin et al. (1997).*
[b] *Per 100,000 woman-years, age-adjusted using the world standard.*

Table 4.3 International Prostate Cancer Rates (1990)

	Incidence	Mortality
Europe		
Denmark	30	25
Spain	25	12
United Kingdom	25	12
Slovakia/Poland	25	10
Americas		
Canada	70	
US (W)	100	15
US (E)	175	30
Columbia	25	
Costa Rica		20
Cuba		20
Asia		
China	2.5	
India	5	
Japan	10	5
Singapore	7	5
Oceania		
Australia	50	20
New Zealand	40	20

(26.5–38.5) and that the rates strongly increase with acculturation in the United States (55.2–72.9).

The data presented in Table 4.3 on international prostate cancer incidence and mortality show a similar effect of acculturation on disease incidence. Thus, the incidences of prostate cancer are much lower in Asian men (2.5–10) than in men from Europe, Oceania, and the United States (25–175). The extraordinarily high incidences of prostate cancer in Caucasian and African American US men, 100 and 175, respectively, are noteworthy and of considerable concern, but remain unexplained. Fortunately, the mortality rates for men in the United States are similar to mortality rates from this disease of men in Europe, which has a much lower prostate cancer incidence rate than the United States.

DIETARY GUIDELINES FOR CANCER PREVENTION

In an effort to provide recommendations based on the most current and reliable scientific evidence to reduce chronic diseases, the US Department of Agriculture (USDA) and the Department of Health and Human Services (DHHS) have jointly published the "Dietary Guidelines for Americans" since 1980. The Dietary Guidelines, which provide advice for healthy Americans over 2 years of age about food choices that promote health and prevent disease, serve as the basis for federal nutrition policy and nutrition education activities. The Dietary Guidelines are reviewed and updated every five years by a Dietary Guidelines Advisory Committee of experts in nutrition and health. The recommendations are based on the most current peer-reviewed scientific studies, which initially was heavily dependent on studies with experimental animals. More recent recommendations depend more heavily on the growing body of clinical and epidemiological studies with humans.

The National Cancer Institute (NCI) has long considered that the potential for dietary changes to reduce the risk of cancer is considerable and that the existing scientific data provide evidence that is sufficiently consistent to warrant prudent interim dietary guidelines that will promote good health and reduce the risk of some types of cancer. The six NCI Dietary Guidelines are consistent with other dietary recommendations from the USDA and DHHS, the American Cancer Society, and the American Heart Association:

Reduce fat intake to 30 percent of calories or less. Studies with both experimental animals and with humans indicate that high intake of fat in the diet is associated with increased tumor growth and mortality. Dietary fat can function as a tumor promoter.

Increase fiber intake to 20 to 30 grams/day with an upper limit of 35 grams. Studies with experimental animals and with humans show that increased dietary fiber is associated with decreased cancer of the colon, which apparently is a result of decreased residence time of carcinogens in the gut.

Include a variety of fruits, vegetables, beans and whole grains in the daily diet. According to information from the American Cancer Society, a large body of scientific work has demonstrated that dietary habits affect cancers at many sites. Diets high in fruits and vegetables have protective effects against cancers of the lung, mouth, esophagus, stomach, and colon.

Avoid obesity. According to information available from the NCI, obesity and physical inactivity may account for as much as 25 to 30% of cancers of the colon, breast, endometrium, kidney, and esophagus. Some studies have also reported links between obesity and cancers of the gallbladder, ovaries, and pancreas.

Consume alcoholic beverages in moderation, if at all. According to recent information available from the American Cancer Society, many research studies have established a relationship between alcohol consumption and cancer. Men who consume two alcoholic drinks a day and women who have one alcoholic drink a day have increased chances of developing cancers of the mouth, esophagus, larynx, breast, and liver.

Minimize consumption of salt-cured, salt-pickled, and smoked foods. High-level consumption of sodium chloride can damage cells lining the gastrointestinal tract and can function as a cancer promoter. Smoked foods often contain carcinogens from the smoke and are often heavily salted. Populations that consume diets that are rich in salt-cured and smoked food have higher incidences of cancers of the stomach.

Although each of these recommendations is supported by considerable experimental and epidemiologic information, published attempts are rare to substantiate the overall effectiveness of compliance with all the recommendations in reducing cancer incidence. One recent study of this type concluded simply that adherence to the nutrition-related behaviors included in the dietary guidelines for Americans *may* be associated with a lower risk of cancer. Consistent with findings for all cancers, the incidences of cancers of the colon, bronchus and lung, breast, and uterus were found to be significantly lower with greater compliance with the dietary guidelines. Relative risks were suggestive of an inverse association between compliance with the guidelines and cancers of the upper digestive tract, rectum, lymph, and hematopoietic system. In contrast, however, the same study found that the incidence of ovarian cancer *increased* with increased compliance with the dietary guidelines. The authors suggest that this surprising latter positive association appears to be attributable to the positive influence on ovarian cancer incidence of physical activity, which has been reported in other studies.

It is important to point out, however, that much of the epidemiologic data on which the dietary guidelines are based recently has been called into question. Most of the human studies depended on retrospective case-control studies, in which the recalled diets of individuals with a specific cancer (i.e., the cases) were compared to the recalled diets of healthy individuals of similar age and circumstances (i.e., the controls). The results of several prospective, so-called cohort studies do not show a general protective effect for all cancers of diets rich in plant-based foods. As these studies have become increasingly sophisticated and specific for certain types of cancers and for specific food plants, however, protective effects are being documented; for example, for vegetable consumption and benign prostate hyperplasia, and for fruit consumption and colorectal cancer. Furthermore, that diet is a source of both carcinogens and anticarcinogens is becoming more fully appreciated, leading some investigators to consider that it is exposure to a proper ratio of cancer inhibitors from vegetables to the cancer initiators, for example in meat, that is responsible for cancer prevention. This is an area of great importance for the maintenance and improvement of human health and will be the subject of intense scientific scrutiny well into the future.

ADDITIONAL READING

Butrum, R.R., Clifford, C.K., Lanza, E. (1988). NCI dietary guidelines: Rationale. *Am. J. Clin. Nutr.* 48:888-895.

Dass, C.R., Choong, P.F. (2008). Cancer angiogenesis: Targeting the heel of Achilles. *J. Drug Target* 16:449-454.

Heber, D., Blackburn, G.L., Go, V.L.W., Milner, J. Eds. (2006). "Nutritional Oncology," 2nd ed. Academic Press.

Michels, K.B., Giovannucci, E., Chan, A.T., Singhania, R., Fuchs, C.S., Willett, W.C. (2006). Fruit and vegetable consumption and colorectal adenomas in the Nurses' Health Study. *Cancer Res.* 66:3942-3953.

Parkin, D.M., Whelan, S.L., Ferlay, J. (1997). Cancer Incidence in Five Continents, Vol. VII, Scientific Publication No. 143. International Agency for Research on Cancer, Lyon.

Rohrmann, S., Giovannucci, E., Willett, W.C., Platz, E.A. (2007). Fruit and vegetable consumption, intake of micronutrients, and benign prostatic hyperplasia in US men. *Am. J. Clin. Nutr.* 85:523-529.

Shannon, A.M., Bouchier-Hayes, D.J., Condron, C.M., Toomey, D. (2003). Tumour hypoxia, chemotherapeutic resistance and hypoxia-related therapies. *Cancer Treat. Rev.* 29:297-307.

CHAPTER 5

Natural Toxins in Animal Foodstuffs

CHAPTER CONTENTS

NATURAL TOXINS IN LAND ANIMAL FOODSTUFFS

Since humans first evolved, the search for food has been one of their primary activities. It is hypothesized that people were originally vegetarians and only gradually adapted to foods from animal sources. In particular, mastery of the use of fire, which occurred about 75,000 years ago, dramatically increased the variety of foods. Over time, people learned that heat treatment detoxified certain poisonous foods. As more and better ways of processing foodstuffs have been learned, the variety of foods available for human consumption has increased.

The animal liver, a large glandular organ, is a nutritious protein-rich food in which important enzymes are concentrated. Beef, calf, pork,

Food Science and Technology

chicken, and lamb livers are commonly sold in Western markets. Toxins occurring in the livers of livestock include bile acids, and excessive levels of essential nutrients (Vitamin A).

Bile Acids

Some livers—bear, beef, sheep, goat, and rabbit—produce toxic acids called bile acid. Bile acids are produced in the liver by the oxidation of cholesterol. In dried bear liver, the bile acid acts as a suppressant (both as a tranquilizer and as a pain killer) on the central nervous system. The Chinese and Japanese discovered this long ago and used dried bear livers in folk medicines over the centuries. Major bile acids are cholic acid, deoxycholic acid, and taurocholic acid, which are shown in Figure 5.1.

Vitamin A

Vitamin A is an important nutrient found in plants and animals. Table 5.1 shows typical food sources of vitamin A. Vitamin A is necessary for normal growth and vision in animals. A deficiency of vitamin A can lead to night blindness, failure of normal bone growth in the young, and diseases of the membranes of the nose, throat, and eyes. Vitamin A is a fat-soluble, light yellow substance that crystallizes as needles (MP 63 °C).

β-Carotene, which breaks down into two molecules of vitamin A in the intestinal mucosa (Figure 5.2), is manufactured in plants and stored in the liver and fat of animals as the palmitate ester. Therefore, it is referred to as provitamin A. Vitamin A is a generic term for a large number of related compounds. Retinol (an alcohol) and retinal (an aldehyde) often are referred to as preformed vitamin A.

It is difficult to define the toxicity of excessive levels of essential nutrients because any food can cause a toxic reaction if it is ingested in excessive amounts. Vitamin A, therefore, is not classified as toxic material even though excessive consumption results in a toxic response.

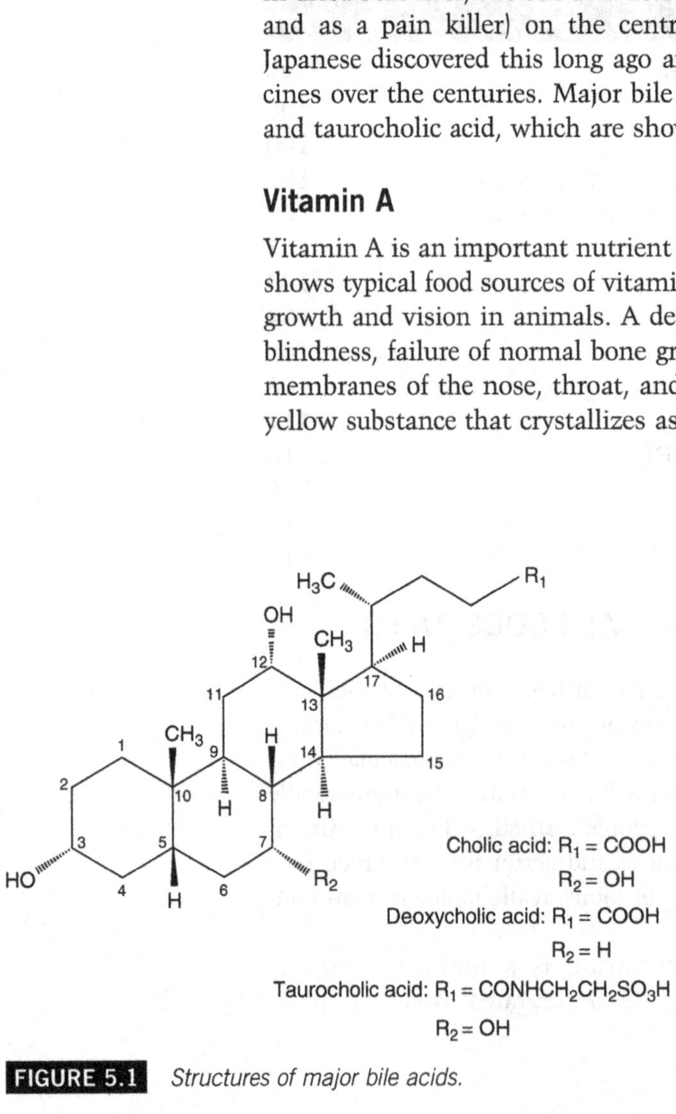

Cholic acid: R_1 = COOH, R_2 = OH
Deoxycholic acid: R_1 = COOH, R_2 = H
Taurocholic acid: R_1 = $CONHCH_2CH_2SO_3H$, R_2 = OH

FIGURE 5.1 *Structures of major bile acids.*

Table 5.1 Typical Food Sources of Vitamin A

Food	Serving	Amount of Vitamin A µg (IU)
Egg	1 large	91 (303)
Butter	1 tablespoon	97 (323)
Whole milk	1 cup (8 oz)	68 (227)
Sweet potato	½ cup mashed	555 (1,848)
Pumpkin	½ cup	953 (3,177)
Carrot	½ cup chopped	538 (1,793)
Cantaloupe	½ medium	467 (1,555)
Spinach	½ cup	472 (1,572)
Broccoli	½ cup	60 (200)
Kale	½ cup	443 (1,475)

Acute Toxicity of Vitamin A

Generally, vitamin A is toxic to humans at the level of 2–5 million IU (International Unit is a standardized measure of vitamin's biological activity). One IU corresponds to 0.3 µg of pure crystalline vitamin A (retinol) or 0.6 µg of β-carotene. The condition caused by vitamin A toxicity is called hypervitaminosis A. Polar bear liver is very rich in vitamin A and at least one case of acute intoxication due to the consumption of the vitamin A-rich liver was reported among Arctic explorers and their dogs. The ingestion of the liver of polar bears produced a painful swelling under the skin. In other reported instances, ingestion of polar bear livers caused joint pain and irritation, dry lips, lip bleeding, and ultimately death. There is also a report that a fisherman who ingested nearly 30 million IU of vitamin A in halibut liver, which contained up to 100,000 IU/g of vitamin A, complained of severe headache centered in the forehead and eyes, dizziness, drowsiness, nausea, and vomiting in 12 to 20 hours followed by redness and erythematous swelling and peeling of the skin.

Table 5.2 shows the vitamin A content in the livers of various animals. Ingestion of approximately 111 to 278 g of polar bear liver causes acute vitamin A toxicity.

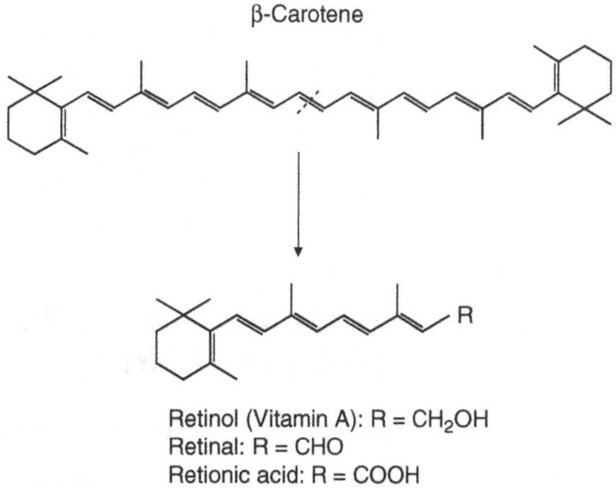

β-Carotene

Retinol (Vitamin A): R = CH_2OH
Retinal: R = CHO
Retionic acid: R = COOH

FIGURE 5.2 *Formation of vitamin A from β-carotene.*

Table 5.2 Animal Livers Containing High Levels of Vitamin A

	Content (IU/100 g fresh)
Animal	
Polar bear	1.8 million
Seal	1.3 million
Sheep or cattle	4,000–45,000
Food	
Butter*	2,400–4,000

*Used as comparison.

Table 5.3 Tolerable Upper Level of Intake for Preformed Vitamin A

Age Group	Amount (IU)/day
Infants–Children (1 ~ 3 years)	2,000
Children (4–8 years)	3,000
Children (9–13 years)	5,677
Adolescents (14–18 years)	9,333
Adults (19 years and older)	10,000

Chronic Toxicity of Vitamin A

Chronic vitamin A toxicity is induced by 1000 μg (approximately 3000 IU)/kg body weight/day. Values vary from source to source but this is the value that commonly is recognized. Table 5.3 shows the tolerable upper level of intake for preformed vitamin A. Signs of chronic toxicity include dry itchy skin, desquamation, loss of appetite, headache, cerebral edema, and bone and joint pain. Severe cases of hypervitaminosis A may result in liver damage, hemorrhage, and coma. Generally, signs of toxicity are associated with long-term consumption of vitamin A (25,000–33,000 IU/day).

TRANSMISSIBLE SPONGIFORM ENCEPHALOPATHIES (TSEs) AND PRIONS

Discovery of Bovine Spongiform Encephalopathy (BSE)

BSE is a relatively new disease found in cattle. BSE is included among transmissible spongiform encephalopathies (TSEs), which are a family of diseases of humans and animals characterized by spongy degeneration of the brain with severe and fatal neurological signs. Symptoms include change in mental state and abnormalities in posture, movement, and sensation. The clinical disease usually lasts for several weeks and it is invariably progressive and fatal. BSE was found to be a transmissible neurodegenerative fatal brain disease, found first in cattle but transmissible to humans with an incubation period of 4 to 5 years. It was first recognized and defined in the United Kingdom in November 1986. BSE reached its peak in 1992, when 36,680 cases were confirmed. BSE occurs in adult animals of both sexes, typically in animals aged 5 years or more.

Discovery of Toxic Principle, Prion

In the 1960s, another type of TSE was reported to be caused by an infectious agent made solely of protein. This theory had been developed to explain the mysterious infectious agent causing the diseases scrapie and Creutzfeldt-Jakob Disease (CJD), which resisted ultraviolet radiation, yet responded to agents that disrupt proteins. This toxic protein was named as "prion" by combining the first two syllables of the words *proteinaceous* and *infectious*. Table 5.4 shows diseases believed to be caused by prion. All known prion diseases, collectively called TSEs, are untreatable and fatal. A vaccine has been developed in mice that may provide insight into providing a vaccine in humans to resist prion infections.

Table 5.4 Diseases Believed To Be Caused by Prions

In humans

- Creutzfeldt-Jakob disease (CJD) and its varieties
- Gerstmann-Sträussler-Scheinker syndrome (GSS)
- Fatal familial insomnia (FFI)
- Sporadic fatal insomnia (SFI)
- Kuru
- Alpers syndrome

In animals

- Scrapie in sheep and goats
- Bovine spongiform encephalopathy (BSE) in cattle
- Transmissible mink encephalopathy (TME) in mink
- Chronic wasting disease (CWD) in elk and mule deer
- Feline spongiform encephalopathy (FSE) in cats
- Exotic ungulate encephalopathy (EUE) in nyala, oryx, and greater kudu
- Spongiform encephalopathy of the ostrich

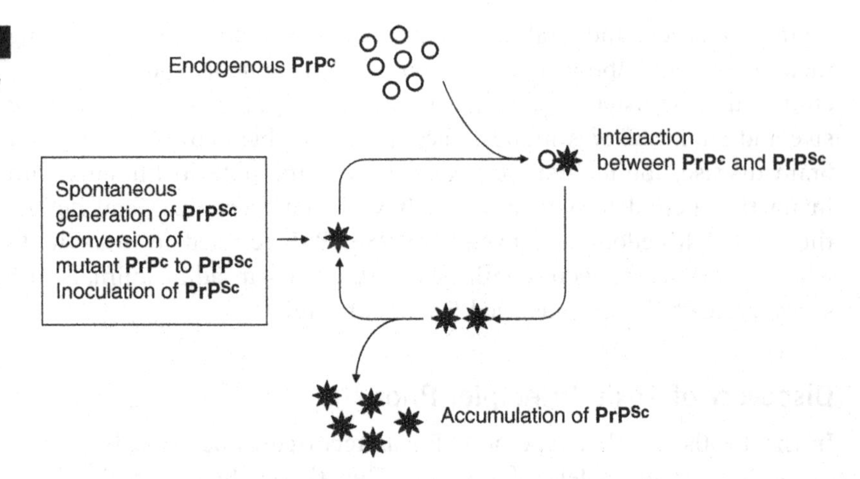

FIGURE 5.3

Proposed mechanism of prion propagation.

Mode of Action of Prion

Prion protein is found throughout the body, even in healthy people and animals. The prion protein found in infectious material has a different structure and is resistant to proteases, enzymes in the body that can normally break down proteins. Prions cause neurodegenerative disease by aggregating extracellularly within the central nervous system to form plaques known as amyloids, which disrupt the normal tissue structure. This disruption is characterized by "holes" in the tissue with resultant spongy architecture due to the vacuole formation in the neurons. Figure 5.3 shows the proposed mechanism of prion propagation.

The normal form of the protein is called PrP^{c} (structurally well-defined), and the infectious form is called PrP^{Sc} (polydisperse and defined at a relatively poor level)—the *C* refers to "cellular" or "common" PrP, and *Sc* refers to "scrapie," a prion disease occurring in sheep. The infectious isoform PrP^{Sc} is able to catalyze the formation of other normal PrP^{c} proteins into the infectious isoforms by changing their conformation and then the infection progresses.

NATURAL TOXINS IN MARINE FOODSTUFFS

Fish are important sources of both food and income for maritime countries. On a global scale, 19% of animal protein for human consumption is derived from fish, and more than one billion people rely on fish as an important source of animal protein, with some small island countries depending on fish almost exclusively. The fishery industry employs nearly 200 million people, with international trade of fisheries products reaching over $50,000,000,000 per year.

The recent rise in Western per capita demand for fish seems to be associated with the recognition of fish as a health food. On the eve of WWII, yearly fish consumption in the Western world rarely exceeded 10 kg per capita. In the recent decades, however, yearly fish consumption in the West has risen to the present level of about 22 kg per capita.

Many fisheries are in crisis, however. The global catch of fish is declining by about 500,000 tons per year from a peak in the 1980s. Furthermore, the UN FAO reports that 75% of the world's commercially important marine fish stocks are either fully fished, overexploited, depleted, or slowly recovering from a collapse. The major cause of this decline is overfishing by humans.

In addition to this severe stress placed on major fisheries, recent decades have been marked by an extraordinary expansion in the nature and extent of harmful marine algal blooms. Some blooms are associated with potent toxins in the causative algae, while others cause problems to other organisms simply because of high algal biomass. Some are of concern at exceedingly low cell densities. Adverse and sometimes devastating effects of some of these algal blooms have been well documented as toxic effects on humans and marine mammals and high mortalities in farmed fish and shellfish species.

Although there are many species of toxic and poisonous marine organisms, this chapter is limited to a discussion of the major natural toxins that affect the marine food supply. Many of these toxicities have been long known to originate from toxins produced by alga that contaminate marine fish or shellfish. We have come to realize in recent years, however, that all the toxicities to be discussed result from contamination of the marine food organism with alga or bacteria. In the case of one of the oldest known toxicities—puffer fish poisoning—toxicity recently has been shown to result from a symbiotic association of the fish with a highly toxic marine bacterium.

Tetrodotoxin—Puffer Fish Poison

Tetraodontidae is a family of primarily marine and estuarine fish. The family includes many familiar species that are variously called puffers, balloonfish, blowfish, bubblefish, globefish, swellfish, toadfish, and toadies. As their various names indicate, these fish have in common the ability to inflate their bodies with water when they sense predatory pressure. In contrast to their cousins, the porcupine fishes, which have large conspicuous spines, the tetradons have small, almost sandpaper-like spines. The family is named for the four large teeth, fused into an upper and lower plate. Puffer fish are the second most poisonous animal in the world (the first is the Golden Poison Frog). The skin, ovaries, eggs, and liver of many *Tetraodontidae* are highly toxic to humans, but the muscle meat is considered a delicacy that is prepared by specially trained and licensed chefs in Japan, Korea, and elsewhere (Figures 5.4, 5.5).

The toxic agent tetrodotoxin, however, is widely distributed in nature, in both marine and terrestrial animals in four different phyla, which include species as diverse as frogs, newts, starfish, crabs, octopus, and marine snails. It has been reported to occur in about 40 species of puffer fish, the most important food source of the toxin. The high toxicities of puffer fish have been known for thousands of years and were reported in Egypt and China as early as 2000–3000 BCE. Between 1888 and 1909 there were 2090 deaths reported in the world, mostly in Japan, which is about 67% mortality in the total poisonings. From 1956 to 1958, of the 715 reported cases, 420 people died. Finally, from 1974 to 1979, there were 60 cases reported with 20 deaths. More recent statistics from Japan indicate a continued high (60%) mortality rate with an incidence rate of about 20 deaths each year.

Following consumption of toxic puffer fish, symptoms usually appear very rapidly between 5 to 20 minutes after ingestion, depending on the dose of toxin. Although each individual may experience symptoms differently, the following are the most common symptoms of puffer fish poisoning.

FIGURE 5.4

Takifugu puffer fish (Takifugu rubripes) *preferred in Japanese cuisine.*

FIGURE 5.5
Fugu sashimi prepared in paper-thin slices for an elegant meal.

Symptoms usually include:

- Initial numbness, tingling of the lips, tongue, and inside of the mouth
- Gastrointestinal disturbances, including vomiting, abdominal pain, and diarrhea
- Difficulty walking and extensive muscle weakness
- General weakness, followed by paralysis of the limbs and chest muscles (cognition is usually clear during these symptoms)
- Drop in blood pressure and a rapid and weak pulse
- Death, which can occur within 30 minutes

With severe cases there is gradual onset of respiratory distress, and death from respiratory failure occurs within 6 hours of exposure. Convulsions can occur after up to 24 hours. If the person can survive 18 to 24 hours, the prognosis for complete recovery is good and no chronic manifestations of puffer fish poisoning have been reported. There is no known antidote for tetrodotoxin, and treatment of poisoning is to relieve symptoms.

Although the toxicity of toxic puffer fish has been the subject of much concern and study for centuries, the pure toxin, tetrodotoxin, was not isolated until 1950. The compound was crystallized and structurally

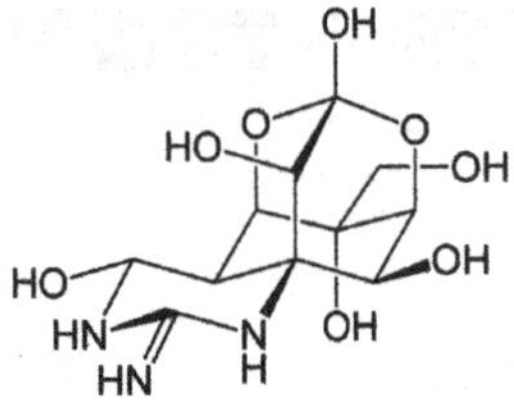

FIGURE 5.6 *Structure of tetrodotoxin.*

characterized in 1964. Studies of this pure material then established tetrodotoxin (Figure 5.6) as one of the most toxic nonprotein natural products known. Tetrodotoxin has an LD_{50} of 8 μg/kg in mice. Ingested tetrodotoxin binds to specific sites in voltage-gated sodium channels and inhibits the influx of sodium ions into neurons. This leads to the inhibition of action potential generation and axonal transmissions in motor, sensory, and autonomic nerves in the peripheral nervous system, which results in the prevention of cell depolarization and subsequent neurotransmitter release.

Recent studies of the ecological effects of tetrodotoxin suggest multiple functions including venom toxin to stun prey, a defense mechanism in the skin to ward off predators, and for the puffer fish, a pheromone to attract males to gravid females. It is interesting to note that tetrodotoxin is not made by fish or other higher organisms, but is made by relatively common marine bacteria, specifically *Pseudoalteromonas haloplanktis tetraodonis*, that associate with these animals. This association of the toxin-producing bacteria with many higher organisms has evolved over millennia to benefit the host and guest.

A relatively recent discovery is the chemical analyses of a powder used in Haiti to induce a semiconscious, so-called zombie state in adherents. This powder was shown to contain high levels of tetrodotoxin, suggesting that this marine toxin might be responsible for these induced effects associated with the practice of Voodoo.

Paralytic Shellfish Poisoning (PSP)

Although shellfish was a food staple for coastal populations since prehistoric times, the importance of this source of food is greatly reduced with the development of more diverse food supplies in modern times (Figure 5.7). Nevertheless, shellfish are enjoyed, primarily as a delicacy, by millions of people worldwide, resulting in a lucrative shellfish industry.

The potential hazards of consumption of shellfish are well established. Outbreaks of typhoid fever attributed to consumption of raw shellfish had been known for centuries. Indeed, the outbreak in Dublin, Ireland, in 1860 is commemorated by the well-known Irish folk song tribute to Molly Malone, a fishmonger who is thought to have contracted yellow fever as a result of sampling her wares. The refrain from the song, "Cockles and mussels, alive, alive-oh," apparently attests to her practice of selling shellfish that were indeed fresh. Improved, standardized, and

FIGURE 5.7

Mussels in the intertidal near Newquay, Cornwall, England.

enforced practices for commercial production, harvesting, and storage of shellfish have eliminated commercially available shellfish as a source of yellow fever and greatly reduced the incidence of any illness from consumption of shellfish. However, the principal public health concerns related to shellfish result from contamination with wastewater-derived viral pathogens and with bacterial agents of an environmental origin. Molluscan shellfish, including mussels, oysters, scallops, and clams have been identified by the US FDA as the source of a majority of seafood-borne illnesses in the United States. As a result, the FDA has given high priority to provide the public greater assurance of the quality and safety of these products.

In addition to the potential bacterial and viral hazards associated with consumption of uncooked shellfish, the sporadic occurrence of acute neurotoxicity—paralytic shellfish poisoning (PSP)—even in cooked shellfish can be a major concern. Occasionally, in usually cooler local waters, where shellfish have been consumed for decades without associated health problems, the shellfish may suddenly become poisonous for periods of only a few weeks. These toxic outbreaks often are associated with algal blooms of single-celled dinoflagellates, which can cause a red-brown discoloration of the water. This proliferation of toxic dinoflagellates, known as red tide, is favored by warmer weather. Thus, regulatory agencies in the United States restrict the taking of shellfish to the cooler months from September to April. Although many species of algae can produce a range of colors in water, PSP is associated primarily with the proliferation of only a few species, namely *Alexandrium catenella* (formerly *Gonyaulax catenella*) or *A. tamarense-excavatum* (formerly *G. tamarensis*).

Although the incidence of poisonings from PSP generally is decreasing worldwide, the occurrence of toxic red tides and the associated adverse effects on the environment are of increasing concern. In the United States, the most recent cases of PSP have occurred along the northeast Atlantic Coast, northwest Pacific Coast, or in Alaska. Most cases have involved recreational shellfish collectors and not commercial vendors. Since 1927, a total of 500 cases of PSP and 30 deaths have been reported in California. Sporadic and continuous outbreaks of PSP occur along the Gulf Coast from Florida to Texas. Sporadic outbreaks have been reported in Europe, Asia, Africa, and the Pacific Islands. Red tide and its resultant massive kills of various birds and marine animals have become an enormous recent concern in Europe, prompting numerous international congresses to address the problem.

As was true for the notorious tetrodotoxin, there has been intense interest in the chemical identity of saxitoxin for centuries. The compound was obtained in purified preparations since the early 1900s. It was isolated in pure form in 1957 and was named in recognition of the source from which it was isolated, the Alaskan butter clam (*Saxidomus giganteus*). The structure of the compound was not fully established, however, until 1975, and was synthetically prepared very shortly thereafter, in 1977 (Figure 5.8). Since that time, the characterizations of over 30 naturally occurring analogues of saxitoxin have been reported.

The toxic characteristics of saxitoxin are similar in many respects to the very potent neurological effects of tetrodotoxin. Thus, the rapid onset of symptoms of paraesthesia, increasing paralysis, and eventual death by respiratory failure is the same for the two toxins. In mice, the lethal potencies of the saxitoxin and tetrodotoxin are also similar with about 8 μg/kg b.w. (via intravenous injection) and 260 μg/kg (via ingestion), respectively. Not surprisingly then, saxitoxin also blocks influx of sodium in neurons by binding to a specific site in voltage-gated sodium channels. Nevertheless, the finding that the puffer fish is sensitive to saxitoxin but not to tetrodotoxin provided the initial evidence that the two toxins bind to different sites in the sodium channel, a notion that is now well established (*vid infra*).

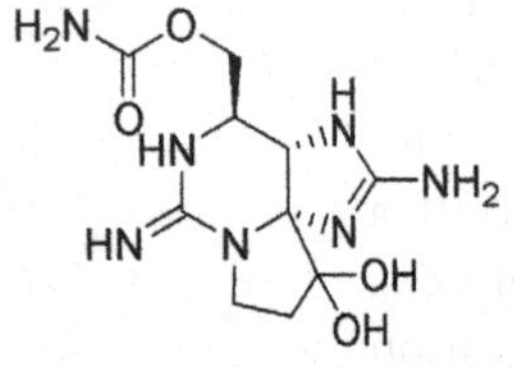

H_2N, O, HN, H, N, NH_2, HN, N, N, OH, OH

Saxitoxin

FIGURE 5.8 *Structure of saxitoxin.*

Ciguatera

A somewhat variable but distinct human toxicity that can result from ingestion of many species of primarily reef-dwelling tropical and subtropical fish is known as ciguatera. The name for this toxicity was derived from the Spanish term for a marine turban snail, cigua, which was thought to be the cause of illness in settlers in Cuba. Documented reports of the poisoning date back to the sixteenth century in tropical areas of the Pacific and Indian Oceans.

Ciguatera poisoning occurs sporadically and is unpredictable. It is thought to affect 50,000 to 500,000 people per year globally. Areas of the United States with endemic ciguatera are Hawaii, Florida, Puerto Rico, Guam, Virgin Islands, and the Pacific Island territories. Occasional outbreaks of ciguatera also have been reported in North Carolina, South Carolina, Louisiana, and Texas. Ciguatera poisoning is the most common nonbacterial, fish-borne poisoning in the United States. Ciguatera poisoning also is endemic in Australia, the Caribbean, and the South Pacific Islands.

Although ciguatera is rarely fatal, its impact in local areas can be significant. In some regions, a loss of work productivity due to ciguatera poisoning is a major concern with an estimated incidence as high as 500 cases/100,000 population and an average recovery time in bed of about three days. A further impact is the loss of fish markets because of fear of toxicity from the locally caught fish. In ancient times, this concern is said to have motivated migrations of native peoples in the Pacific Islands. This has been documented as an important contributor to out-migration from more isolated outer islands to more populated areas, due to lack of markets for fish and to difficulties in obtaining alternative foods. Moreover, in some areas of the Caribbean, imported fish are sold in restaurants in preference to potentially ciguatoxic fish caught locally.

Ciguatera toxicity results from the association of fish species with certain toxic algae. More than 400 species of fish have been implicated in ciguatera poisoning, starting with herbivores and then climbing up the food chain to the larger carnivorous fish. For as yet not fully explained reasons, these fish can accumulate high levels of certain toxic marine dinoflagellate species, most notably *Gambierdiscus toxicus*, which produce potent lipophilic neurotoxins, including the characterized natural product, ciguatoxin (Figure 5.9). Species of fish most frequently implicated include groupers, amberjack, red snappers, eel, sea bass, barracuda, and Spanish mackerel. Since the toxin is lipophilic and quite stable in the fish, it undergoes bioaccumulation, leading to higher toxicity in the larger fish. Fish larger than 2 kg contain significant

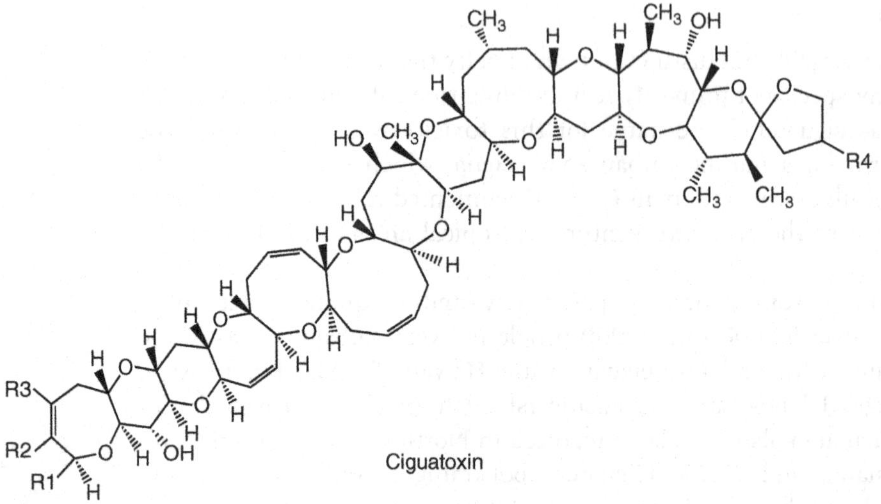

FIGURE 5.9 *Structure of ciguatoxin (P-CTX-1).*

amounts of toxin and readily produce toxic effects when ingested. The presence of toxin does not affect odor, color, or taste of the fish and the toxin is quite stable to cooking.

Ciguatera poisoning is a variable syndrome that encompasses a range of adverse biological effects. The effects usually include gastrointestinal symptoms such as nausea, vomiting, and diarrhea, which often are followed by neurological symptoms such as headaches, muscle aches, paraesthesia, numbness, ataxia, and, in some cases, hallucinations. Severe cases of ciguatera can also result in symptoms known collectively as cold allodynia, which is a burning sensation on contact with cold. There is some evidence that ciguatera poisoning can be transmitted sexually and in mother's milk. Onset of symptoms may be within 15 minutes or, rarely, as late as 24 hours after ingestion of the toxin. Generally, symptoms are noted within 6 to 12 hours after ingestion of the toxic fish. Symptoms increase in frequency and severity over the subsequent 4 to 6 hours. Although the symptoms usually last for only several days, there are reports of the effects lasting from weeks to years. Ciguatera poisoning seldom is lethal. The typical mortality rate is 0.1%, although rates as high as 20% have been reported. Death usually is attributed to cardiovascular depression, respiratory paralysis, or from shock due to low blood volume (hypovolemia).

The complex and variable toxicology that has been ascribed to ciguatera suggests an equally complex etiology. Indeed, more than 20 precursor ciguatoxins, and the closely related gambiertoxins, have been identified in *G. toxicus* and in herbivorous and carnivorous fish. The toxins become more polar as they undergo oxidative metabolism and pass up the food chain. The major Pacific ciguatoxin (P-CTX-1) causes ciguatera at levels of about 0.1 μg/kg in the flesh of carnivorous fish. The LD_{50} of this substance is in the range of about 0.25 to 4 μg/kg by intraperitoneal injection in mice. Of particular interest are the effects of the combined mixture of the various analogs of the toxin that occur in the fish. These studies are yet to be conducted. One key mode of action of ciguatoxin is that it binds to and activates sodium ion channels, causing cell membrane hyperexcitability and instability. It seems unlikely, however, that this single mode of action can explain all the diverse toxic effects that are characteristic of ciguatera poisoning.

Neurotoxic Shellfish Poisoning (NSP)

Neurotoxic shellfish poisoning (NSP) is an additional type of poisoning produced by consumption of shellfish that are contaminated with toxic species of phytoplankton. Although the toxic phytoplankton *Gymnodium breve* can bloom in sufficient numbers to color the water, thus producing a red tide, this toxicity, in contrast to PSP, is restricted to warmer tropical waters, and lethal human poisonings are rare. Massive fish kills and the seasonal occurrences of red tides have been long known to Native Americans and were documented off the west coast of Florida since 1844. Shellfish toxicity was documented in 1880 and aerosol-related respiratory symptoms in human inhabitants were described in 1917. Since 1946, when the causative toxic dinoflagellate was first discovered, extensive red tides have been observed nearly every year in Florida and elsewhere in the Gulf of Mexico. Although the duration of red tides can be as short as a few weeks, algal blooms that have persisted for many months have been documented in the southeastern United States. The surface area covered by prolonged red tides can be up to 10,000 km^2, and in some cases, as revealed by satellite images, have involved the entire western coast of Florida.

Florida red tides can affect humans, wildlife, fishery resources, and the regional tourist-related economy. As *G. breve* cells die and break up, they release a suite of powerful neurotoxins, known collectively as brevetoxins (Figure 5.10). Although some illness related to shellfish consumption occurs occasionally in this area, efforts of monitor and control consumption of toxic shellfish have been quite effective in preventing NSP in humans. However, fish kills, bird kills, and occasional invertebrate kills are common sights during red tides. For example, a massive red tide that occurred in 1988 along the

FIGURE 5.10

Structure of brevetoxin and analogs.

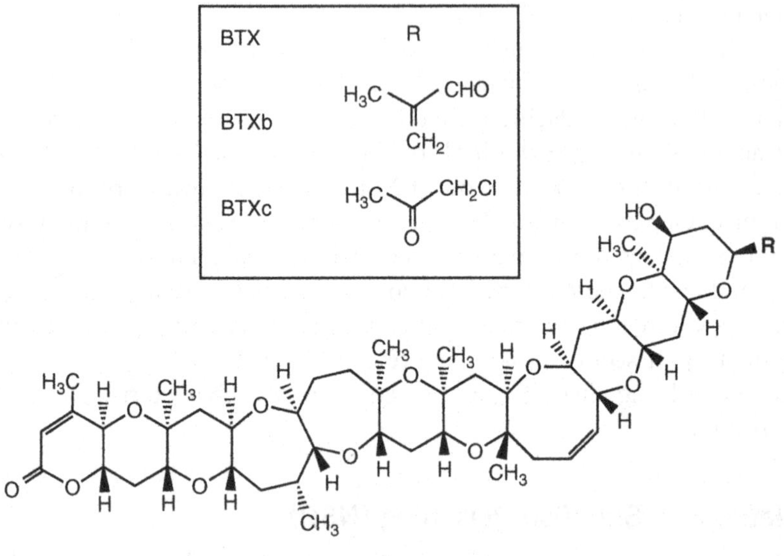

eastern coast of the United States led to the deaths of 740 bottlenose dolphins due to brevetoxin poisoning. In 1996 more than 150 manatees, an endangered species, died due to brevetoxin exposure during a prolonged red tide along the southwest coast of Florida.

Although the toxic effects of NSP can be rapid in onset, they are less dramatic than the effects of PSP. The time of symptom onset can range from 15 minutes to 18 hours following ingestion of toxic shellfish, and the duration of toxicity is usually less than 24 hours but can range from 1 to 72 hours. The symptoms of NSP include: numbness; tingling in the mouth, arms, and legs; incoordination; and gastrointestinal upset. Some patients report reversal of sensitivity to temperature. Aerosolized toxin in the surf can produce an allergic response characterized by runny nose, conjunctivitis, bronchospasm, and cough in sensitive individuals along the shore. Recovery normally occurs in two to three days and is usually complete.

The mechanism of NSP action has been the subject of extensive studies. Brevetoxin and its over 10 different congeners were first isolated and chemically characterized in 1981. The acute toxicity of brevetoxin was found to be similar to the toxic potency of ciguatoxin with an intravenous LD_{50} = 0.5 μg/kg in mice. The pure compounds were shown to bind and depolarize nerve membranes by causing a shift in sodium channel activation toward negative membrane potentials and by inhibiting normal inactivation. Like ciguatoxins, brevetoxins can bind to a specific site in the sodium channel, which results in the uncontrolled influx of sodium ions into the neuron (*vid infra*).

Amnesic Shellfish Poisoning (ASP)

A relatively recently discovered toxicity that results from consumption of contaminated shellfish is called amnesic shellfish poisoning (ASP). Although the toxic agent, domoic acid (DA) (Figure 5.11), was first identified in 1960 in a type of seaweed in Japan, the toxic effects of DA became well known during a serious seafood poisoning incident in Canada in 1987. The illness occurred following the consumption of mussels (*Mytilus edilus*) when more than 100 people were hospitalized with at least four fatalities. Since that time there has been an increasing frequency of reports of DA-related incidents in the world, including in Spain, Mexico, New Zealand, and the United States. In 1998, an El Niño year, there was an outbreak of sea lion poisonings and mortalities in Monterey Bay, California, and reports of high levels of DA in razor clams collected from beaches in Washington and Oregon.

In 2007, reports in the popular media appeared, indicating that DA was detected in record quantities in shellfish on the Southern California Coast, and that this toxin was responsible for the poisonings of hundreds of marine animals in the area that year. DA also was blamed for the poisonings of thousands of marine animals and birds since a major outbreak in 2002. California State health officials stated that about 50 dolphins, a minke whale, and scores of sea birds also were killed by that season's toxic algal bloom.

The primary sources of ASP toxin in shellfish are species of diatoms (*Pseudo-nitzschia australis, P. multiseries, P. pungens*), which can proliferate to high concentrations in the water. Filter feeding shellfish then accumulate the diatoms and associated toxins as a result of their normal feeding practice. In one study of DA production by these organisms, DA levels were found to be as high as 1% or more of the dry weight of the plankton. Contaminated mussels had digestive glands that were engorged with *P. pungens*. Studies conducted in Monterey Bay showed that blooms of this phytoplankton occur in high enough densities to contaminate animals several times a year. In addition, the studies revealed that fish and krill that fed on plankton contained toxin levels above the FDA regulatory limit (20 μg DA/g of tissue) even though animal kills were not reported. The findings of these studies suggested that blooms of DA-producing diatoms are a regular occurrence in Monterey Bay, and probably elsewhere, and that wildlife may be affected more often than was previously suspected.

As is the case for other forms of marine poisoning caused by toxic plankton, the incidence of ASP seems to be increasing in recent years. The DA levels in several areas and outbreaks of DA poisonings also have been rising. The main algae that produce DA are most prominent off the California Coast and high water temperatures mostly between March and June trigger the blooms. With water temperatures predicted to become even higher as climate change warms the world's oceans, the threat of DA poisoning to humans and wildlife is likely to increase.

The toxicology of DA is fairly well understood. Following consumption of toxic levels of DA of about 200 μg/g shellfish, patients first experience gastrointestinal distress within 24 hours after eating the contaminated shellfish. Other reported symptoms have included dizziness, headache, and disorientation. A unique symptom of this poisoning is a permanent short-term memory loss, which is the basis for the name of this toxicosis. In severe poisoning, seizures, weakness or paralysis, violent head weaving, and death may occur. DA shows only moderate acute toxicity in mammals with an LD_{50} in mice of 3.6 mg/kg by intravenous injection. This poisoning is seldom life threatening for humans, however, and recovery is usually complete within two or three days. DA is heat-stable and similar in structure to the endogenous neurotransmitter, glutamate, and to another algal toxin, kainic acid (Figure 5.11).

DA binds to receptors in neurons with a 3-fold and up to 100-fold greater affinity, respectively, than kainate and glutamate. The binding of DA to glutamate and kainate receptors, which results in the depolarization and then the firing of the neuron. DA also hyperactivates nerve signaling, which can result in toxic lesions in the hippocampus, a major site in the brain for memory. Some studies have shown, however, that DA might not fully account for the toxic effects of the shellfish. Furthermore, an impure domoic acid preparation isolated from mussels is more neurotoxic for cultured human neurons than purified domoic acid. This increase in toxicity is believed to be due to potentiation of domoic acid effects by glutamic acid and aspartic acid, which are naturally present in high concentrations in mussel tissue. This synergy might well be the basis for the pronounced and consistently observed mammalian toxicity of DA-contaminated shellfish, in spite of the relatively low toxicity (high LD_{50}) of pure DA in mice.

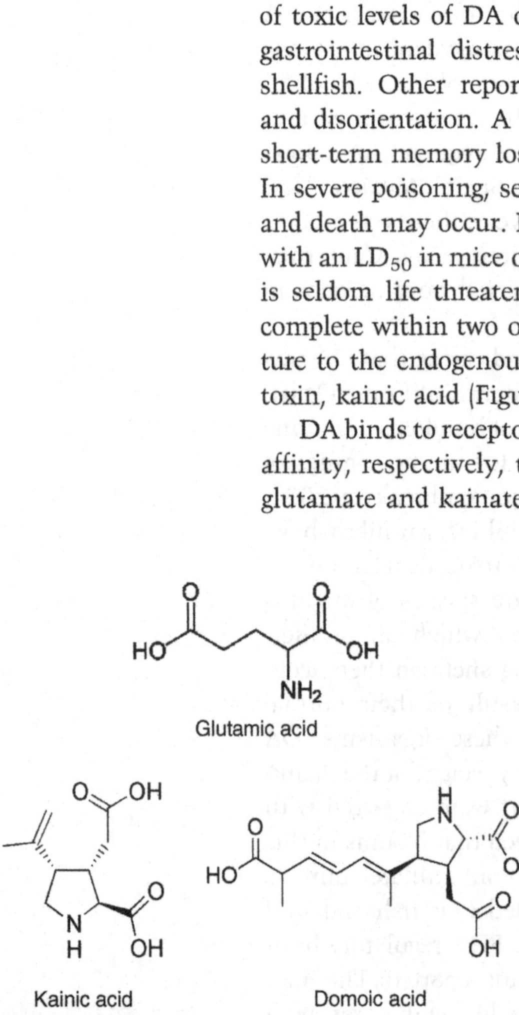

Glutamic acid

Kainic acid

Domoic acid

FIGURE 5.11 *Structure of domoic acid and the neurotransmitter glutamic acid and kainic acid.*

Voltage-Gated Na^+ Channels

Voltage-gated sodium channels function to regulate the influx of sodium ions into neurons required for normal conduction of the nerve impulse. Intense study of these channels over many years, often using some of the marine neurotoxins described in this chapter, has led to a detailed understanding of the mechanisms by which these important channels function. The molecular structure of the sodium channel from the electric organ of the electric eel recently was determined using single particle image analysis derived from ultra-high resolution cryo-electron microscopy. The structure was found to be significantly different from the proposed structure visualized for many years as a simple cylindrical shape with a central pore through which ions flow. The channel was found to be a bell-shaped molecule pierced by a network of channels arranged like the diagonals of a cube intersecting in the center that is presumed to be the ion pore (Figure 5.12). This structure indicates that the channel actually has four entry points converging onto a central cavity that then branches out to four exit pathways into the interior of the cell. Four additional channels are hypothesized to carry the gating charges that underlie the conformational response by the

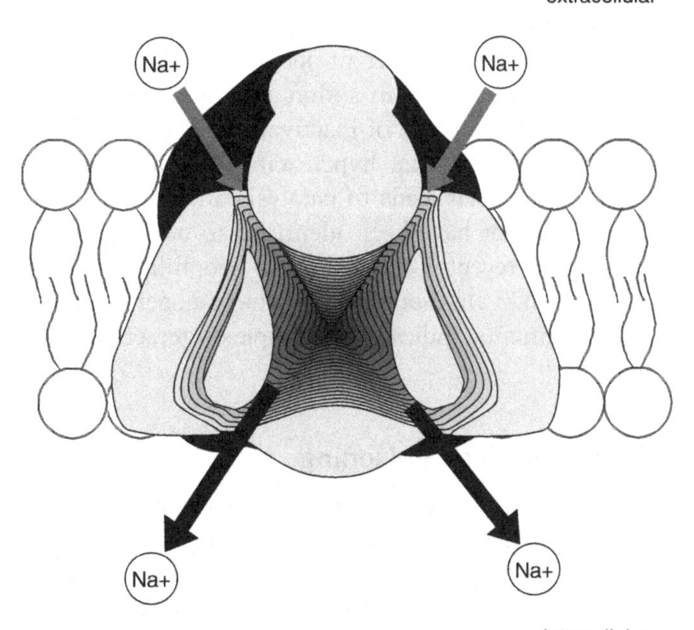

FIGURE 5.12 *Schematic representation of the voltage-gated sodium channel.*

channel to changes in the transmembrane electrical potential difference. The channel is proposed to function by the movement of the central external mass in and out, in a piston-like motion, to open the pathways for the ions into the intracellular cavity.

Sodium channels are the molecular targets for several potent toxins, including tetrodotoxin (TTX), saxitoxin (STX), brevetoxin, and ciguatoxin, that were discussed in this chapter, as well as other toxins such as anemone toxin, batrachotoxin, veratridine, conotoxin, and certain insect toxins that were not discussed here. As a group, these toxins act at six or more distinct receptor sites on the sodium channel protein.

The toxic actions of several of the marine toxins discussed in this chapter are now well understood in terms of their interactions with the voltage-gated sodium channels. Tetrodotoxin and saxitoxin are pore-blocking neurotoxins. The amino acid residues that form neurotoxin receptor sites for these toxins are located in the central pore of the receptor. The toxicological effects of closed sodium channels are decreased excitability of neurons, leading to respiratory paralysis and death. Selective pressure from the presence of STX in the natural environment or TTX in prey has selected for mutations in the ion selectivity filter that cause resistance to these toxins in some marine organisms. Also, a nonaromatic residue (asparagine in fugu) in the pore confers TTX resistance in the puffer fish.

Brevetoxin and ciguatoxin are pore-activating neurotoxins. They bind to neurotoxin receptor sites of open channels and prevent the closing of the channel. This results in a shift in activation to more negative membrane potentials and a block of inactivation. The toxicological effects of the open channels are neuronal hyperexcitability, followed by a desensitization of the receptors that leads to paralysis and death. Transmembrane segments of the receptor have been identified to participate in the formation of the neurotoxin receptor site for these lipophilic toxins. Binding determinants for brevetoxin are thought to be widely dispersed throughout the transmembrane segments, indicating a complex interaction of various domains in the protein.

Scombroid Fish Poisoning

Scombroid fish poisoning is the most common cause of ichthyotoxicosis worldwide. In the United States, such poisoning represents one of the major chemical food-borne illnesses reported to the Centers for Disease Control and Prevention. From 1968 to 1980, 103 incidents involving 827 people were reported to the CDC. From 1973 to 1986, 178 outbreaks affecting 1096 people were reported.

Scombroid poisoning results from the consumption of certain foods, most often fish, that contain unusually high levels of histamine. Spoiled fish from the families *Scombridae* and *Scomberesocidae*, such as tuna, mackerel, and bonito, are most commonly implicated in incidents of scombroid poisoning, but other nonscombroid fish, such as mahi-mahi, bluefish, and sardines, have been implicated. On rare occasions, scombroid poisoning has also been reported following consumption of Swiss and other cheeses.

The symptoms of scombroid poisoning are generally those of histamine poisoning and resemble an acute allergic response. The symptoms include nausea, vomiting, diarrhea, an oral burning sensation or peppery taste, hives, itching, red rashes, and hypotension. The onset of the symptoms usually occurs within a few minutes after ingestion of the implicated food, and the duration of symptoms ranges from a few hours to 24 hours. Mortality is rare. Antihistamines can be used effectively to treat this toxicosis. Histamine is formed in foods by certain bacteria that are able to decarboxylate amino acids, including histidine. However, foods containing unusually high levels of histamine may not appear to be spoiled. Foods with histamine concentrations exceeding 50 mg per 100 g of food generally are considered to be hazardous. Proper handling and refrigerated storage can prevent histamine formation in fish.

The etiology of scombroid poisoning has been the subject of some controversy over the years. Under controlled conditions, histamine consumed orally at a level that is present in toxic fish shows very low toxicity because various amine oxidases that occur in the gut are capable of deactivating histamine before it is absorbed. Early investigations of the toxic agent in the fish suggested the presence of a histamine-like substance, called "saurine," with enhanced toxicity. Subsequent studies in other laboratories were unable to identify a separate toxin in the fish, but noted that different salts of histamine can behave differently on chromatography. The suggestion was made that saurine is simply a conjugate acid of histamine. Further studies showed, however, that certain other products of amine decarboxylation found in the fish, most notably, cadaverine and putrescine, were strong competitive inhibitors of the amine oxidases that detoxify histamine. Thus, it is reasonable to suggest that the unusual oral toxicity of histamine in the fish is a result of the inhibitory effects of additional bacterial products such as cadaverine and putrescine on histamine deactivation in the gut (Figure 5.13).

There is considerable concern by scientists and regulatory agencies over the fact that the impact of toxic algae on human health and the environment has been increasing. Thus, the number of documented incidences of blooms of algae that produce toxins has increased in frequency, intensity, and geographical distribution in recent decades. In the past, toxic algal

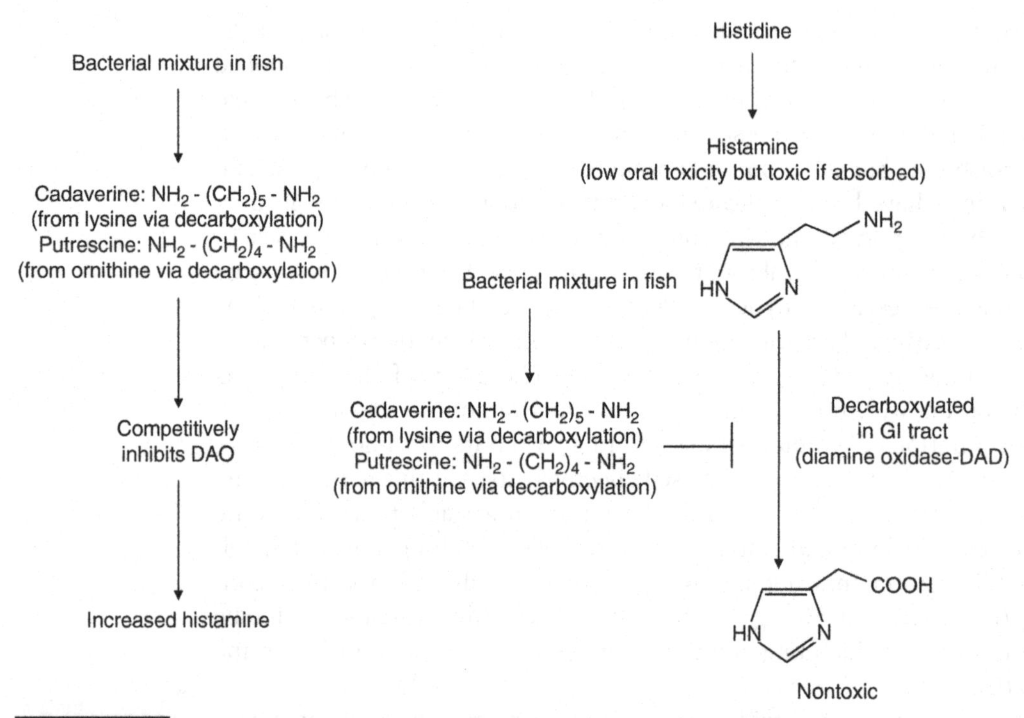

FIGURE 5.13 *The mechanism of scombroid poisoning involves the production of histamine and other amine products of amino acid decarboxylation by contaminating bacteria of histidine-rich fish. The other amines can inhibit the detoxification of histamine in the gut.*

blooms in the United States have been observed in limited geographic areas, such as the west coast of Florida and the coastlines of Alaska and Maine. Currently, however, toxic blooms have been reported in every coastal state and toxicities are being documented from an increasing number of species. Although the data on the worldwide occurrence of toxic algal blooms are less complete, reports of increased incidence of algal blooms in some areas are striking. For example, an eight-fold increase in algal blooms was reported between 1976 and 1986 in Hong Kong. The causes of these increases are far from clear. Suggested possibilities include increased surveillance, the development of aquaculture, accumulation of nutrients from run-off, transport of cysts in ship ballast, decreased silica from damming waters, diminished growth of safer competing diatoms, and upwelling of nutrient-rich water, possibly due to global warming. It is clear that although the occurrence of toxic algal blooms have been recognized for millennia, their true significance to the threatened marine environment and to human health will be subjects of intense worldwide concern well into the future.

SUGGESTED FURTHER READING

Hwang, D.F., Noguchi, T. (2007). Tetrodotoxin poisoning. *Adv. Food Nutr. Res.* 52:141-236.

Isbister, G.K., Kiernan, M.C. (2005). Neurotoxic marine poisoning. *Lancet Neurol.* 4:219-228.

Jeffery, B., Barlow, T., Moizer, K., Paul, S., Boyle, C. (2004). Amnesic shellfish poison. *Food Chem. Toxicol.* 42:545-557.

Lehane, L., Olley, J. (2000). Histamine fish poisoning revisited. *Int. J. Food Microbiol.* 58:1-37.

Llewellyn, L.E. (2006). Saxitoxin, a toxic marine natural product that targets a multitude of receptors. *Nat. Prod. Rep.* 23:200-222.

CHAPTER 6

Toxic Phytochemicals

CHAPTER CONTENTS

Members of the Plant Kingdom are rich sources of products of vital importance to members of the Animal Kingdom. Plants provide oxygen, energy, vitamins, and shelter that are essential to animal life. Animals are important in plant reproduction and seed dispersal, and assist in the breakdown of dead plant material. Plants also are rich sources of small molecular weight, nonnutritive natural products, often called phytochemicals, that can exhibit beneficial (drugs and hormones) or negative (phytotoxins) effects in animals, including humans. In some cases, the plant products play roles in the plant's defense against other organisms. For example, an allelochemical is a chemical substance that is released by a plant that inhibits the growth or germination of another plant. The process is called allelopathy. In contrast, a phytoalexin is a metabolite that is produced by a plant usually in response to damage by plant pathogens and that can inhibit the growth of invading organisms. Some phytochemicals can strongly influence insect reproduction. In most cases, however, the functions of these phytochemicals in the plants are not known.

In this chapter we discuss a selected group of phytotoxins that are important as examples of natural chemicals that adversely affect human health or that have been widely used as approved therapeutic agents. We also provide brief discussions of phytoalexins and of the interactions of certain phytochemicals in the diet with established drugs to produce adverse diet–drug interactions.

PHYTOTOXINS

Goitrogens

Goiter is defined as a noncancerous and noninflammatory enlargement of the thyroid gland and has been recognized since ancient times. The first description of thyroid goiter dates back to China around 2700 BCE. In many endemic goiter areas of the world, the inhabitants became so accustomed to goiter that these growths became marks of beauty in women. For example, portraits from the seventeenth century Dutch court portray upper-class women with small goiters. Goiters were not considered to be attractive in men, however. The fashionable use of high collars during the early nineteenth century in Europe had its origins in the effort by male royalty to conceal prominent goiters. Until the institution of iodide supplementation of table salt in the 1924, goiter was endemic in Great Lakes, Appalachian, and Northwestern regions of the United States, with prevalence in children in these "goiter belt" regions of 26 to 70%. Furthermore, prior to the institution of iodide supplementation in Nigeria in 1993, the goiter incidence rate was as high as 67%. An assessment of hypothyroidism in this region in 2005 found that goiter prevalence had decreased to only 6%.

Although the incidence of goiter in the United States has decreased considerably since the initiation of iodide supplementation, goiter and hypothyroidism continue to be serious problems in some areas even with iodide supplementation. Studies conducted in Maine, Kentucky, and many other states have identified significant incidences of goiter even with adequate iodide consumption. Furthermore, in recent years, iodide deficiency has been a growing problem. For example, a US national health survey conducted in the early 1970s found a moderate to severe iodine deficiency incidence of only about 2.5% in the population. A similar survey conducted in the early 1990s found at least a moderate iodine deficiency rate of 11.1% of the population. This dramatic increase in iodine deficiency is thought to result from an aggressive effort by public health bodies to restrict sodium consumption as a means of reducing hypertension and the risk of cardiovascular disease. These concerns over excess sodium consumption have led to striking increases in goiter prevalence in many other countries, as well. World Health Organization data indicate that between 1993 and 2004 the prevalence of goiter has grown 81.4% in Africa, 80.7% in Europe, and 62.9% in the Eastern Mediterranean region, with a worldwide increase of 31.7%. The causes of these increases in goiter prevalence during a period when iodide supplementation has increased are the subject of intense interest by public health agencies.

Goiter is an adaptive response to insufficient levels of thyroid hormones and, in the adult, is usually a nonlethal but readily diagnosed symptom of hypothyroidism. Other more serious effects of hypothyroidism include thyroid dysfunction, loss of energy, increased cardiovascular disease risk, impaired mental and physical development (cretinism), and prenatal and infant mortality. The adverse effects of hypothyroidism on fetal development are a particular concern since even mild maternal iodine deficiency can result in a range of serious intellectual, motor, and hearing deficits in offspring. Researchers in this field estimate that early maternal hypothyroxinemia (weeks 8–12) may potentially affect up to 5% of newborns in the United States. Moreover, the prevention of neurological deficits in newborns by iodine supplementation was identified as a major goal of the historic World Summit for Children that convened in 1990. An emerging concern is a possible role of hypothyroidism in the growing incidence of autism.

Accumulating evidence indicates that iodine deficiency often occurs in conjunction with other antithyroid factors to produce endemic goiter. Environmental goitrogens when low in concentration may normally be ineffective but may become significant when iodine supply is restricted. In some cases, the goitrogen might be sufficiently potent by itself to cause goiter or thyroid insufficiency despite an abundance of iodine.

The environmental goitrogens to be discussed in the following sections are classified on the basis of their mode of hypothyroid action. As indicated in Figure 6.1, thyroid hormone metabolism is regulated by the hypothalamic/pituitary axis at several levels.

Iodide Trapping

The pituitary gland is stimulated to produce thyroid-stimulating hormone/thyrotropin (TSH), which activates most of the steps of thyroid hormone synthesis. Iodine uptake occurs through a sodium-iodide symporter, NIS. This active transport mechanism allows thyroid cells to concentrate iodine

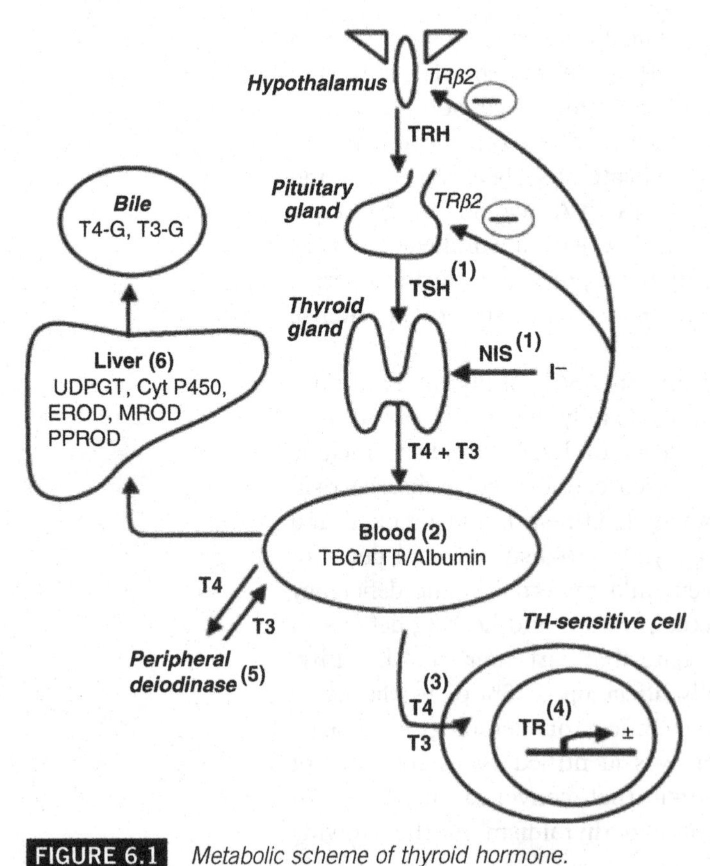

FIGURE 6.1 *Metabolic scheme of thyroid hormone.*

at 20- to 40-fold above its level in the extracellular space. The same transporter is found in the breast, and inhibition at that site has been proposed as another factor that can lead to iodine deficiency in breastfed infants. NIS is inhibited by thiocyanate and nitrate ions.

Organification

In the presence of thyroperoxidase (TPO) and hydrogen peroxide, iodide is oxidized to iodine and incorporated into mono- and diiodotyrosyl (MIT and DIT) residues of thyroglobulin. TPO also couples DIT and MIT to form T4 and T3. TSH stimulates TPO activity whereas moderate concentrations of iodide reduce it. Organification is inhibited by reducing substances such as goitrin from *Brassica* plants and certain flavonoids, as well as insufficient or excess iodide which inhibits hydrogen peroxide synthesis.

Proteolysis

Lysosomal degradation of thyroglobulin hydrolyzes the iodothyronines T4 and T3 from their protein support. The lysosomal enzymes cathepsins B, L, and D are active in thyroglobulin proteolysis. This process can be blocked by excess iodide and lithium.

Deactivation

The major routes of deactivation of T3 and T4 are by glucuronidation and sulfate formation. Glucuronidation of thyroxine is the rate-limiting step in the biliary excretion of thyroxine. In contrast to T3 itself and the stable glucuronide, T3 sulfate is rapidly degraded by successive deiodination of the tyrosyl and phenolic rings. Phenobarbital and dioxin affect thyroid function by altering hormone disposition due to induction of hepatic thyroxine glucuronyltransferase. PCBs can displace T4 from specific binding proteins (transthyretin, thyroxine binding globulin) to increase clearance.

Peripheral Activation

T3 may be synthesized and secreted by the thyroid gland or it can be produced by deiodination of T4 in the liver and other tissues. This deiodination of T4 to T3 is regulated by deiodinases, which are selenium-containing enzymes. T4 deiodination is inhibited by the thyrostatic drug, propylthiouracil, and by the food additive, FD&C Red #3. Cadmium and methylmercury can inhibit specific deiodinases, as well.

Environmental Antithyroid Substances

Many plants in the mustard family (Cruciferae) contain thyrotoxic substances such as goitrin, allylthiourea, and thiocyanate. These substances are not present in the undamaged plant, however, but are produced by the enzymatic conversion of thioglucoside precursors, called glucosinolates, when the required thioglucosidase, myrosinase, comes into contact with the substrate following disruption of the cell's integrity. Upon enzymatic hydrolysis, the glucosinolate called progoitrin is rapidly converted to goitrin (Figure 6.2). Progoitrin is present most notably in swedes and turnips and at low levels in other plants of the *Brassica* genus, including cabbage, broccoli, cauliflower, kale, kohlrabi, Brussels sprouts, and rutabaga. The highest levels occur in the seeds of these plants. Other sources of goitrin-like compounds have been detected in various species of herbs and shrubs of the Barbarea and Residea families.

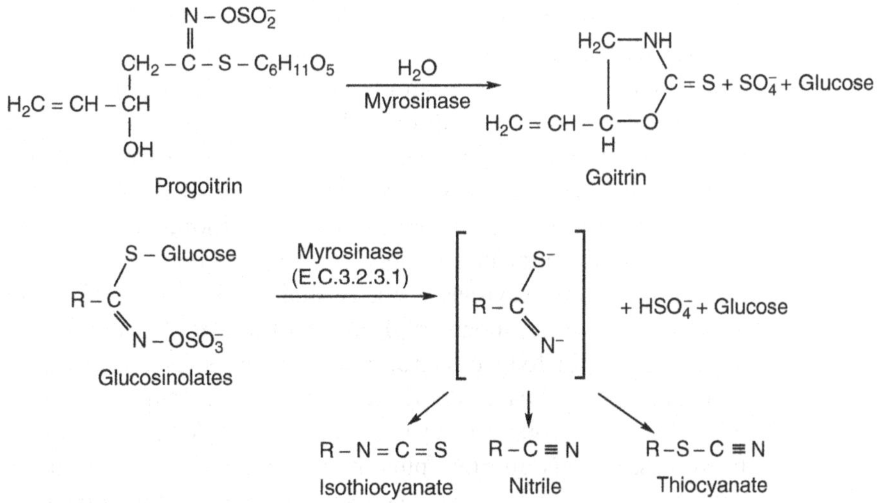

FIGURE 6.2 *Goitrin production in Brassica plants.*

The goitrogenic activities of *Brassica* plant components have been the subjects of many studies. Studies conducted as early as 1928 showed that a high level of cabbage consumption produced goiter in rabbits. Subsequent studies of the effects of dietary vegetables on radioiodine uptake in the thyroid gland of human subjects found that raw rutabaga was consistently the most active food plant. That activity was lost by normal cooking suggested the requirement for the enzymatic conversion of a precursor in the plant (progoitrin) to the active substance(s) (goitrin), and that this enzymatic activity was not present in the gastrointestinal tract. Goiter can be most consistently induced in animals when the seeds of *Brassica* species, especially rapeseed, are included in the feed. The goitrogenic effects of the leafy portion of the plants are generally variable and less pronounced than the effects of seeds, especially if the diet is deficient in iodide. It is unlikely, therefore, that consumption of *Brassica* plants, as a normal part of a nutritionally adequate diet, will induce thyroid enlargement. However, it seems plausible that if iodide intake is marginal, consumption of high levels of *Brassica* vegetables may produce some symptoms of hypothyroidism.

The goitrogenic effects of goitrin are well established. Administration of goitrin to rats for 20 days induces an enlargement of the thyroid gland, decreases iodine uptake by the gland, and decreases thyroxine synthesis. Goitrin also has been implicated in the development of endemic goiter in school children of Tasmania. This goiter endemia could not be prevented by iodine supplementation. Goitrin has also been involved as an environmental goitrogen in Finland. Goitrin is secreted in the milk of cows fed on goitrin-rich rapeseed meal at a level of about 0.1% the level in the feed. Studies with rodents showed that goitrin administered to the dams produced symptoms of hypothyroidism in the nursing pups.

Thiocyanate is an additional dietary goitrogen of considerable importance. Thiocyanate is produced as a by-product of glucosinolate hydrolysis and as the principal detoxification product of cyanide. Thiocyanate interferes with active uptake and concentration of inorganic iodide by the thyroid and inhibits the enzyme thyroperoxidase, thereby preventing the incorporation of iodine into thyroglobulin. Several staple foods in the tropics contain large amounts of cyanogenic glycosides that are detoxified as thiocyanate. These plants include cassava, millet, yam, sweet potato, corn, bamboo shoots, and lima beans. Poorly refined cassava (Manihot esculenta) (Figure 6.9) is a major source of cyanide in parts of Africa. Thiocyanate toxicity from cassava consumption, plus the combined effect of iodine and selenium deficiency, is thought to cause endemic goiter and to contribute to endemic cretinism observed in parts of Africa.

A further important group of naturally occurring antithyroid substances are the plant polyphenols. Chemicals in this large class of natural products have diverse roles in the plants ranging from protection against ultraviolet light and pathogens, to providing colors that attract pollinators. Plant polyphenols such as quercetin, resveratrol, and epigallocatechin produce many positive effects in humans and, as will be discussed in a later section, are under intense study as remedies for a range of disorders including cancer and cardiovascular disease. In addition, however, many polyphenols exhibit antithyroid activities primarily as a result of their inhibitory effects on organification. Inhibitors in this group include genistein and daidzein from soy; catechin from traditional tea (*Camellia sinensis*); quercetin found in apples, onions, red grapes, citrus fruits, broccoli, cherries, and berries; kaempherol from grapefruit; rutin found in buckwheat; tannins in many kinds of nut; and apigenin and luteolin in millet. The latter substances, along with thiocyanate, are thought to contribute to the high prevalence of goiter in an iodine deficient population in West Africa for whom millet is a dietary staple. These substances may exacerbate the effects of endemic or sporadic low iodine-induced hypothyroidism in other populations, as well.

A less well-defined source of hypothyroid substances is contaminated drinking water. Studies in India and elsewhere showed a clear association of goiter prevalence with contaminated water supplies. By simply providing an uncontaminated water supply in this region, the incidence of hypothyroidism in schoolchildren was decreased significantly. Although specific antithyroid substances were not identified in these studies, many possible pollutants have this activity. Thus, powerful antithyroid products such as resorcinol, methoxy-anthracene, phthalate esters, and phthalic acid are known to contaminate water in areas rich in coal and shale. Certain antithyroid disulfides, which are similar to substances formed in onions and garlic, are present in high concentrations in aqueous effluents from coal-conversion processes. Furthermore, 60% of all herbicides, notably, 2,4-dichlorophenoxyacetic acid (2,4-D) and thioureas, exhibit antithyroid activities. Other antithyroid agents include polychlorinated biphenyls (PCBs), perchlorate, and mercury. Thus, normal thyroid function may be compromised by any of a host of pollutants found in the water supplies of some regions.

Favism

Favism is a hemolytic disorder in genetically susceptible populations exposed to pro-oxidant chemicals including those found in fava beans (*Vicia faba*) (Figure 6.3). Symptoms of favism include hemolytic anemia,

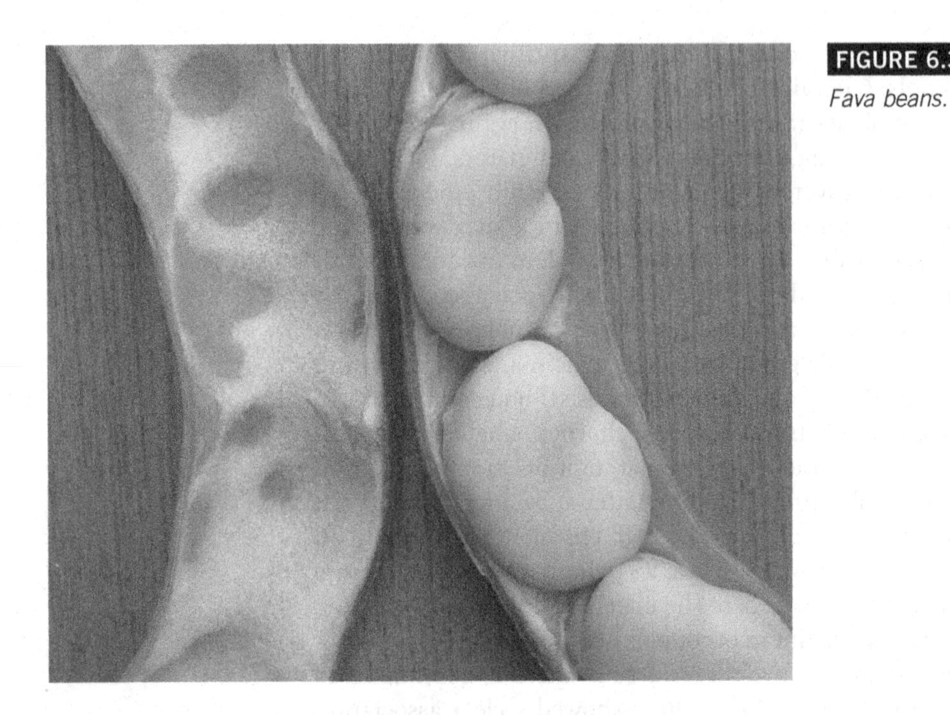

FIGURE 6.3
Fava beans.

jaundice, and hyperbilirubinemia. In susceptible populations, the annual incidence of favism ranges from less than one to around nine cases per 10,000 persons. Nearly all these cases are children under 15 years old, with 85 to 95% under 6 years old with boys in the large majority. Mortality is currently confined to children less than 5 years of age but is rare if treatment is available.

The genetic susceptibility for the disorder is a deficiency of erythrocyte glucose-6-phosphate dehydrogenase (G6PD). G6PD deficiency, which is a result of mutations in the X-linked G6PD gene, is the most common human enzymopathy and is present in approximately 400 million people throughout the world. G6PD deficiency is most prevalent, however, in areas of Africa, India, around the Mediterranean Sea, China, and the Tropics. Although most individuals with this deficiency are asymptomatic, fully 10 to 20% of the population experience favism at some time in their lives. The high prevalence of G6PD deficiency in these areas is thought to have evolved in response to the malaria parasite, *Plasmodium falciparum*, against which G6PD deficiency plays a protective role. In addition, however, with recent population migrations, the prevalence of G6PD deficiency is growing in other parts of the world where malaria is not prevalent, notably North America and northern Europe.

The fava bean is one of the earliest domesticated food plants. Cultivation of the fava bean has been dated back to about 7000 BCE in the Middle East. Favas have been found in 6000-year-old archeological sites in Spain and in 4000-year-old Egyptian tombs. The bean was cultivated in Massachusetts by some of the first European settlers to the New World. Fava beans, which also are known as broad beans or horse beans, are common in the Mediterranean diet in general and are the most common food in the Egyptian diet.

The mode of action of fava bean in the induction of hemolytic anemia has been the subject of considerable inquiry. Analysis of the bean for substances that decrease the levels of reduced glutathione in isolated erythrocytes led to the identification of two related glycosides, vicine and convicine, which occur in very high levels (1–2 g/100 g dry wt.) in fresh beans. Cooking does not appreciably reduce the levels of the glycosides in the beans. Hydrolysis of the glycosides by exposure to simulated gastric juice or to β-glucosidase, which is present at high levels in ripe beans, produces the active aglycones, divicine and isouramil. A further phenolic substance that has been implicated in the cause of favism is L-DOPA (3,4-dihydroxy-L-phenylalanine). Although L-DOPA apparently is not active by itself in producing the effects associated with favism, its combination with divicine and isouramil produces a synergistic increase in hemolytic activity. The principal effect of these substances is to produce an oxidative stress in cells that leads to the loss of reduced glutathione in treated erythrocytes from G6PD-deficient subjects. The mechanism of induction of oxidative stress involves the production of quinones and quinone imines from the parent hydroquinones. Redox cycling of the oxidized products via high capacity, single electron reductive processes (e.g., NADPH-CYP450 reductase), as indicated in Figure 6.4, can result in catalytic production of reactive oxygen species (ROS). The increased levels of ROS in turn induce, among other products, increased levels of oxidized glutathione, protein-bound mixed disulfides, cellular protein aggregates, and cross-linked structural proteins in the erythrocyte. These and other abnormalities of the erythrocytes, including the disruption of cellular calcium homeostasis, lead eventually to their destruction by phagocytosis in the spleen.

The G6PD-deficient cells are unable to counter the effects of the pro-oxidant substances because of the central role of G6PD in the control of cellular redox status. As indicated in Figure 6.5, G6PD is responsible for mediating the oxidation of G6P to 6-phosphogluconolactone, which is the first reaction in the pentose phosphate pathway of glucose metabolism. A major function of this pathway is to provide cellular reducing power in the form of NADPH. NADPH, in turn, is central to the cellular metabolic defenses against oxidative stress, especially as a cofactor for reductive

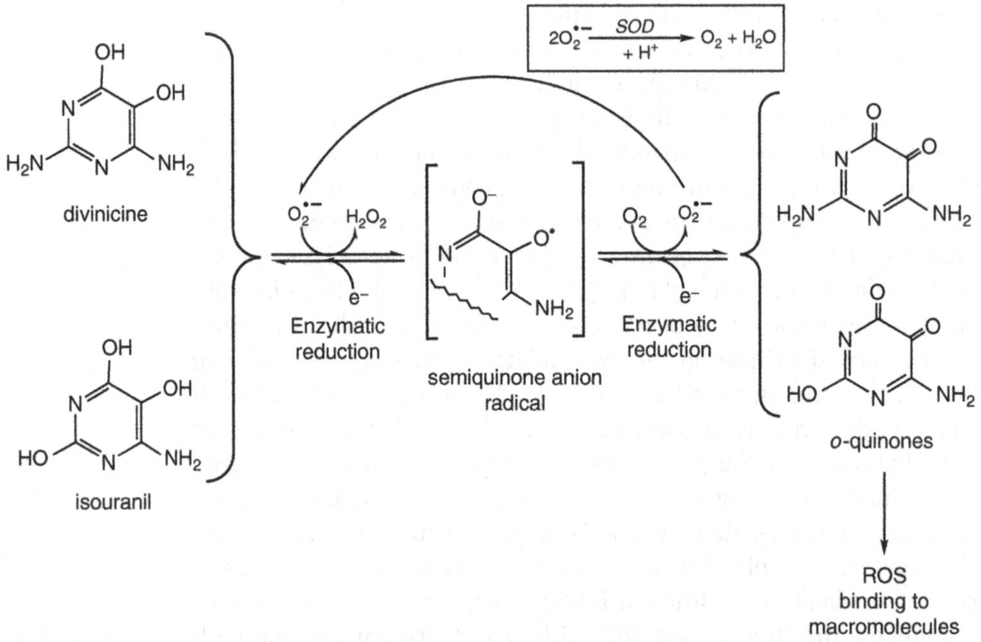

FIGURE 6.4 *Redox cycling of oxidized products via high capacity, single electron reductive processes.*

enzymes such as catalase, glutathione peroxidase, glutathione reductase, and thioredoxine reductase. Under conditions of oxidative stress in normal cells, the capacity of the cellular defenses is increased by increased expression of G6PD. Thus, the inability of G6PD-deficient cells to increase the supply of NADPH leaves them highly vulnerable to induced oxidative stresses.

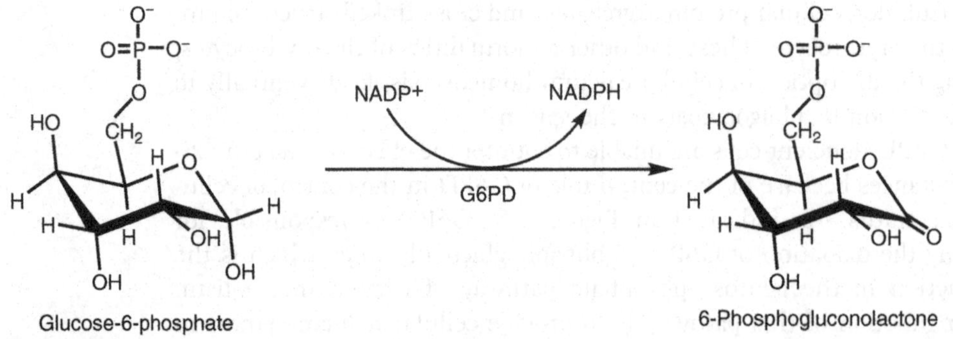

FIGURE 6.5 *G6PD reaction.*

Neurolathyrism

Neurolathyrism is an ancient disease that is caused by the consumption of peas of the genus *Lathyrus* (*L. sativus*), commonly known as vetch pea, chickling pea, or grass pea, and by many other names in various languages. The disease apparently was first described in the time of Hippocrates (circa 400 BCE), and occurred in epidemic proportions in parts of North Africa, the Middle East, Asia, and India until the twentieth century. Very high incidences of neurolathyrism also occurred in prisoners and in peasants during World War II in Europe. Chronic *Lathyrus* poisoning continues to affect large populations in Ethiopia, India, and Bangladesh. The toxicity of *Lathyrus* is well known to regulatory agencies and the sale of the peas is illegal in most regions. Nevertheless, *Lathyrus* is a protein-rich, hardy crop that thrives under conditions of poor soil, drought, and even flooding, and thus is used as an emergency crop for feed and food in many areas. As expected, it is during times of drought or after floods that the prevalence of neurolathyrism is highest.

Several aspects of the etiology of neurolathyrism are quite well understood. Neurolathyrism is known to occur in humans following consumption of at least 300 g/day of the *L. sativus* peas for a period of at least three months. Neurolathyrism is a progressive, neurodegenerative condition of motor neurons. It is characterized in humans, primarily men, by a stiffening of the muscles in the legs that predominantly targets a type of neuron called a Betz cell in the spinal tract. The initial symptoms are increasing rigidity of calf muscles and loss of control over the legs. The toxin in the peas that causes the disorder is the amino acid derivative, β-N-oxalyl-α,β-diaminopropionic acid (ODAP) (Figure 6.6), which occurs as 0.1 to 2.5% of the dry weight of the peas. Administration to male mice for 40 days of a dose of ODAP that would likely result from consumption of the toxic peas (i.e., 5 mg/kg body weight) produced decreases in neuronal activities characteristic of human neurolathyrism.

Studies with several model systems have shed light on the mechanism of toxic action of ODAP. Early studies showed that the toxic effects of ODAP in rodents could be inhibited by administration of antagonists of the AMPA subclass of glutamate receptors. ODAP also has been shown to facilitate release of glutamate from presynaptic elements and to inhibit glutamate uptake, which results in a large increase in glutamate concentration at the synapse. This increase in glutamate concentration produces toxic effects in the neurons primarily as a result of the

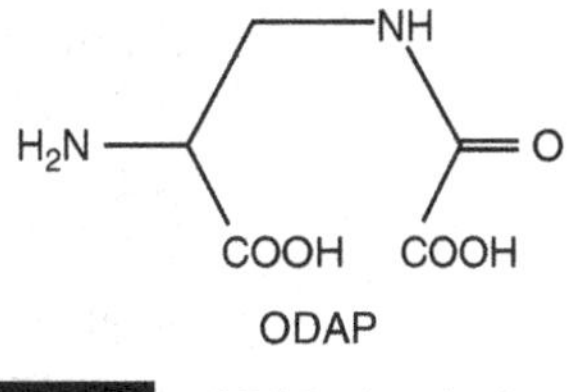

NH
H_2N
O
COOH COOH
ODAP

FIGURE 6.6 *ODAP chemical structure.*

release of ROS from inhibition of complex I of mitochondria electron transport chain by glutamate. This role of ROS in the effect of ODAP is consistent with the reported protective effects against neurolathyrism of antioxidants such as vitamin C, as well as a nutritionally adequate diet. It is hypothesized, as well, that neurolathyrism is rarely seen in women because of the protective effects of estrogen against cellular damage by ROS.

The toxic effects of high concentrations of glutamate in neurons involve both receptor mediated and receptor independent mechanisms. Under normal conditions of neuronal activation, glutamate activates glutamate receptors that induce calcium ion influx, which is coupled to a large variety of biological responses, including the release of neurotransmitters and growth factors and modulation of neurotransmission. This neuronal activation by glutamate is followed by a concerted deactivation of the pathway by a process that includes a decrease in glutamate concentration at the synapse. Uncontrolled activation of the receptor, however, leads to a prolonged increase in intracellular calcium ion concentration, which triggers ROS production and activates cell death pathways. Glutamate also functions by a receptor independent mechanism to induce cellular oxidative stress that has been called glutamate-induced oxytosis. Cells are sensitive to this type of toxicity because high levels of extracellular glutamate can block the glutamate–cysteine antiporter in the plasma membrane, which results in the inhibition of glutathione synthesis. Glutamate-induced oxytosis in cultured cells, then, involves depletion of GSH, increased ROS production, inhibition of complex I of the mitochondrial electron transport chain, increased calcium influx, and subsequent cell death. The observation that antioxidants can block the lathyrogenic effects of ODAP indicates that oxidative stress induced by both receptor independent and receptor dependent mechanisms are essential to the full development of the disorder.

Cyanogenic Glycosides

Cyanogenic glycosides are a group of water-soluble substances that occur naturally in over 2,500 plant species. Common plant products that are rich sources of cyanogenic glycosides include certain types of lima beans, sorghum, bitter almonds, choke cherry seeds, apple seeds, bamboo shoots, and cassava. Hydrolysis of these glycosides yields a ketone or aldehyde, a sugar, and the highly toxic hydrogen cyanide. Cyanogenic glycosides can function in the defense of many of these plants against invading pathogenic organisms and against many herbivores, including humans. Indeed, there are many well-documented reports of serious illness and deaths resulting from the consumption of cyanide-rich plants by humans, farm animals,

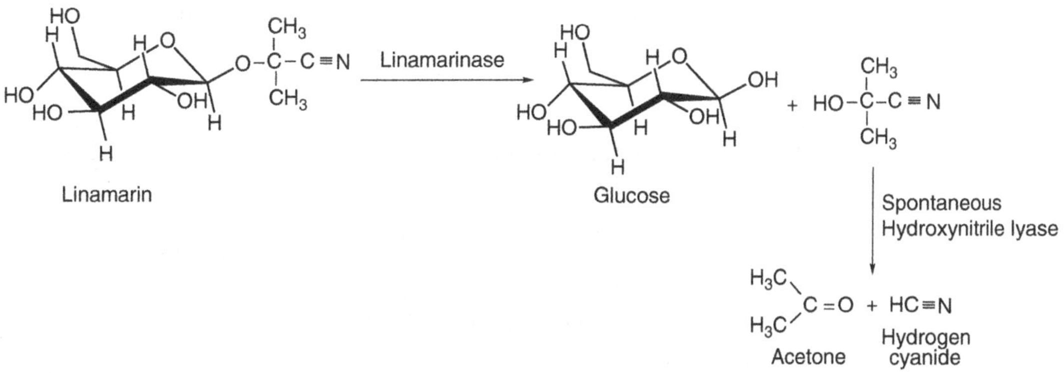

FIGURE 6.7 *Cyanide liberation from linamarin.*

insects, and other organisms. Hydrogen cyanide is released from cyanogenic glycosides in chewed or chopped plants by an enzymatic process involving two reactions (Figure 6.7).

The first step is hydrolysis by a β-glucosidase to yield the cyanohydrin and a sugar. Most cyanohydrins are relatively unstable in aqueous solutions and decompose spontaneously to the corresponding ketone or aldehyde and hydrogen cyanide. This decomposition is accelerated, however, by the action of the enzyme, hydroxynitrile lyase. The cyanogenic glycoside and the enzymes necessary for release of hydrogen cyanide are all present but separated in the plant. When fresh plant material is damaged, for example, by chewing, cutting, or during insect attack, cell wall structures are broken down sufficiently to allow the enzymes and the cyanogenic glycoside to come into contact and to allow the reaction to proceed.

The development of simple methods to detoxify cyanide-rich plant products has allowed their use as important food sources. For example, cassava root is a major source of dietary carbohydrate for hundreds of millions of people in South America and Africa. The levels of cyanide available from unprocessed cassava are high enough to produce serious effects of cyanide poisoning on chronic consumption. Detoxification of cassava is traditionally a multistep process that includes chopping and grinding in running water, which induces enzymatic production of hydrogen cyanide and removes both the freed cyanide and the glycoside in the wastewater. Fermentation and boiling processes are also used in some traditional procedures for production of cassava flour. However, in spite of well-developed processing procedures that can greatly reduce the levels of hydrogen cyanide available from cassava products, the hazard of chronic cyanide poisoning from cassava remains significant in some areas. A major problem continues to be the increased cost of more highly processed (and detoxified) flour. Efforts by public health and governmental organizations are ongoing to promote the production and consumption of adequately processed and safe cassava products.

Cyanide is considered to be a highly toxic substance with both acute and chronic effects. Cyanide produces multiple immediate biochemical effects in biological systems, including inhibition of the antioxidant defense, alteration of critical cellular ion homeostasis, and inhibition of cellular respiration. Overt symptoms of acute poisoning include mental confusion, muscular paralysis, and respiratory distress. The minimal lethal oral dose of hydrogen cyanide is estimated to be 0.5 to 3.5 mg/kg body weight. Cyanide exerts its acute toxic effects by binding to the ferric ion of cytochrome oxidase, the oxygen-reducing component of the mitochondrial electron transport chain. Recent studies have shown that this inhibitory effect of cyanide involves the low level production of nitric oxide, which augments the effect of cyanide by binding to the cupric ion present in cytochrome oxidase. The overall result is cessation of cellular respiration.

Cyanide ion can be metabolized to several products as indicated in Figure 6.8. The major excretion product of cyanide is thiocyanate, the production of which is catalyzed by the mitochondrial enzyme, thiosulfate sulfurtransferase, also known as rhodanese, and mercaptopyruvate sulfurtransferase (MST). Rhodanese is concentrated in the liver and not in mammalian targets of cyanide lethality, including the heart and central nervous system. MST, however, occurs ubiquitously in both the cytoplasm and mitochondria of cells of all tissues. The cosubstrate for the rhodanese reaction is thiosulfate, which is produced in a multistep process from cysteine that involves desulfuration of cysteine and oxidation of the released hydrogen sulfide. The cosubstrate for MST, mercaptopyruvate, is produced from cysteine as well, but by a direct deamination reaction. In the presence of decreased levels of thiosulfate, as may occur with nutritional deficiencies, cyanate can become a significant metabolite of cyanide. Cyanate can increase the toxicity of cyanide, as discussed next. An additional metabolic pathway for cyanide involves complexation with hydroxocobalamin, which is derived from vitamin B_{12}. This pathway, which

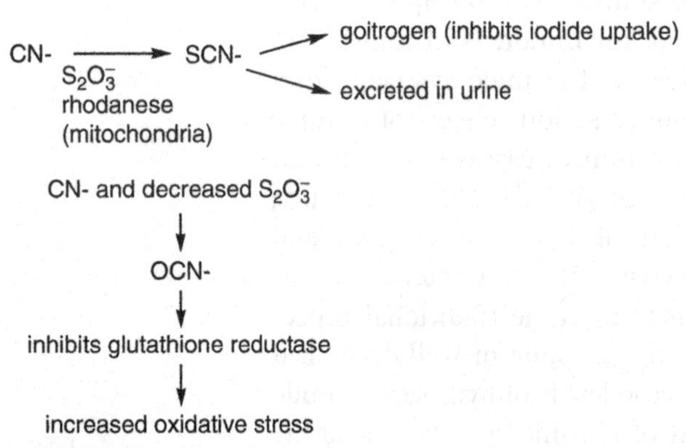

CN- → SCN- → goitrogen (inhibits iodide uptake)
SCN- → excreted in urine
$S_2O_3^-$
rhodanese
(mitochondria)

CN- and decreased $S_2O_3^-$
↓
OCN-
↓
inhibits glutathione reductase
↓
increased oxidative stress

FIGURE 6.8 *Metabolic pathways for cyanide ion.*

produces cyanocobalamin as an end product, may be important in low-level cyanide exposures.

Antidotes for acute cyanide poisoning are an important addition to first aid kits, especially in laboratory or industrial situations where exposure to high levels of cyanide is possible. Antidotes in current use include the cobalt complexes, hydroxocobalamin and dicobalt ededate, and the methemoglobin forming oxidants, nitrites and 4-dimethylaminophenol (DMAP). The cobalt atoms of the former substances form strong complexes with cyanide, which are then excreted in the urine. The heme oxidizing agents convert the ferrous form of heme to the ferric form, which, if present in sufficient concentration, removes cyanide from the ferric ion of cytochrome-*c* oxidase and allows respiration to resume. A volatile form of nitrite is amylnitrite, which can be administered to unconscious victims by inhalation. Amylnitrite is a comparatively weak heme-oxidizing agent, however, and its activity as a vasodilator is now understood to be a major contributor to its well-established antidotal properties. DMAP is a potent heme-oxidizing agent and is administered to patients by intravenous injection. The cobalt compounds are administered by intravenous injection, as well, in some cases along with thiosulfate to facilitate the conversion of cyanide to thiosulfate.

Of considerably greater importance than acute cyanide poisoning, in terms of the numbers of people affected, are the effects of chronic, low-level exposure to cyanide. Heavy consumption of cassava in a nutritionally marginal diet in parts of Africa and South America is associated with at least two disorders that occur with very low incidence in areas where cassava consumption is low, is part of a nutritionally adequate diet, or where cassava is cyanide free. Tropical ataxic neuropathy (TAN) is a neurological syndrome characterized by optic atrophy, ataxia, and deafness, which was first observed in the 1950s in West Africa. A related neurological disorder that results from prolonged consumption of cyanide-rich cassava is known as tropical amblyopia, and is characterized by atrophy of the optic nerves and blindness. A very similar syndrome, known as tobacco amblyopia, was identified in heavy smokers who also consumed a nutritionally deficient diet. The source of cyanide in this case was the tobacco smoke.

Individuals with these disorders have very low concentrations of sulfur-containing amino acids in the blood and elevated levels of plasma thiocyanate. Symptoms of the early stages of the diseases subside when patients are placed on cyanide-free and nutritionally adequate diets, especially including adequate levels of sulfur-containing amino acids and vitamin B_{12}. Goiter may be prevalent regardless of the nutritional quality of a cassava-based diet because of the goitrogenic effects of chronic exposure to

FIGURE 6.9 *Cassava* (Manihot esculenta).

thiocyanate in these populations. Again, as with the incidence of goiter, the prevalence of TAN and tropical amblyopia depend on economic factors as indicated by a rise in TAN incidence in Nigeria in the 1990s during an economic downturn for the region, compared to the more prosperous 1960s for this region.

The mechanism of the chronic toxicity of cyanide is an extension of the acute effects of this toxin. Thus, incomplete inhibition of cytochrome-*c* oxidase in the mitochondria results in the increased production of reaction oxygen species (ROS) and decreased production of ATP. Under these conditions, cells continue to respire but with decreased efficiency and with increased exposure to oxidative stress. Since adequate ATP production is required for the cell to mount its cellular defenses against oxidative stress, this combination of effects of cyanide is particularly hazardous to the cell. A further exacerbating factor is the production of cyanate from cyanide under conditions of low sulfur nutriture. Cyanate is an inhibitor of glutathione reductase, the inactivation of which further reduces cellular defenses against ROS. Although cyanide is no doubt toxic to all cells, the optic neurons are particularly sensitive. It is thought that small caliber axons and especially the long axons of the optic nerve are particularly sensitive to cyanide poisoning because of their normally high respiratory rate and relatively large lipid membrane surface area that is susceptible to ROS.

Lectins

Lectins are a remarkably complex group of widely occurring proteins and glycoproteins that can bind to certain carbohydrates. The lectins contribute to the defense of the plants against insects and herbivory by other organisms including humans since they possess cytotoxic, fungitoxic, insecticidal and antinematode properties either in vitro or in vivo, and are toxic to higher animals. Lectin binding to carbohydrate components of the cell wall can cause clumping or agglutination of the cells. This property is used as a basis for assays of red blood cell types. When lectins bind to carbohydrate components of intestinal epithelial cells, the result may be decreased absorption of nutrients from the digestive tract. In the case of the most highly toxic lectins, the binding is to the ribosome with very pronounced toxic consequences.

Lectins are widely distributed in nature and exhibit significant toxicities. Extracts of over 800 plant species and from numerous animal species show agglutinating activity. Of particular interest here are the lectins that occur in a wide variety of legumes used as food or feed, such as black beans, soybeans, lima beans, kidney beans, peas, and lentils, to name a few. Lectins isolated from black beans cause growth retardation when fed to rats at 0.5% of the diet, and the lectin from kidney beans produces death in rats fed on lectin at 0.5% of the diet for two weeks. Soybean lectin, a less toxic lectin, fed at 1% of the diet to rats produces only growth retardation, with an estimated LD_{50} of about 50 mg/kg. Although thorough cooking detoxifies lectins in legumes, one of the most nutritionally important features of uncooked plant lectins is their ability to survive digestion by the gastrointestinal tract following consumption. This allows the lectins to bind to membrane glycosyl groups of the cells lining the digestive tract. As a result of this interaction, a series of harmful local and systemic reactions is triggered. Locally, lectins can affect the turnover and loss of gut epithelial cells, damage the luminal membranes of the epithelium, interfere with nutrient digestion and absorption, stimulate shifts in the bacterial flora, and modulate the immune state of the digestive tract. Systemically, lectins can disrupt lipid, carbohydrate, and protein metabolism, promote enlargement and/or atrophy of key internal organs and tissues, and alter hormonal and immunological status. At high intakes, lectins can seriously threaten the growth and health of consumers.

Although the toxicities of the large majority of plant lectins are characterized as chronic antinutritional disorders, a few lectins are notorious as highly potent acute toxins. Most notable in this group is ricin, a lectin from castor bean. The seed of the castor bean (*Ricinus communis*, see Figure 6.10) has been recognized for its toxicity for millennia, with clear references to its use in folk medicine in India dating to the sixth century BCE. As few as five castor beans are said to contain enough ricin to kill an adult human. The toxic effects of ricin are seen within two to three hours

FIGURE 6.10 *Castor bean plant* (Ricinus communis).

following consumption of the castor beans or ricin-contaminated material, and include diarrhea, nausea, vomiting, abdominal cramps, internal bleeding, liver and kidney failure, and circulatory failure. Rapid heartbeat can also occur. If the castor beans are swallowed whole, the poisoning will be less severe than if the beans are chewed because little ricin is released from the whole bean. Breathing dust that contains ricin causes cough, weakness, fever, nausea, muscle aches, difficult breathing, chest pain, and cyanosis. Breathing the dust can result in respiratory and circulatory failure. The oral LD_{50} for ricin in mice is 30 mg/kg, which is approximately 1000-fold higher than by injection or inhalation. LD_{50} estimates for humans are in the range of 1 to 20 mg of ricin/kg of body weight following oral exposure, and presumably much lower following inhalation.

Ricin was isolated in purified form in 1 to 5% yield from the castor bean in the late nineteenth century as a relatively heat-stable glycoprotein of approximately 60 kDa molecular weight. The steam distillation process used to prepare castor oil is sufficient to deactivate this protein, which renders this often used folk remedy safe for consumption. Subsequent studies showed that ricin is resistant to digestion. As indicated in Figure 6.11, ricin is composed of two protein chains, A and B, linked by a disulfide bond. The B chain is a lectin and binds to galactose-containing glycoproteins and glycolipids expressed on the surface of cells, facilitating the entry of ricin into the cytosol. The A chain inhibits protein synthesis by irreversibly inactivating eukaryotic ribosomes through removal of a single adenine residue from the 28S ribosomal RNA loop contained within the 60S subunit. This process prevents chain elongation of polypeptides and leads to cell death because of inhibition of protein synthesis. Additional mechanisms of toxicity have been reported as well, including activation of apoptosis pathways, direct cell membrane damage, alteration of membrane structure and function, and release of cytokine inflammatory mediators. A broad group of bacterial and plant toxins have similar A- and B-chain protein components, including diphtheria toxin.

Since ricin is a readily available and highly potent toxin, it has been put to a range of negative and positive uses in modern times. Ricin's properties as a chemical warfare agent were examined and developed in the early

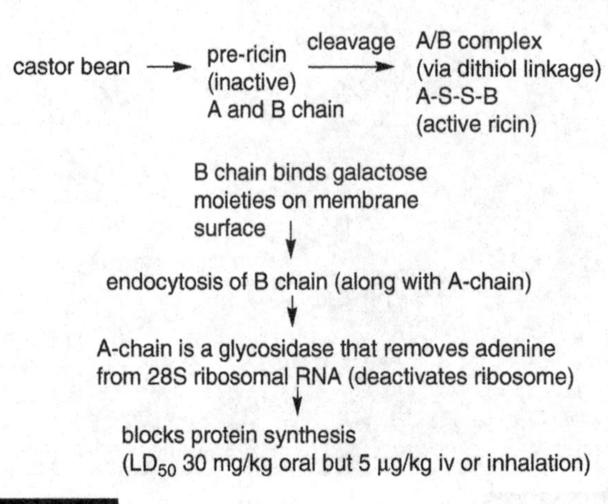

FIGURE 6.11 *Mechanism of toxic action by ricin.*

twentieth century in the United States and elsewhere. More recently, ricin is thought to have been used in the assassination of an international journalist and in an attempted assassination of a US politician. On the more positive side, however, ricin is being studied for use in cancer chemotherapy, in bone marrow transplantation, and in cell-based research. Malignant cells apparently are more susceptible to ricin toxicity than nonmalignant cells because the former express more carbohydrate containing surface lectin binding sites than do nonmalignant cells. Antibody-conjugated ricin can target cancer cells and has been investigated as an immunotherapeutic agent.

Vasoactive Amines

Vasoactive amines are small molecular weight basic organic compounds that can elicit the sympathetic, so-called "flight-or-fight" response in higher organisms. In response to an appropriate stimulus, neurons in this system release primarily norepinephrine and epinephrine, also called noradrenaline and adrenaline, respectively, which bind to adrenergic receptors in peripheral tissues. As shown in Figure 6.12, these substances, along with dopamine, are structurally related catecholamines. Also included in the group of vasoactive amines are serotonin and tyramine. Activation of the adrenergic pathways by these substances induce the characteristic responses that include pupil dilation, increased sweating, increased heart rate, occasional vomiting, and increased blood pressure. If this response is improperly controlled, life-threatening symptoms can result, including heart arrhythmia, hypertensive crisis, and death.

Because exogenous vasoactive amines, such as those that occur in certain foods, can be highly toxic, an effective means of detoxification of these substances has evolved. This fact is clear from the observation that these substances are far more toxic when administered by intravenous injection than by ingestion. Indeed, circulating levels of exogenous vasoactive amines are carefully controlled by the action of monoamine oxidases, MAO-A and MAO-B. MAO-A deaminates serotonin in the central nervous system, as

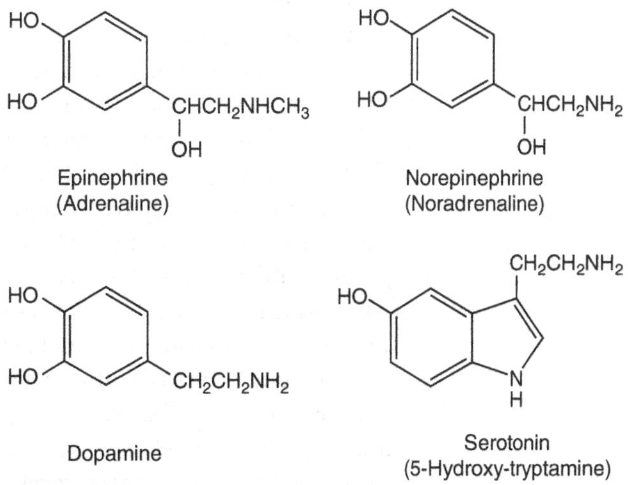

Epinephrine (Adrenaline) — Norepinephrine (Noradrenaline) — Dopamine — Serotonin (5-Hydroxy-tryptamine)

FIGURE 6.12 *Chemical structures of vasoactive amines.*

well as dietary monoamines in the gastrointestinal system. MAO-B is found predominantly in liver and muscle and deaminates dopamine and phenylethylamine. Bioactive amines may also be deaminated by diamine oxidase (DAO) activity in the gut that can provide protection from small amounts of amines that are normally present in foods. Because of rapid metabolic conversion of amines by MAO and DAO, oral administration of vasoactive amines to normal mammals generally has little effect on blood pressure. Marked effects are seen, however, when patients or experimental animals also are treated with MAO inhibitors.

With the development of MAO inhibitors for therapeutic purposes, it became clear that dietary vasoactive amines could present a significant threat to human health. MAO inhibitors are used to inhibit the actions of monoamine oxidase, especially in the central nervous system, to serve as antidepressants or as anti-Parkinsonian agents.

The first generation of MAO inhibitor drugs, including Isocarboxazid, Marplan, Nardil, and Parnate, are nonspecific, inhibit both isoforms of MAO, and the inhibition is considered irreversible. The second generation of drugs, termed RIMA for reversible inhibitor monoamine, which include L-Deprenyl, Rasagiline and selegiline, are selective in the inhibition of MAO-B, reversible, and carry little risk of a hypertensive effect in low dosage as in the treatment of Parkinson disease. To be effective in depression treatment, however, the higher dosages that are required begin to inhibit all isoforms, and the potential for a hypertensive crisis increases.

The levels of the vasoactive amines have been examined in a wide range of food items. The results of these analyses indicate that low levels (0–28 μg/g) of several of these substances, including serotonin, tyramine, dopamine, and norepinephrine, occur naturally in banana pulp, tomato, avocado, potato, spinach, and orange. Much higher levels of specifically tyramine (20–2170 μg/g) were detected primarily in food products that were subjected intentionally or as a result of spoilage to microbial action. Foods that are most consistently in this group and, therefore, are considered as significant hazards for patients under treatment with MAO inhibitors include aged cheeses, aged and cured meat, improperly stored or spoiled meats, poultry and fish, Marmite (yeast extract), sauerkraut, soy sauce and other soy condiments, and tap beer. Lower levels of vasoactive amines, and lower hazard, are associated with red and white wine, and bottled or canned beers.

Any foods with free amino acids, especially tyrosine and phenylalanine, are subject to vasoactive amine formation (Figure 6.13) if poor sanitation and low quality foods are used, or if the food is subjected to storage conditions suitable for bacterial growth. An estimated 80% of all hypertensive events are associated with consumption of cheese by patients treated with

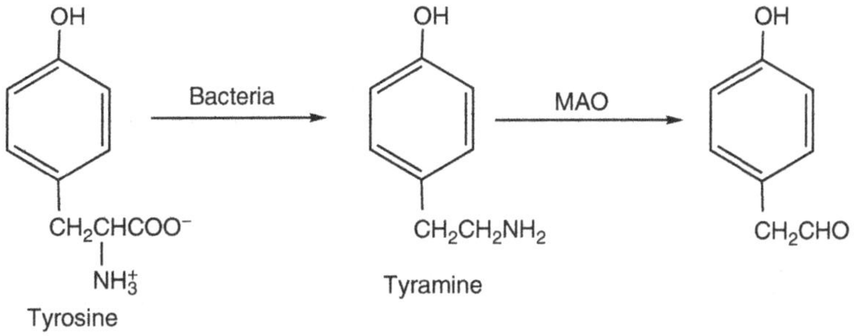

OH Bacteria OH MAO OH

CH_2CHCOO^- | NH_3^+ Tyrosine

$CH_2CH_2NH_2$ Tyramine

CH_2CHO

FIGURE 6.13
Vasoactive amine formation.

MAO inhibitors. These patients should not consume a few foods such as aged cheeses, especially English Stilton, Cheshire, and Danish bleu, because a clinically significant level of tyramine may be present in one ounce or less.

Caffeine

Caffeine is a methylated xanthine derivative that occurs naturally in highest levels in coffee, tea, and chocolate products. As indicated in Table 6.1, the levels of caffeine in coffee range from 40 to 180 mg/5-oz. cup and from 20 to 110 mg/cup for traditional teas made from leaves of *Camellia sinensis*. Baker's chocolate and dark chocolate contain caffeine levels of 5 to 35 mg/oz. with levels ranging from 2 to 30 mg/cup for chocolate drinks. Significant amounts of caffeine are added to certain soft drinks and over-the-counter headache medications, as well. It is interesting to note, further, that the caffeine content of very large coffee portions that are sold in specialty outlets is nearly equal to the caffeine in three over-the-counter caffeine pills. Because of the widespread availability and use of caffeine and caffeine-containing products, caffeine is considered to be the most widely self-prescribed and used drug in modern society.

The discoveries of caffeine-containing plants are ancient and the associated legends are part of the folklore of many cultures. Archeological evidence indicates that our human ancestors may have consumed caffeine-containing plants as far back in time as 3500 BCE. Early peoples found that chewing the seeds, bark, or leaves of certain plants had the effects of easing fatigue, stimulating awareness, and elevating mood. A popular Chinese legend claims that an emperor, who reigned in about 3000 BCE, accidentally discovered that when some leaves, presumably tea leaves, fell into boiling water, a fragrant and restorative drink resulted. The history of coffee has been recorded as far back as the ninth century CE. A popular legend traces

Table 6.1 Caffeine Levels in Various Beverages

Item	Average (mg)	Range
Coffee (5-oz. cup)		
Brewed, drip method	115	60–180
Brewed, percolator	80	40–170
Instant	65	30–120
Decaffeinated, brewed	3	2–5
Decaffeinated, instant	2	1–5
Teas (5-oz. cup)		
Brewed, major U.S. brands	40	20–90
Brewed, imported brands	60	25–110
Instant	30	25–50
Iced (12-oz. glass)	70	67–76
Some soft drinks (6 oz.)	18	15–30
Cocoa beverage (5 oz.)	4	2–20
Chocolate milk beverage (8 oz.)	5	2–7
Milk chocolate (1 oz.)	6	1–15
Dark chocolate, semi-sweet (1 oz.)	20	5–35
Baker's chocolate (1 oz.)	26	26
Chocolate-flavored syrup (1 oz.)	4	4
NoDoz max strength (1 pill)	200	
Mountain Dew (12 oz.)	55	
Excedrin (1 pill)	35	
Espresso (1 oz. Starbucks)	35	
Coffee (8 oz. Starbucks)	250	
Coffee, grande (16 oz. Starbucks)	550	

its discovery to an Ethiopian goat herder who observed goats that became elated and sleepless at night after browsing on coffee shrubs. On trying the berries that the goats had been eating, the goat herder experienced the same vitality. The legend suggests further that the wife of the goat herder inadvertently dry roasted the berries and used them to make an aqueous decoction that we call coffee. The use of cola nut and cocoa are also ancient, with evidence for the use of the latter dating to 600 BCE in Mayan culture.

Caffeine is a neurological stimulant that produces biological effects in almost all organs of the body. The major cellular action of caffeine is to block receptors of adenosine, which act as negative regulators of function throughout the body. Blockage of these receptors by caffeine and the related purine, theophylline, therefore stimulates activities of the associated organs and tissues, including the central nervous system. At low doses of around 200 mg for an adult, caffeine produces central nervous system stimulation, diuresis, relaxation of smooth muscles, cardiac muscle stimulation, and increased gastric secretion. The centuries-old belief that caffeine can improve physical performance in fatigued individuals has been substantiated by many studies, but the physical performance of rested subjects is not improved by caffeine.

The authors of one such study that measured the effects of caffeine on the performance of members of the US Navy Seals Special Forces indicated "even in the most adverse circumstances, moderate doses of caffeine could improve cognitive function, including vigilance, learning, memory, and mood state." The authors of another study of the effects of caffeine on stress from test-taking in students found that "caffeine had significant effects on blood pressure and heart rate in habitual coffee drinkers that persisted for many hours during the activities of everyday life" and that "caffeine may exaggerate responses to the stressful events of normal daily life." There is considerable current interest, as well, in the observation that caffeine consumption shows a strong negative correlation with the incidence of Alzheimer's disease. Results of studies with a rodent model of this disease also show a strong protective effect of caffeine for this disorder.

Excessive consumption of caffeine can produce many adverse reactions, as well. The most consistently observed toxic effects of caffeine include nervousness, irritability and cardiac arrhythmias. Adverse effects can also include vomiting, abdominal pain, agitation and seizures. The oral LD_{50} for caffeine is approximately 200 mg/kg, or about 12 g/60-kg person. Thus, a lethal dose for an adult of this moderately toxic substance would require rapid consumption of about 50 cups of coffee or 50 caffeine tablets. Death occurs from ventricular fibrillation with blood caffeine levels over 100 μg/ml.

The consumption of caffeine by pregnant and nursing women has been of considerable recent concern. There was general agreement until recently that low caffeine intake (<150 mg/day) during pregnancy is not likely to harm the fetus, and only high caffeine intake (>300 mg/day, equivalent to more than 3 cups of coffee/day) should be avoided during pregnancy due to an association with increased incidence of birth defects. The results of an important recent study conducted with a large population of patients associated with a major health organization and controlled for many variables resulting from pregnancy, however, suggest that these allowances for caffeine consumption during pregnancy may be too generous. This recent study showed that women who consumed only 200 mg or more of caffeine per day had fully twice the miscarriage risk as women who consumed no caffeine. The results showed further that even women who consumed less than 200 mg of caffeine daily had more than a 40 percent increased risk of miscarriage compared to women who did not consume caffeine. The increased risk of miscarriage appeared to be due to the caffeine itself, rather than other possible chemicals in coffee because caffeine intake from noncoffee sources such as caffeinated soda, tea, and hot chocolate showed a similar increased risk of miscarriage. If these findings are reproduced by other investigators in other settings, they are likely to lead to recommendations for greater restrictions on caffeine consumption by pregnant women. In the meantime, it is prudent from women who are pregnant or about to become pregnant to strongly curtail or discontinue their consumption of caffeine.

Curare

Native peoples throughout the world have employed natural plant materials as poisons for hunting and warfare since prehistoric times. For centuries prior to the arrival of Europeans, the indigenous human populations of the Amazon basin in South America used extracts primarily of the climbing shrubs or vines *Strychnos toxifera* and *Chondrodendron tomentosum* as sources of a potent arrow and dart poison known as curare, which is translated into English as "the flying death."

According to published reports on curare that first appeared in 1516, tribes in the Amazon region had developed a variety of methods for the preparation of the poison. All the methods involved crushing and cooking the roots, bark, and stems of the selected plant, along with the addition of other plants and venomous animals. A light syrup was produced by repeated boiling of the mixture and was applied to the tips of arrows and darts. The latter were carefully heated near a flame to produce a hardened, black, or dark brown, tar-like coating that was resistant to damage during storage or use.

Continuing studies since the discovery of curare active compounds have well established their toxic mode of action. Crystalline d-tubocurarine (Figure 6.14), the primary active compound in curare, was isolated in 1935 in the classic studies by Harold King. d-Tubocurarine was shown subsequently to block nicotinic acetylcholine receptors at neuromuscular junctions to produce paralysis in muscles. This alkaloid first affects the muscles of the toes, ears, and eyes, and then muscles of the neck, arms, and legs. Finally, with a high enough dose, muscles of the lungs are paralyzed and the victim dies by asphyxiation. The victim remains awake and aware that he is

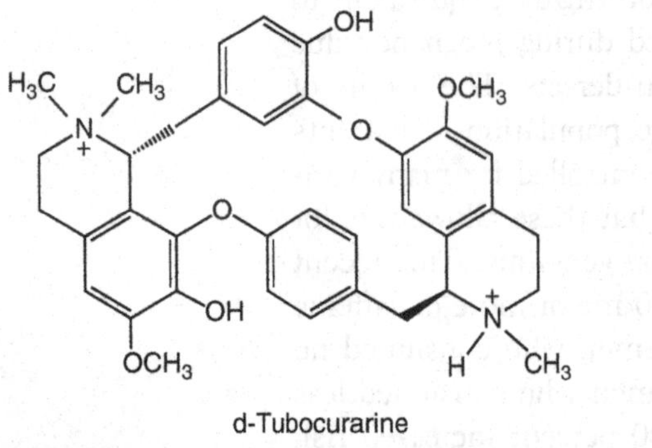

FIGURE 6.14 Chemical structure for d-tubocurarine.

unable to move and is losing the ability to breathe. The LD_{50} for d-tubocurarine was found to be only 0.5 mg/kg by intravenous injection into rabbits, which classifies this alkaloid as an extremely toxic agent. Recovery is possible, however, if respiration is maintained, which allows for the metabolism, detoxification, and excretion of the toxin. Curare alkaloids are not toxic following ingestion, apparently because their high degree of polarity inhibits their absorption for the gastrointestinal tract. This explains why game taken with curare-coated arrows and darts is safe to eat and why natives could safely test the potency of their poison preparations by tasting the mixture for bitterness!

There have been many mostly unsubstantiated uses of curare in folk medicine of indigenous tribes. Thus, curare has been used as a diuretic and for madness, dropsy, bruises, edema, fever, and kidney stones. d-Tubocurarine has been put to very effective use in Western medicine, however, starting in the 1940s as a muscle relaxant during surgical anesthesia. The medical uses of the curare compounds, however, have been superseded by a number of curare-like synthetic drugs that have a similar pharmacodynamic profile but with fewer side effects.

Strychnine

As with curare, the first uses of strychnine were as a mixture with other natural products in a plant extract preparation applied to the tips of arrows and darts by native hunters since before recorded history. The principal source of strychnine is *Strychnos nux-vomica*, an evergreen tree native to Southeast Asia, especially India and Myanmar, and cultivated elsewhere (Figure 6.15). Its dried seeds or beans and sometimes its bark (called nux-vomica) were used as ordeal poisons by natives and continue to be used by some groups in herbal remedies. The seeds were first brought to Europe in the fifteenth century, likely as a poison for game and rodents. The seeds were first used in European medicine in 1640 as a stimulant.

Because of its potent biological effects, studies of the active principles of *S. nux-vomica* were initiated very early in the history of the development of organic chemistry as a science. The potent emetic activity of the extracts of this plant were well known to western science by the time the species was named with the Latin designation for "vomiting nut" by Linnaeus in the eighteenth century. Subsequent studies starting in 1818 by French chemists Pelletier and Caventou were guided by this activity and resulted in the isolation of strychnine as the active principle (Figure 6.16). The structural complexity of the compound was beyond the analytical capabilities of

FIGURE 6.15 *The principal source of strychnine* (Strychnos nux-vomica).

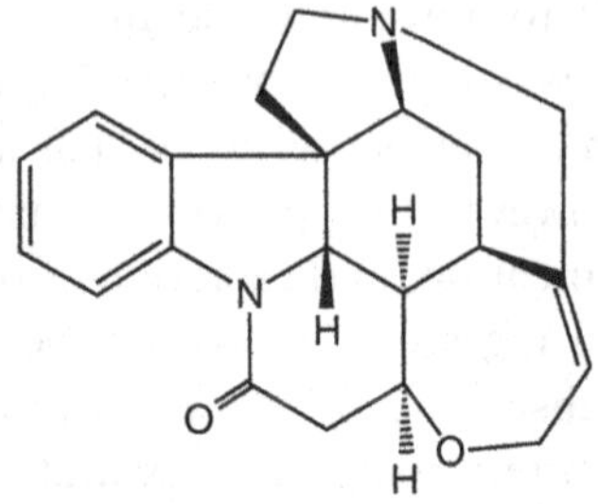

FIGURE 6.16 *Chemical structure for strychnine.*

the time, however, and the structure of strychnine was elucidated over 100 years later in 1947 by the English chemist, Sir Robert Robinson, and coworkers. The total synthesis of the compound was achieved in 1963 by the American chemist, R.B. Woodward, and coworkers. Their successful studies with strychnine were crowning achievements of very productive careers for Robinson and Woodward, and both received Nobel Prizes in Chemistry shortly after the publication of their work on this famous natural product.

The biological activity and mode of action of strychnine have been the subjects of many studies since the compound was purified at the beginning of the nineteenth century. Strychnine is considered to be one of the bitterest substances known, with a detectable limit of as low as 1 ppm by weight in water. By stark contrast, the taste threshold for sucrose is nearly 2,000 ppm for humans. This very strong bitterness is thought to contribute to its emetic properties and may account for its use in ordeal trials. Rapid consumption of the *nux-vomica* preparation by a fearless innocent individual leads to the rapid and efficient expulsion of the stomach contents before the toxin can be absorbed in sufficient quantities to cause death. A slower, more tentative rate of consumption by a guilty person, however, would avoid the vomit reflex with lethal consequences. The major toxic effects of strychnine progress from flexed muscles up to painful immobility, convulsions, and finally to complete muscle paralysis. Spasm of the chest muscles and diaphragm result in hypoxia and respiratory failure. Lethal doses for an adult person are in the range of only 0.2 to 0.4 mg/kg body weight, which makes strychnine an extremely toxic agent.

Studies of the mode of action of strychnine indicate that this alkaloid is a highly selective neurotoxin. Strychnine is readily absorbed following ingestion and inhalation and can accumulate in lipid storage sites in the body. Thus, chronic exposure to doses of strychnine that are not acutely toxic can eventually produce a toxic effect. Strychnine blocks postsynaptic receptors of the inhibitory neurotransmitter, glycine, in the spinal cord and motor neurons. Studies suggest that glycine and strychnine bind to partially overlapping sites on this receptor. Strychnine-sensitive glycine receptors are the predominant carriers of fast inhibitory transmission at synapses in the vertebrate spinal cord and brain stem. These receptors belong to the family of ligand-gated ion channels, which also includes acetylcholine and GABA receptors. Activation of this kind of receptor by glycine or other agonists induces the opening of the anion-selective channel of the receptor, thereby allowing influx of chloride into the cytoplasm. The resulting hyperpolarization of the postsynaptic membrane stabilizes the resting potential of the cell and thus inhibits neuronal firing.

Atropine

Atropine is a neuroactive natural product with a long and intriguing history of both illicit and licit uses. Atropine occurs with several other tropane alkaloids prominently in plants of the Solanaceae family, primarily *Mandragora officinarum*, formerly called *Atropa mandragora*, and *Atropa belladonna* (Figure 6.17). *M. officinarum* (also called love apple, love plant, and devil's apple) is a low-growing, large-leafed, flowering plant with deep and large roots that is native to the Mediterranean area and southern Europe, and is a common plant throughout the Middle East.

A. belladonna, also called belladonna, deadly nightshade, herb of Circe, or witch's berry, is a flowering bush that is native to Europe, North Africa,

FIGURE 6.17 Atropa belladonna.

and Western Asia. The term belladonna, meaning beautiful lady in Italian, usually is thought to be derived from the practice of fashionable women, dating back to Cleopatra in Egypt in the first century BCE, to use plant extracts to dilate their pupils for cosmetic purposes. Other interpretations suggest that the name may have referred to a magical or beautiful lady (or spirit) of the forest, and the term may have been used as a substitute for witch, suggesting an association with the use of herbal remedies and mysterious potions.

The use of these plants is thought to date back to prehistoric times, and written records indicate that the plants were put to rather sophisticated uses in ancient Greece and Egypt. For example, the Book of Genesis in the Bible (circa 1400 BCE) describes the use of mandrake to promote conception. Artifacts from ancient Egypt dating from about 1370 BCE suggest the use of mandrake as an aphrodisiac and analgesic agent. The early Greek scholar, Theophrastus (370–285 BCE), stated that mandrake is useful for wounds, skin infection, gout, and sleeplessness. A mixture of mandrake and wine is reported to have been used by a Carthaginian general around

200 BCE to immobilize an invading army. In 184 BCE, Hannibal's army used belladonna plants to induce disorientation in the enemy troops. Around 35 BCE, Mark Anthony's army is said to have been poisoned by *A. belladonna,* and Cleopatra used atropine containing plant extracts to dilate her pupils. The Greek herbalist Dioscorides (ca. 40–90 CE) apparently was the first to record the use of *A. belladonna* as an anesthetic agent.

Large doses of belladonna were used in medieval Europe by witchcraft and devilworship cults to produce hallucinogenic effects. Belladonna also was used during the Middle Ages for psychological torture to gain confessions, both true and false, from stubborn victims. A preparation of the root of *A. belladonna* began to be used as a topical pain reliever in 1860. Belladonna became a major item of world commerce in the succeeding decades for pain relief.

Studies of the isolation of the active principle and mode of action of belladonna and mandrake proceeded relatively quickly following the establishment of these plants as bona fide medicinal usefulness. Using the mydriatic activity as a guide, the primary active principle, atropine, apparently was first isolated in pure crystalline form from dried belladonna root in 1831 by a German pharmacist, A. Mein (Figure 6.18). The substance first was synthesized by Nobel Prize winning German chemist Richard Willstätter in 1901. Subsequently, it was established that the natural product is a racemic mixture of D- and L-hyoscyamine and that most of its physiological effects are due to L-hyoscyamine.

Atropine produces many dramatic effects in humans and experimental animals. The toxic symptoms of atropine include dry mouth, drowsiness, dizziness, constipation, and nausea. Other effects include pupil dilation, blurred vision, fever (due to the inability to perspire), inability to urinate, heart arrhythmia, and excessively dry mouth and eyes. Higher doses can produce sensations of a burning throat, delirium, restlessness and mania, hallucinations, difficulty breathing, and flushed skin that is hot and dry. Death will ensue with constriction of the airways and suffocation. Whereas the therapeutic doses of atropine are in the range of only about 30 μg/kg or less, lethal effects are reported in humans following doses of about 1.5 mg/kg and greater. Atropine has relatively little effect on most domestic animals and birds, but is quite poisonous to dogs and cats. The effects result from atropine's ability to lower the "rest and digest" activity of all muscles and glands regulated by the parasympathetic nervous system. Atropine is a potent competitive antagonist of the human muscarinic

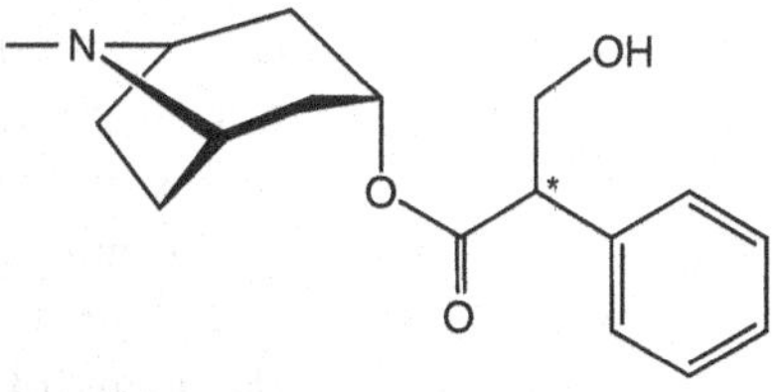

FIGURE 6.18 Chemical structure of atropine.

acetylcholine receptors. Of the many recognized activities exhibited by atropine including narcotic, diuretic, sedative, antispasmodic, and mydriatic, the most important uses currently are in the treatment of eye diseases and as an antidote to anticholinesterase nerve agents that have been used in recent times in chemical warfare and terrorism.

PHYTOALEXINS

Phytoalexins are plant metabolites that are produced by a plant in response to environmental stresses. Invading organisms such as bacteria, viruses, fungi, and nematodes can induce the production of phytoalexins in plants, often with the result that the growth of the invading organism in inhibited. This phenomenon is a general response to stress, however, since exposure to cold, ultraviolet light, physical damage, and certain chemical compounds such as metal salts, polyamines, and certain pesticides also can elicit the production of the same phytoalexins in a given plant. A classic example of production of these so-called stress metabolites, or natural pesticides, occurs in potatoes inoculated with the blight fungus *Phytophthora infestans*. When administered on the surface of a potato slice, certain strains of this fungus will initially grow rapidly, and then their growth rates will gradually decrease. If an extract of the infected potato slice is placed in contact with a pure culture of the same fungus, the fungus will not grow. This response to a variety of fungi has been observed in many other plants such as peas, green beans, broad beans, soybeans, carrots, sugar beets, cabbage, and broccoli, to list just a few. The plant responses in many of these cases are triggered by certain polysaccharide components of the fungal cell walls. The quantities of phytoalexins produced by a plant can be quite remarkable. For example, soybeans infected with the fungus *Phytophthora megasperma* can produce glyceolin I at levels as high a 10% of the dry weight of the infected tissue within just a few days.

The chemical structures of the phytoalexins are generally modifications of structures of metabolites produced in the unstressed plants. This indicates that the compounds are produced by modifications of the plant's normal metabolism. Representative structures of several well-studied phytoalexins are presented in Figure 6.19.

Because of their chemical diversity, abundance, pesticide-like activities, and their presence in the feed and food chains, the phytoalexins have been the focus of considerable concern primarily because of their well-demonstrated effects in livestock and their potential effects in humans. Sweet potato phytoalexins provide an example of the considerable toxic potential

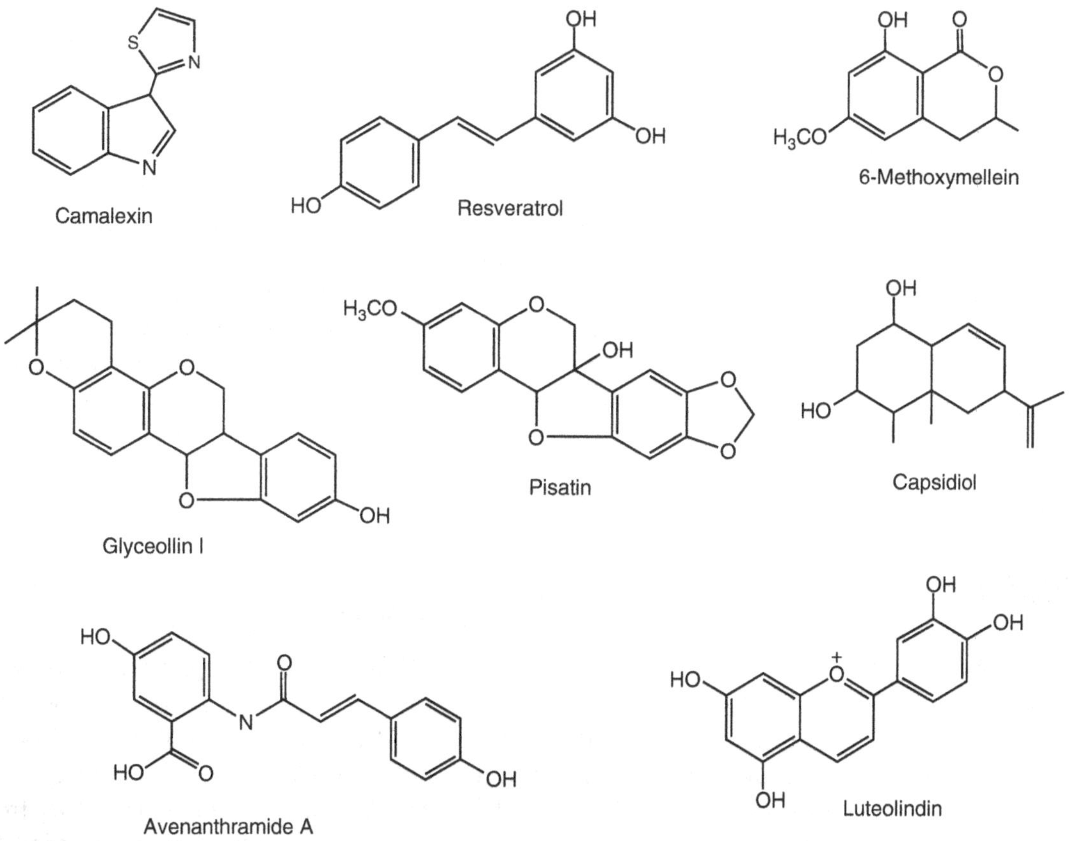

FIGURE 6.19 *Chemical structures of phytoalexins.*

of these metabolites. Consumption of moldy sweet potatoes in the feed is known to produce severe respiratory distress, pulmonary edema, congestion, and death in cattle. The toxic substances present in the moldy sweet potatoes, but not in the uninfected sweet potatoes, are a group of structurally related furans (Figure 6.20). Two of the compounds, ipomeamarone and ipomeamaronol, cause liver degeneration in experimental animals with LD_{50} = 230 mg/kg. The lung edema factors from infected tubers include 4-ipomeanol, 1-ipomeanol, ipomeanine and 1,4-ipomeadiol, with ipomeanine being the most toxic of the group to mice with an oral LD_{50} = 26 mg/kg. Each of these substances can produce an acute toxic response in mice that is indistinguishable from the acute response produced by the administration of the crude sweet potato extract from infected tubers.

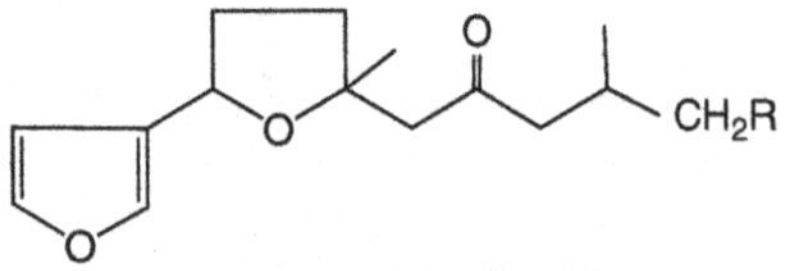

Ipomeamarone: R = H
Ipomeamaronol: R = OH

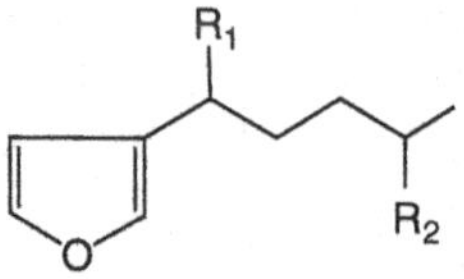

	R_1	R_2
4-Ipomeanol	H	OH
1,4-Ipomeadiol	OH	OH
Ipomeanine	H	O
1-Ipomeanol	OH	H

FIGURE 6.20 *Toxins from damaged sweet potatoes.*

The toxic terpenes from damaged sweet potato can occur in only slightly damaged tubers used for food. The presence of these substances is associated with a darkened, transverse ring below the skin of the tuber. The liver toxin, ipomeamarone, was shown to be present in commercially available sweet potatoes at levels ranging from 0.1 to 7.8 mg/g of sweet potato. Thus, human exposure to toxic phytoalexins from sweet potato and other plant products is likely to be quite common. Nevertheless, the levels of exposure appear generally to be below the toxic range since there seems to be no documented case of human poisoning from metabolites of stressed plants of any genus. On the contrary, as described in Chapter 12 of this volume, some phytoalexins such as resveratrol from grapes, exhibit potentially important beneficial effects in humans.

HERB-DRUG INTERACTIONS

Herbal products have been used in traditional medicine for millennia and in some cases have proved to be sources of currently used medicines. In recent decades there has been a resurgence of interest in traditional and new herbal products even in advanced societies for the prevention and treatment of many human maladies from memory loss to cancer. Although there are many claimed benefits for the use of these remedies with little established adverse effects, progress in testing of this very large group of substances for safety and efficacy has been slow. An increasingly recognized problem with the use of many of these products, however, is their interaction with the effects of the established therapeutic agents used in Western medicine.

Although a full discussion of the many reported herb–drug interactions is beyond the scope of this discussion, a few examples of important interactions will be presented as an introduction to this field of study.

St. John's wort is one of the most commonly used herbs for the self-treatment of mild to moderate depression in Western countries. St. John's wort produces several polyphenolic substances, most notably hypericin, that can decrease the beneficial anticoagulant effect of warfarin, apparently by inducing the expression of CYP1A2 and CYP3A4 in the liver. These enzymes are largely responsible for the metabolism of warfarin. St. John's wort and cyclosporine coadministration after organ transplantation may result in therapeutic failure of cyclosporine and transplant graft rejection, also because of increased metabolism of cyclosporine.

Grapefruit juice contains naringin and bergamottin, which are potent inhibitors of CYP3A4 and CYP1A2. Thus, in contrast to St. John's wort, grapefruit consumption can decrease the metabolism and increase the potency of several drugs including cyclosporin, estrogens, and others. This effect can be very strong as indicated by an observed nine-fold increase in serum concentration of a cholesterol-lowering drug simvastatin following grapefruit juice consumption.

Watercress was reported to contain unidentified phytochemicals that are potent inhibitors of CYP2E1, a phase I metabolic enzyme that contributes to the metabolism of many drugs including acetaminophen and the muscle relaxant, chlorzoxazone. A single treatment with 50 g of watercress delayed the clearance of chlorzoxazone by over 50% in human volunteers.

Tomatoes, eggplant, and potatoes contain solanaceous glycoalkaloids that can delay recovery from anesthesia due to acetylcholinesterase inhibition.

Calcium-containing dairy products (milk, yogurt, cheese) and supplements can block gut absorption of tetracycline antibiotics.

Charbroiled meat products contain polycyclic aromatic hydrocarbons that can induce CYPs and increase metabolism of antiasthmatic drugs, warfarin, steroid hormones, and other drugs.

Other herbs with established adverse effects on various medications include garlic, ginseng, ginkgo, kava, milk thistle, licorice, *Echinacea*, valerian, *Ephedra*, bitter melon, *Lycium*, papaya, borage, and fenugreek as indicated in Table 6.2.

Table 6.2 Herbs with Established Adverse Effects on Various Medications

Medical Plant	Vernacular Name	Drugs	Herb-Drug Interactions
Hypericum perforatum L.	St. John's wort	Cyclosporine, Midazolam, Tacrolimus, Amitriptyline, Digoxin, Indinavir, Warfarin, Phenprocoumon, Theophylline, Irinotecan, Alprazolam, Dextromethorphan, Simvastatin	Decreases blood concentrations of these drugs
		Oral contraceptives (ethinyl estradiol/desogestrel)	Breakthrough bleeding and unplanned pregnancies
		Sertraline, paroxetine, and nefazodone	Serotonin syndrome
		Antidepressants or serotonergic drugs	Gastrointestinal disorder, allergic reactions, fatigue, dizziness, confusion, dry mouth, photosensitivity
Allium sativum L.	Garlic	Saquinavir	Decreases plasma concentration of saquinavir
		Warfarin sodium	Alters bleeding time
		Paracetamol	Pharmacokinetic variables of paracetamol changes
		Chlorpropamide	Hypoglycemia
Panax ginseng	Ginseng	Phenelzine sulfate	Induction of mania and blood concentration reduction of alcohol (ethanol) and warfarin; headache; trembling
		Estrogens or Corticosteroids	Additive effects
Salvia miltiorrhiza (Lamiaceae)	Danshen	Warfarin	Enhances anticoagulation and bleeding
Ginkgo biloba L.	Ginkgo	Warfarin, aspirin, ticlopidine, clopidogrel, dipyridamole	Bleeding
		Thiazide diuretic	Raises blood pressure
		Trazodone	Coma
		Levodopa	Increases "off" periods in Parkinson patients
Piper methysticum Forst. f.	Kava	Alprazolam	Semicomatose state or coma
		Cimetidine and terazosin	Lethargy and disorientation
		Benzodiazepines	Coma
Silybum marianum (L.) Gaertner (Asteraceae)	Milk thistle	Indinavir	Decreases trough concentrations
Glyeyrrhiza glabra L. (Fabaceae)	Licorice	Spironolactone	Pharmacological effect offset

Continued

Table 6.2 Herbs with Established Adverse Effects on Various Medications *(continued)*

Medical Plant	Vernacular Name	Drugs	Herb-Drug Interactions
Echinacea purpuvea; Piper methysticum Forst. f.; *Panax ginseng; Allium sativum* L.; *Hypericum perforatum* L.	Echinacea; kava	Anticancer drugs	Pharmcokinetic interactions
Echinacea purpurea (Asteraceae)	Echinacea	Anabolic steroids, Amiodarone, Methotrexate, and Ketoconazole	Hepatotoxicity
Tanacetum parthenium (L.) Sch. Bip. (Asteraceae); *Allium sativum* L.; *Ginkgo biloba* L.; *Zingiber officinate Roscoe; Panax ginseng*	Feverfew; garlic; ginkgo; ginger; ginseng	Warfarin sodium	Alteration of bleeding time
Valeriana officinalis L.	Valerian	Barbiturates	Excessive sedation
		Central nervous system depressants	Increased drugs effect
Ephedra sinica Stapf.	Ma Huang	Caffeine, decongestants, and stimulants	Hypertension, insomnia, arrhythmia, nervousness, tremor, headache, seizure, cerebrovascular event, myocardial infraction
Momordica charantia L. and *Panax ginseng*	Karela or bitter melon and ginseng	Diabetes mellitus drugs	Blood glucose level effect
Lycium barbarum L. (Solanaceae), *Mangifera indica* L. (Anacardiaceae) and *Carica papaya* L. (Caricaceae)	Lycium, mango and papaya	Warfarin	Anticoagulant effect increased
Trigonella foenum graecum L.	Fenugreek	Glipizide, Insulin and other drugs that may lower blood sugar levels	Excessive decrease of blood sugar levels
		Heparin, Ticlopidine and Warfarin	Bleeding
Cyamopsis tetragonoloba (L.) Taub. *(Fabacae)* and *Triticum* spp. (Poaceae)	Gum guar and wheat bran	Digoxin	Plasma digoxin concentration decreased
Borago officinalis L. and *Oenothera bniennis* L. *(Onagraceae)*	Borage and evening primrose oil	Anticonvulsants	Seiizure threshold lower

SUGGESTIONS FOR FURTHER READING

Audi, J., Belson, M., Patel, M., Schier, J., Osterloh, J. (2005). Ricin poisoning: A comprehensive review. *JAMA* 294:2342-2351.

Cauli, O., Morelli, M. (2005). Caffeine and the dopaminergic system. *Behav. Pharmacol.* 16:63-77.

Hammerschmidt, R. (1999). Phytoalexins: What have we learned after 60 years? *Ann. Rev. Phytopathol.* 37:285-306.

Ho, H.Y., Cheng, M.L., Chiu, D.T. (2007). Glucose-6-phosphate dehydrogenase—From oxidative stress to cellular functions and degenerative diseases. *Redox Rep.* 12:109-118.

Philippe, G., Angenot, L., Tits, M., Frédérich, M. (2004). About the toxicity of some Strychnos species and their alkaloids. *Toxicon.* 44:405-416.

Ravindranath, V. (2002). Neurolathyrism: Mitochondrial dysfunction in excitotoxicity mediated by L-beta-oxalyl aminoalanine. *Neurochem. Int.* 40:505-509.

Rietjens, I.M., Martena, M.J., Boersma, M.G., Spiegelenberg, W., Alink, G.M. (2005). Molecular mechanisms of toxicity of important food-borne phytotoxins. *Mol. Nutr. Food Res.* 49:131-158.

Román, G.C. (2007). Autism: Transient in utero hypothyroxinemia related to maternal flavonoid ingestion during pregnancy and to other environmental antithyroid agents. *J. Neurol. Sci.* 262:15-26.

Skalli, S., Zaid, A., Soulaymani, R. (2007). Drug interactions with herbal medicines. *Ther. Drug Monit.* 29:679-686.

Sneader, W. (2005). Drug discovery: A history. John D. Wiley and Sons.

Vasconcelos, I.M., Oliveira, J.T. (2004). Antinutritional properties of plant lectins. *Toxicon.* 44:385-403.

Vetter, J. (2000). Plant cyanogenic glycosides. *Toxicon.* 38:11-36.

CHAPTER 7

Toxins from Fungi

CHAPTER CONTENTS

MYCOTOXINS

Fungi produce a multitude of substances consisting of wide-ranging chemical structures and biological activities. Certain fungal metabolites are highly desired components in some foods such as cheese, whereas other metabolites are important antibiotics, such as penicillin and cephalosporin. Some fungi, however, can produce substances that are potent acute or chronic toxins or carcinogens. These toxic agents are called **mycotoxins**, a term that usually is reserved for toxins produced by filamentous fungi. The mycotoxicoses produced by many of the toxic fungi, primarily as a contaminant of animal feed and food, have been known for centuries to adversely affect the health of domestic animals and humans. Mycotoxicoses in domestic animals can result in decreased growth rates, abnormal reproduction, disease, and early death of the animals, against all of which effects the livestock producer must be continually vigilant. The dramatic adverse effects in humans of a few mold-infected foods have been known for centuries also, but the role of mycotoxins as possible human carcinogens has been the subject of intensive study only since the early 1960s.

Ergotism

In early human history, the association of the consumption of certain grains with human disease, especially rye, became known as ergotism. In about 1100 BCE, after the first cultivation of rye in China, writings indicated that this grain could produce a characteristic set of severe toxic effects. Similar reports date back to about 600 BCE in Assyria. In the first century BCE, Julius Caesar's legions suffered an epidemic of what became known as ergotism during one of their campaigns in Gaul. Well-documented epidemics of ergotism, involving many thousands of deaths, continued to occur regularly in Europe during the eleventh to the thirteenth centuries, and then more sporadically up to the sixteenth century. Although in developed countries ergotism continues to be a problem primarily in livestock, conditions of adversity in which the safety of the grain harvest is not monitored have led to outbreaks of human ergotism as recently as 2001 in Ethiopia from consumption of ergotized barley. It is sobering to note that whereas ergot was the cause of widespread suffering in Medieval Europe, Hippocrates was well aware of the adverse and the useful effects of ergot by as early as 370 BCE, and had described an ergot preparation for the inhibition of postpartum hemorrhage.

By the mid-1700s, the cause of ergotism was established, or rediscovered, to be derived from the consumption of grain, usually rye, that is contaminated with the fungus *Claviceps purpurea*. Under cool, damp growing conditions the fungus proliferates, invades the grain kernels and grows to form characteristic sclerotium. The sclerotia apparent in the blighted rye are curved, black to purple masses up to 6 cm long (Figure 7.1). This mass is the resting stage of the fungus that can remain viable under dry conditions and germinate when moistened. The growth of as many as 50 species of *Claviceps* on a variety of food and animal feed crops in the grass family is associated with ergotism.

FIGURE 7.1 *Ergot sclerotium on rye.*

The symptoms of ergotism are of two kinds. The **gangrenous** type of ergot poisoning is characterized by severe pain, inflammation, and a burned appearance of the extremities, which may become blackened. In severe cases, loss of fingers, toes, and even hands and feet can occur. In the Middle

Ages, the burned appearance of the tissues manifested by ergotism was called "Holy Fire" or "St. Anthony's Fire" because of the beliefs that the disease was caused by divine retribution on the sinners and that St. Anthony could provide relief from the suffering. In fact, many believers in the protective powers of St. Anthony did experience relief from the tortuous effects of the poisoning, due, at least in part, to the practice of taking a pilgrimage to Padua, Italy, where St. Anthony's remains were enshrined. Of course, the likelihood is great that during their journey, the supplicants dined on food that was not contaminated with ergot, thus discontinuing their exposure to the toxins. In any case, the strength and influence of the Catholic Church grew considerably during this period due to the demonstrated positive effects of a belief in the protective powers of St. Anthony against a scourge that had run rampant in Europe for centuries (see Figure 7.2).

The effects of the second type of ergot poisoning, **convulsive** ergotism, are quite distinct from the gangrenous type and have occurred at times without the co-occurrence of gangrenous ergotism. The symptoms are primarily neurological in nature and include writhing, tremors, numbness, blindness, convulsions, and hallucinations, often along with temporary or permanent psychosis. Perhaps due to the lack of obvious physical injuries, sufferers of the convulsive form of the disease often were believed to be possessed by demons and under spells. Indeed, a recent analysis of the events surrounding the Salem witch trials that occurred in Salem, Massachusetts in 1692 concluded that the girls who made the condemning accusations against their townsfolk may have been under the influence of convulsive ergot poisoning. No one imagined at that time, that over 200 years later in the twentieth century, ergot would prove to be the source of some of the most potent and enigmatic hallucinogenic agents ever discovered.

The physiological basis for the gangrenous form of ergotism is the constriction of blood vessels. This activity is due to vasoconstriction in peripheral veins and arteries induced by the ergotamine and ergocristine alkaloids from the fungus. The effects of high doses of these alkaloids are most pronounced in the more poorly vascularized distal structures, such as the fingers and toes, and can result in acute ischemia, mummification, and loss of the digit. Ergotamine was isolated as the first purified ergot alkaloid in 1918 by the Swiss chemist, Arthur Stoll. This compound shares structural similarity with neurotransmitters such as serotonin, dopamine, and adrenaline, and can bind to several cell receptors acting both as agonist and antagonist in signal

FIGURE 7.2 *16th century woodcut showing St. Anthony with a victim of ergotism.*

transduction pathways within tissues. The peripheral vasoconstrictive effects of ergotamine are mediated through activation of one of the serotonin receptors (5-HT2) along with activation of α-adrenergic receptors. The activation effects of the alkaloid are prolonged compared to other agonists apparently because of slow dissociation of the alkaloid from the receptors. This activation can stimulate vascular smooth muscle and cause vasoconstriction in both arteries and veins. The antimigraine effect of low doses of ergotamine, for which the compound is used clinically today, is due to constriction of the intracranial extra cerebral blood vessels through the 5-HT1B receptor, and by inhibition neurotransmission by 5-HT1D receptors.

The physiological bases for the convulsive form of ergotism are less well understood. The convulsive symptoms are thought to be caused by the clavine group of alkaloids, including agroclavine, elymoclavine, and lysergol. Animal experiments have shown that agroclavine, elymoclavine, and lysergol exhibit excitatory effects on the central nervous system, as does LSD, suggesting that these substances may well be psychoactive in humans as well. However, the effects of the clavine alkaloids on neurological receptors is likely to be very complex as has been determined for LSD, which can affect the activities of at least six 5-HT receptors. Thus, a full understanding of the psychotropic effects of ergot is likely to require considerable further study. This certainly continues to be the case for LSD, even after over 70 years since its synthesis and serendipitous discovery as a potent hallucinogen in 1938 by Albert Hofmann in Switzerland (Figure 7.3).

The selective occurrence of one form of ergotism over the other also remains an unresolved issue. Several hypotheses have been proposed to explain this well known phenomenon, however. The apparently more reasonable explanations include biosynthetic variability of the *Claviceps* species, species differences in the infecting fungus, and the possibility that another microorganism may be parasitizing the *Claviceps* and converting the vasoconstrictive alkaloids to psychotropic alkaloids more similar in structure to LSD.

Ergonovine was first isolated from *Claviceps* in 1935 and found to be a potent inducer of uterine constriction. Ergonovine induces significant vasoconstriction but does not exhibit the adrenergic blocking action of ergotamine (Figure 7.4). Ergonovine, usually in combination with other drugs, is used in obstetrics in the third stage of labor, principally to decrease postpartum bleeding and to facilitate delivery of the placenta. High doses of ergonovine can produce LSD-like psychotropic symptoms.

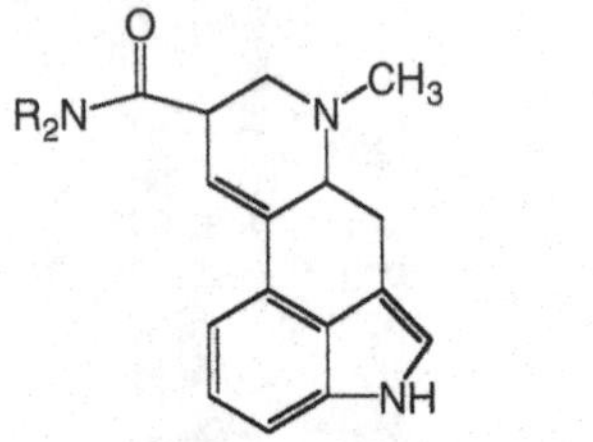

Lysergic acid diethyl amide (LSD) (R=ethyl)

FIGURE 7.3 *Structure of lysergic acid diethyl amide.*

Alimentary Toxic Aleukia

Alimentary toxic aleukia (ATA), also known as septic angina, is another mycotoxicoses that has caused great human suffering. Sporadic occurrences of ATA have been reported to occur primarily in Russia since the nineteenth century. Outbreaks were recorded in 1913, 1932, and toward the end of World War II. Russian descriptions of symptoms include fever, hemorrhagic rash, bleeding in the nose, throat, and gums, destruction of air passages, extreme leukopenia (low level leukocyte), sepsis, and destruction of bone marrow. Outbreaks of the disease were generally sudden, and mortality rates were often in excess of 50% of the affected population.

Ergotamine Ergonovine

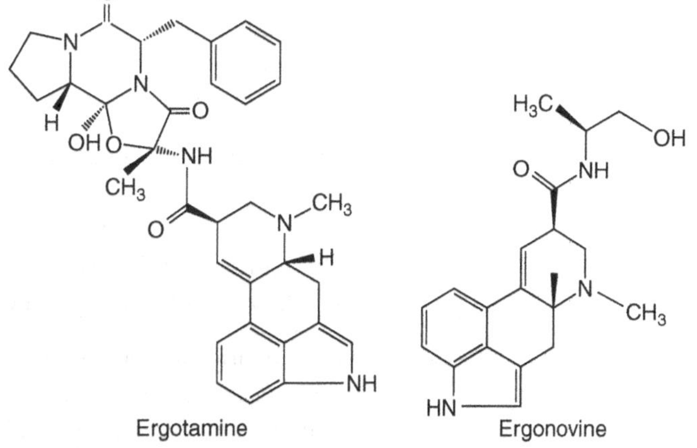

FIGURE 7.4 *Chemical structures of active ergot alkaloids.*

There are four recognized stages of ATA:

Stage 1: Shortly after exposure to the toxins, victims experience burning sensations in the mouth, throat, esophagus, and stomach. These symptoms often are followed by vomiting, diarrhea, and abdominal pain due to inflammation of the gastric and intestinal mucosa. Victims at this stage also experience headaches, dizziness, fatigue, tachycardia, heavy salivation, and fever. The leukocyte count is reduced from normal.

Stage 2: After suffering for three to nine days with the symptoms of Stage 1, Stage 2 begins with a reduction in intensity of the symptoms. Although the victim may begin to feel well and capable of normal activity, the destruction of the bone marrow continues with a further reduction of leukocytes and a worsening anemia. The body's resistance to infection is reduced. A general weakness, acute headache, and reduced blood pressure are apparent. The duration of Stage 2 may be from several weeks to months. If the consumption of the contaminated food is discontinued at this time and the victim obtains medical care, the chances of recovery are good.

Stage 3: This stage is signaled by the appearance of hemorrhages on the skin and on the mucous membranes of the mouth and tongue, as well as in the intestine and stomach. As the toxicosis progresses, the necrotic lesions proliferate along with an increased tendency for infection. Esophageal lesions often occur and the involvement of the epiglottis may cause laryngeal edema leading in many cases to strangulation. Destruction of the hematopoietic system becomes nearly complete, and death ensues.

Stage 4: If the Stage 3 patient receives acute medical treatment including blood transfusions and heavy dose antibiotic treatment, he may survive the ordeal and thus enter Stage 4, the period of convalescence. This period requires several months of treatment for the nearly full recovery of the hematopoietic system.

Several theories about the cause of ATA included vitamin deficiencies or new infectious agents. However investigators eventually discovered that the disease was caused by consumption of grain left in the fields over the winter and harvested in the spring. Examination of the fungal flora of overwintered cereals implicated in the outbreaks of ATA revealed a rich array of fungal species. Toxin production by purified fungi, determined by the application of fungal extracts to the skin of rabbits, was then implicated in the toxicity of moldy cereals to people.

Fungi of the genus *Fusarium* were shown to produce the highest number of toxic isolates in the rabbit skin assay. The species most often associated with ATA outbreaks were *F. poae* and *F. sporotrichoides* (also called *F. tricinctum*). These are fairly common, so-called cryophilic fungi because they have the peculiar capability of producing toxic substances in increased amounts when grown under conditions that include a period of growth at temperatures near 0 °C. Investigators found that field conditions that included relatively high winter temperatures, deep snow cover, frequent freezing and thawing, and light rainfall in autumn were conducive to fungal growth and toxin production in grain and the soil.

The fungal toxins that are the causative agents for ATA are a group of polycyclic sesquiterpene derivatives called trichothecenes (Figure 7.5). T-2 toxin (Figure 7.6), the most heavily implicated toxin of the group, was isolated for toxigenic *Fusarium* species and shown to produce in cats all the symptoms of ATA in humans. Importantly, removal of T-2 toxin from the fungal extract resulted in the loss of the toxicity of the extract. Unfortunately, analyses of T-2 toxin in the grain that caused human ATA were not conducted, leaving unresolved this uncertainty about the role of T-2 in the cause of ATA.

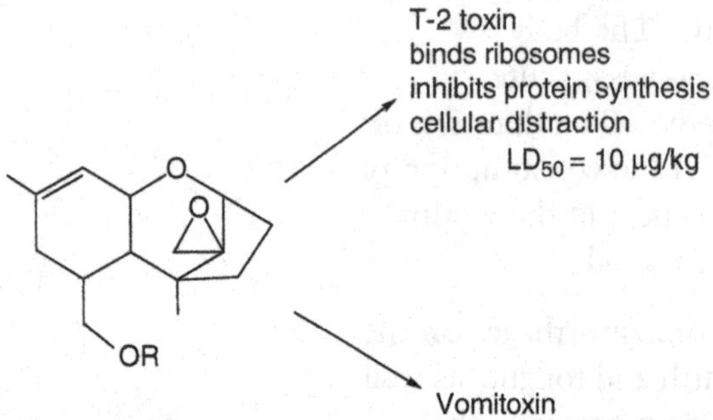

T-2 toxin
binds ribosomes
inhibits protein synthesis
cellular distraction
LD_{50} = 10 µg/kg
O
O
OR
Vomitoxin

FIGURE 7.5 *Trichothecene structure and activity.*

Considerable progress has been made since the discovery of T-2 toxin and the other

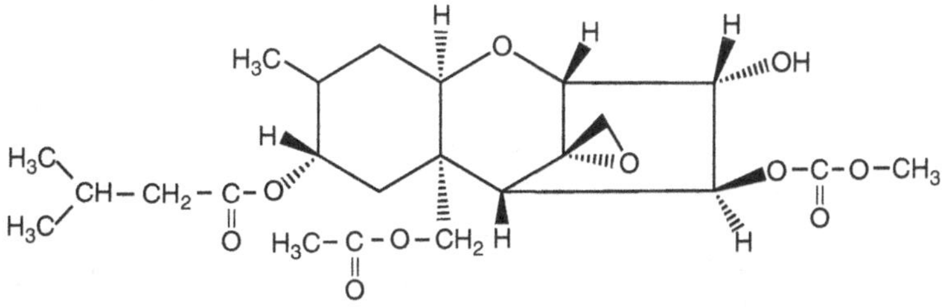

T-2 Toxin

FIGURE 7.6

Structure for T-2 toxin.

trichothecenes in unraveling the modes of action of these substances. The major effect of T-2 toxin and other trichothecenes is that they inhibit protein synthesis by binding to ribosomes, causing ribotoxic stress, which leads to the activation of stress kinases (JNK/p38), the disruption of DNA and RNA synthesis, and inhibition of mitochondrial function. T-2 toxin most strongly affects actively dividing cells such as those lining the gastrointestinal tract, skin, lymphoid, and erythroid cells. Immunosuppression is a further major effect of inhibition of protein synthesis that is caused by a decrease in antibody levels, in immunoglobulins, and in certain other humoral factors such as cytokines.

Fumonisins

A devastating neurological syndrome in horses called equine leukoencephalomalacea (ELEM) has been known since the mid-nineteenth century and has long been associated with the consumption of moldy corn. As the name of the syndrome indicates (i.e., softening of the brain with accumulation of white blood cells), the main target of the toxicity is the central nervous system. The major symptoms occur rapidly following consumption of the moldy corn, and include neurologic overexcitement, blindness, incoordination, and facial paralysis. Death can occur within 24 hours. There is severe edema of the brain with extensive hemorrhaging and necrosis of neurons.

Intensive studies of the etiology of ELEM were begun in South Africa shortly after a serious field outbreak of the disease in 1970. These studies soon led to the identification of the common fungus, *Fusarium verticillioides*, which is also known as *F. monilifome*, as the causative organism. This fungus is one of the predominant fungi associated with corn intended for human and animal consumption worldwide. The fungal metabolites responsible for the toxicity, fumonisins B1, B2, and B3 were identified nearly two decades later (Figure 7.7). In addition to being a cause of ELEM, FB1 was subsequently shown to induce a pulmonary edema syndrome in pigs and liver cancer in rats. FB1 also is a cancer promoter and initiator in

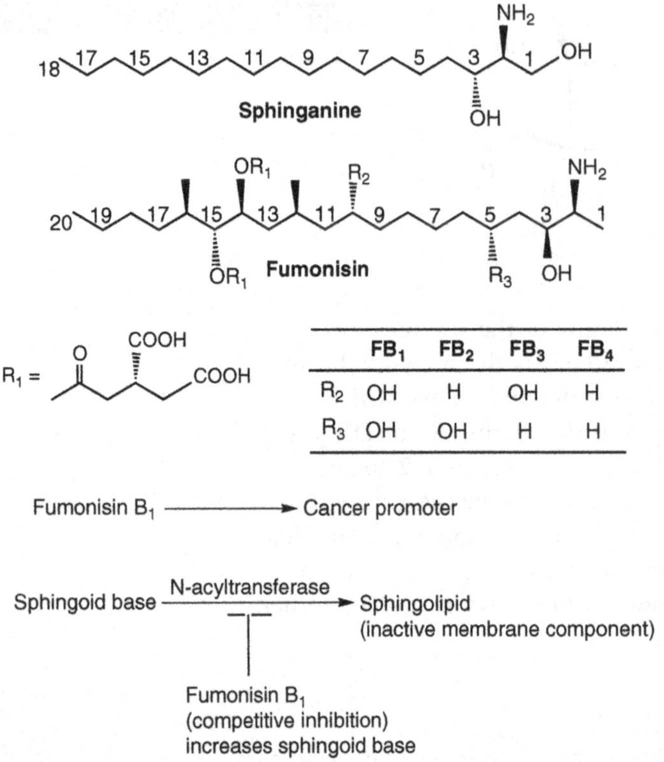

Sphinganine

Fumonisin

R_1 = (COOH, COOH side chain)

	FB_1	FB_2	FB_3	FB_4
R_2	OH	H	OH	H
R_3	OH	OH	H	H

Fumonisin B_1 → Cancer promoter

Sphingoid base —N-acyltransferase→ Sphingolipid (inactive membrane component)

Fumonisin B_1 (competitive inhibition) increases sphingoid base

FIGURE 7.7 *Structures and activities of fumonosins and related endogenous substances.*

rat liver and is hepatotoxic to horses, pigs, rats, and monkeys. The occurrence of fumonisins in homegrown corn has been associated with an elevated risk for human esophageal cancer in the former Transkei region of South Africa and in China. A preferred dietary item in at least one of the Chinese villages with high incidence of esophageal cancer was a kind of corn bread on which a heavy growth of natural mold was promoted to provide a characteristic flavor. The natural occurrence of FB_1, and other fumonisins, has been reported in commercial corn and/or corn-based feeds worldwide. Associations of increased esophageal cancer risk with high levels of FB_1 in corn used for food were reported for the areas of China, South Africa, northern Italy, Iran, and the southeastern United States.

The mechanism of action of FB_1 has been the subject of considerable interest. A substantial body of experimental evidence has supported the hypothesis that the structural similarity between sphinganine and FB_1 may result in the disruption of sphingolipid metabolism. Sphingolipids are important structural components of the cell membrane, especially of neurons. Sphingolipid metabolism is an important part of the cascade of events leading to altered cell growth, differentiation, and cell injury. FB_1 was found to be a competitive inhibitor of sphinganine N-acetyltransferase, the central enzyme that is responsible for the conversion of sphingoid base to sphingolipid. The resulting blockage of sphingolipid biosynthesis and accumulation of sphinganine leads to activation of stress protein kinase activity, activation of tumor cell growth, induction of lipid peroxidation, and tissue necrosis.

An intriguing and as yet unresolved issue related to the toxicity of the fumonisins concerns the very pronounced species differences in sensitivity to these compounds. For example, whereas the oral LD_{50}s of FB_1 in rats and mice are very large (over 5000 mg/kg body weight), toxic effects characteristics of ELEM in horses are seen at oral doses as low as 0.75 mg/kg. Exposure to only 0.3 mg/kg body weight FB_1 in naturally contaminated feed for less than two weeks proved to be lethal in ponies. Moreover, the liver cancer potencies (TD_{50}s) of FB_1 administered orally to mice and rats are only 6.8 mg/kg and 1.5 mg/kg, respectively.

Aflatoxin

Moldy feed has long been associated with diseases and other adverse effects in domestic animals. These maladies generally were thought to be problems of yield of livestock and poultry products for farmers and ranchers and the possible implications for human health were generally of little concern. For example, various liver diseases, collectively called hepatitis-X disease, were identified by veterinarians to occur in swine, cattle, and dogs. Improved methods of feed handling, production, and storage considerably reduced the occurrence of these types of disorders by the middle of the twentieth century. It was not until 1960, however, that the human health implications of disorders in farm animals began to be clearly recognized. At that time, over 100,000 young turkeys in England died of a disease that was dubbed turkey-X disease. The disorder was characterized by extensive necrosis of the liver in the turkeys. At about the same time, attention was drawn to an increasing mortality and liver tumors in hatchery-raised trout in Oregon. It was later shown that both the peanut meal used as a feed supplement for the turkeys and the cotton seed meal used to supplement the trout feed were heavily contaminated by a common fungus known as *Aspergillus flavus* (Figure 7.8). Subsequent efforts to isolate the active toxins in

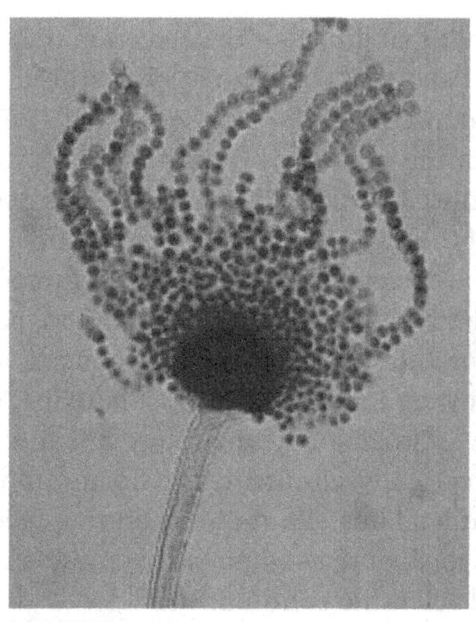

FIGURE 7.8 *A. flavus fruiting body (sporangium) (left) and infected peanuts (right).*

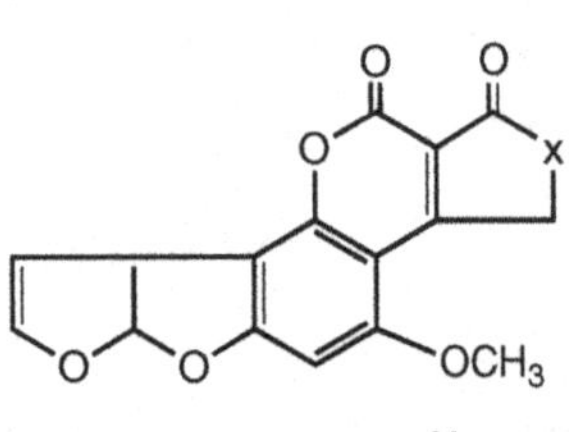

	X
Aflatoxin B_1	CH_2
Aflatoxin G_1	CH_2O

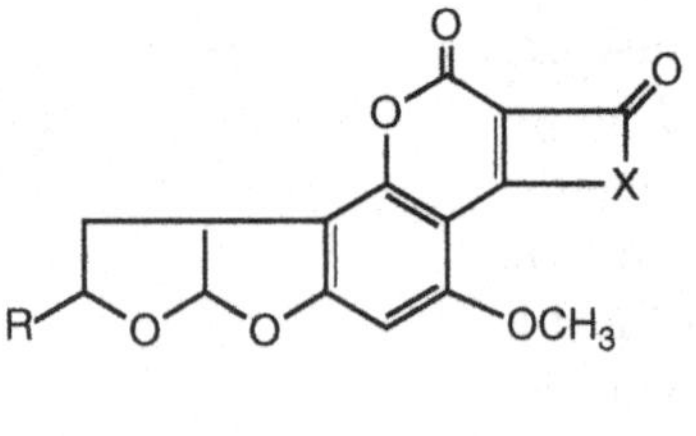

	R	X
Aflatoxin B_2	H	CH_2CH_2
Aflatoxin G_2	H	CH_2CH_2O
Aflatoxin B_{2a}	OH	CH_2CH_2
Aflatoxin G_{2a}	OH	CH_2CH_2O

FIGURE 7.9 *Structures of aflatoxins.*

the fungus led to the characterization of a series of related substances that were named the aflatoxins.

The aflatoxins are a series of bisfuran polycyclic fungal products. Based on their characteristic blue or green fluorescence under ultraviolet light, these compounds were named aflatoxin (AF) B_1, B_2, G_1, and G_2, all of which are mold metabolites (Figure 7.9). Hydroxylated metabolites of AFB_2 and AFG_2 also were isolated from the fungus and were called AFB_{2a} and AFG_{2a}. The compounds are lipid soluble and exhibit considerable stability to most cooking conditions, including during the roasting of peanuts.

Sources

The aflatoxins are produced by several related fungal species, including the agriculturally important species *A. flavus* and *A. parasiticus*. *A. flavus* is a common constituent of the microflora in air and soil throughout the world. This species causes deterioration of stored wheat, corn, rice, barley, bran, flour, and soybeans, as well as peanuts, cottonseed, and other commodities. The fungus generally does not invade the seeds of living grain or intact peanuts. Growth occurs primarily when the products are stored under conditions of relatively low moisture that eliminate or reduce the growth of competing species such as *Penicillium* and *Fusarium* fungi.

With advances in analytical capabilities for the detection of aflatoxins, it quickly became clear that food and feed contamination by these products is common. Assays conducted soon after their discovery in the 1960s showed that aflatoxins were present in the majority of samples of peanuts, peanut meals, and peanut butter produced in the United States. Analyses of samples collected from around the world, particularly from Africa and Asia, showed that aflatoxins could be detected in a wide variety of important food and feed products, including barley, cassava, corn, cottonseed, peas, millet, cowpeas, rice, sesame seed, sorghum, soybean, sweet potatoes, and wheat. Even samples of dry spaghetti were found to contain aflatoxins. Because of the natural occurrence of the aflatoxins and their very broad distribution, it became clear that efforts to totally eliminate the toxins from the food and feed supplies would be futile. Thus, efforts to determine relatively safe levels of exposure to the most toxic of the compounds became an important priority for the regulatory and scientific communities.

Acute Toxicities

The aflatoxins show considerable species variation in their acute and chronic effects. Prominent acute effects in rodents include hepatic lesions with edema, biliary proliferation, and parenchymal cell necrosis. In rhesus monkeys, AFB_1 treatment commonly results in fatty infiltration and biliary proliferation with portal fibrosis.

Ducklings, trout, pigs, rabbits, and adult male rats are some of the more sensitive species to the acute toxic effects of aflatoxins with $LD_{50} < 2.0$ mg/kg body weight for AFB_1 administered by the oral route. Adult mice and hamsters are relatively insensitive with an oral $LD_{50} > 9.0$ mg/kg. Nonhuman primates exhibit an intermediate level of sensitivity with $LD_{50} = 2.2 - 8.0$ mg/kg, for the several species that have been examined. There are significant exceptions to this overall categorization of acute toxicity, however, since very young mice are highly sensitive to AFB_1, whereas female rats are quite insensitive with oral $LD_{50} = 18$ mg/kg. The compounds with the double bond in the furan ring, AFB_1 and AFG_1, are considerably more toxic than the derivatives with saturated furan rings, AFB_2 and AFG_2. The species differences in sensitivities to AFB_1 generally depend on the level of epoxide hydrolase (EH), which mediates the hydrolysis of the epoxide, and of GST (mGST3A-3, GSTT1) activity, which mediates the conjugation of the activated metabolite, AFB_1-epoxide, to form the inactive glutathione adduct. This GST activity is generally low in the sensitive species and high in the less sensitive species.

Carcinogenic Activity

The relative species sensitivities to the chronic effects of the aflatoxins are generally similar to the sensitivities to the acute effects. Thus rats, rhesus monkeys, and cynomolgus monkeys are sensitive to the carcinogenic effects of AFB_1 with $TD_{50} = 3$, 8 and 20 μg/kg body weight, respectively, for the induction of hepatocellular carcinoma. In contrast, mice exhibited no carcinogenic effects in this dosage range. A large body of information of this kind supports the claim that AFB_1 is the most potent liver carcinogen known. As for the acute effects of AFB_1, the carcinogenic effects are attributable to the action of metabolites that are capable of reacting with cellular macromolecules. This activation process is mediated in the liver primarily by CYP enzymes and in the lung primarily by prostaglandin H synthase and lipoxygenase. AFB_1 is bioactivated by epoxidation of the terminal furan ring double bond, which generates the exo- and endo-AFB_1-epoxides electrophilic products (Figure 7.10). AFB_1-exo-epoxide is the more mutagenic of the

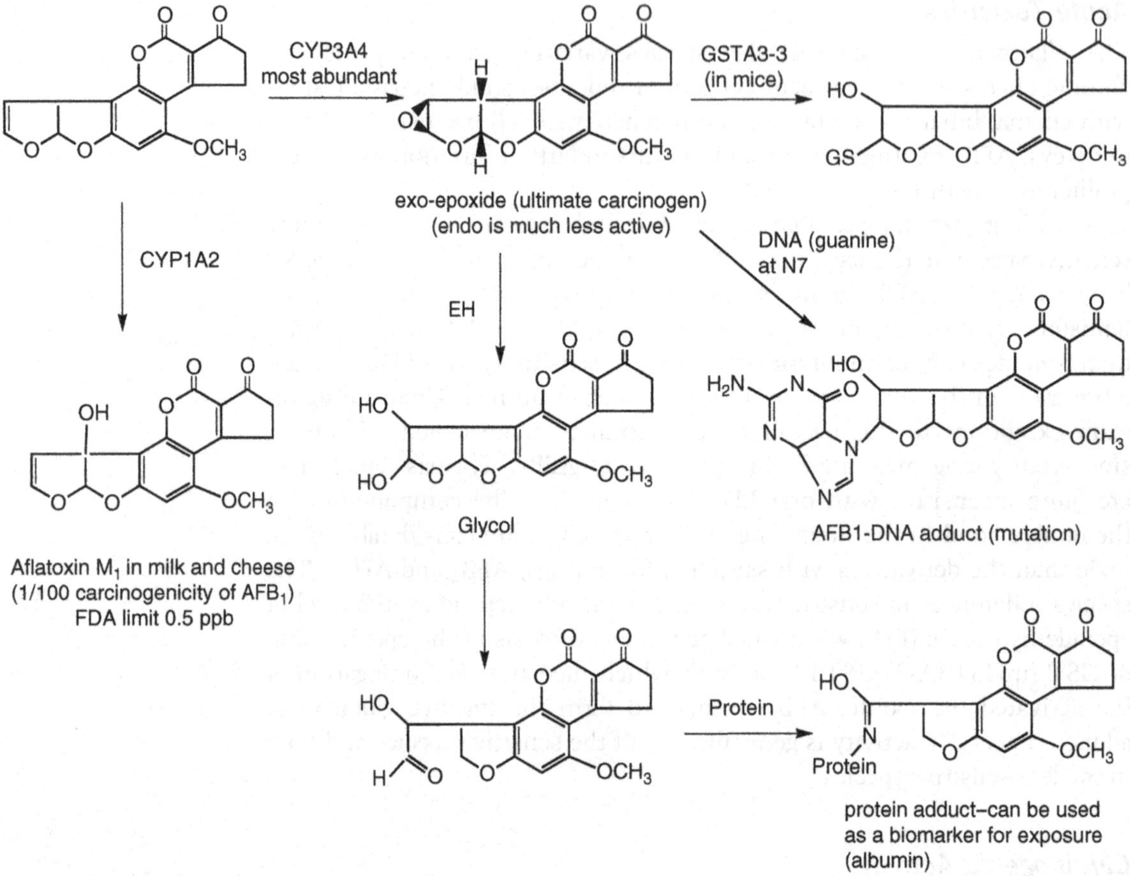

FIGURE 7.10 *Metabolism and DNA binding of AFB_1.*

two isomers and is capable of alkylating nucleic acids and proteins under nonenzymatic conditions.

Although the exo-isomer has not been isolated from metabolic systems, its existence is inferred from the structures of its stable reaction products. Also, synthetic AFB_1-exo-epoxide produces the same macromolecular adducts under nonenzymatic conditions that are produced by this derivative prepared under enzymatic conditions. The reactivity of AFB_1-exo-epoxide with DNA is at least 1000-fold greater than the reactivity of the endo isomer, apparently because intercalation of the furanocoumarin moiety of the epoxide between the bases in DNA orients the epoxide for facile backside attack (SN2) by N7 of guanine to open the epoxide and produce the primary AFB_1-DNA adduct, trans-8,9-dihydro-8-(N7-guanyl)-9-hydroxyaflatoxin B_1 (AFB_1-N7-Gua). Only very low levels of DNA adducts are formed with the

AFB_1-endo-epoxide because the reaction orientation that favors the backside attack and opening of the epoxide is not achievable. This alkylation of guanine by AFB_1-exo-epoxide, if not repaired, results in a G $\rightarrow$T transversion mutation, which can lead to cancer initiation. The level of AFB_1-N7-Gua adducts excreted in the urine is used as a biomarker of exposure to AFB_1 in humans, and is well correlated with tumor production in studies with laboratory animals.

Although AFB_1 is clearly a potent liver carcinogen in experimental animals, and is recognized as a human carcinogen by the International Agency for Research on Cancer (IARC), this compound by itself apparently has little effect on liver cancer rates in the world. Instead, the impact of this mycotoxin in human liver carcinogenesis depends on the potentiating effects of the commonly occurring viral infection with hepatitis B virus. Indeed, as occurs primarily in parts of Asia and Africa, for populations that are exposed to high levels of environmental AFB_1 exposure and are infected with HBV, the relative risk of developing liver cancer is increased 15-fold over the risk of populations that are exposed to high levels of AFB_1 but are not infected with HBV. Furthermore, this risk level increases to 77-fold for subjects who carry inactivating mutations in any of the enzymes that detoxify AFB_1-exo-epoxide. As was discussed more fully in Chapter 4, HBV functions as a tumor promoter to synergistically increase the carcinogenic potency of AFB_1.

Although AFB_1 is best known as a hepatocarcinogen, epidemiological studies also have shown a positive association between human lung cancer occurrence and inhalation exposure to AFB_1. Many studies have reported an association of occupational exposure to inhaled AFB_1, particularly in the form of contaminated grain dust, with increased respiratory cancers. Levels of AFB_1 in respirable grain dust have been reported to be as high as 52 ppm, which is well above the FDA regulatory action level of only 20 ppb for AFB_1 in food. Human lung cancer also is reported to be associated with dietary AFB_1 exposure in humans and in laboratory animals. Although CYP1A2 and CYP3A4 are responsible for activation of AFB_1 in human liver, CYP2A13 mediates this reaction in human lung. CYP2A13 is expressed predominantly in respiratory tissues including the nasal mucosa, lung, and trachea, and is responsible for metabolic activation of the tobacco-specific carcinogen, 4-(methylnitrosamino)-1-(3-pyridyl)-1-butanone (NNK).

With the realization of high carcinogenic potency of AFB_1 and its wide distribution in food and feed, it was quickly realized that regulatory standards for contamination levels of the aflatoxins must be set and enforced. The current action level set by the US FDA for aflatoxins in the general food supply is 20 ppb, but for milk is only 0.50 ppb. It is interesting to note, however, that as of 2001 the European Union established an upper limit for aflatoxins of only 4 ppb for most cereals and peanuts, and of only 0.050 ppb in milk and 0.025 ppb in milk-based food for infants. The maximum level allowed in the United States for aflatoxin contamination in corn used for dairy cattle feed is 20 ppb and for most breeder livestock feed is 100 ppb. The significance of the lower allowable limits for aflatoxins in Europe on trade and human health remains to be established.

Metabolism

There are several metabolites of AFB_1 that are important in its biological activity in addition to the epoxide derivatives discussed previously (Figure 7.11). Aflatoxin M_1 (AFM_1) occurs in milk of cows and other mammals, including humans, fed on AFB_1-containing feed or food. AFB_1

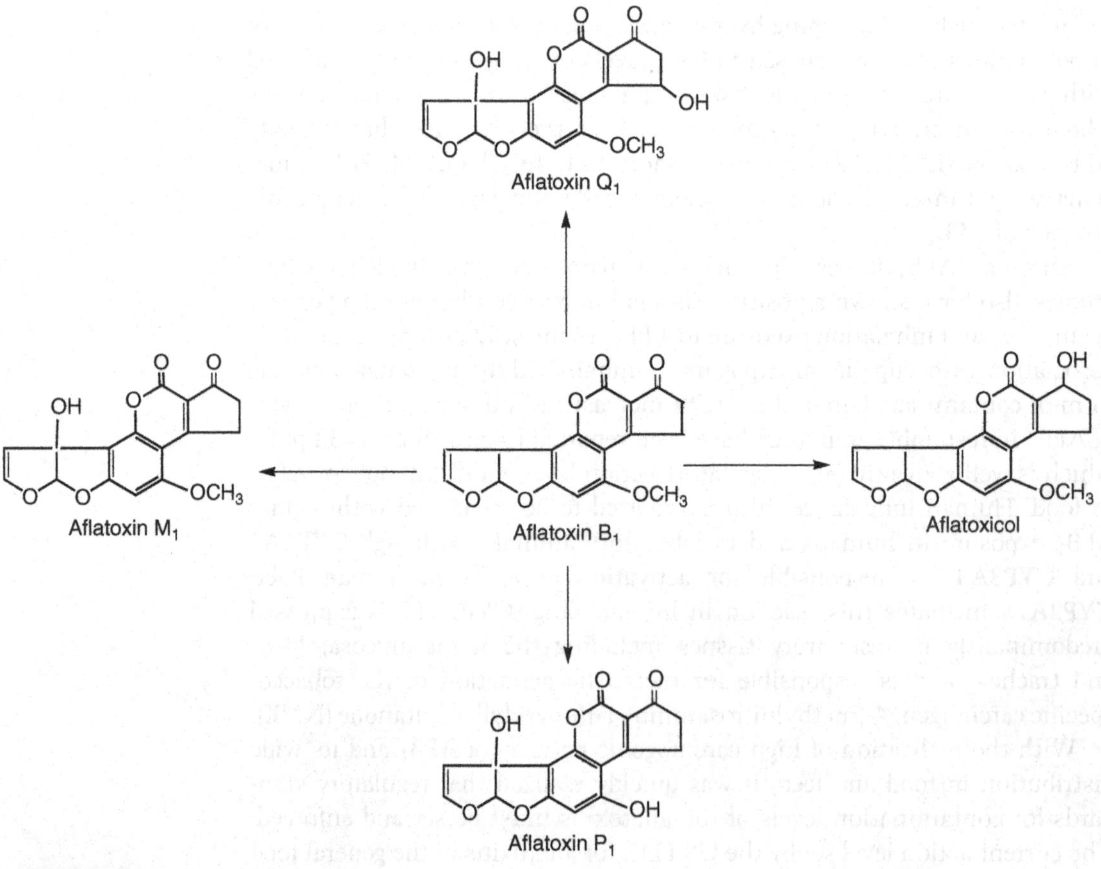

Aflatoxin Q_1

Aflatoxin M_1

Aflatoxin B_1

Aflatoxicol

Aflatoxin P_1

FIGURE 7.11 *Metabolic products of AFB1.*

is converted by human CYP1A2 to AFM_1 and AFB_1-epoxide in a ratio of 1:3. AFM_1 also occurs in association with the protein fraction of some milk derivatives such as yogurt and cheese. This metabolite, which is recognized as a probable human carcinogen by IARC, represents about 0.5 to 6% of the administered AFB_1 in different species. AFM_1 is excreted in human urine and feces as a minor metabolite of AFB_1. The total conversion of AFB_1 to AFM_1 in cow's milk is about 1%, and the carcinogenic potency of AFM_1 is about 2 to 10% of AFB_1. However, the acute toxicities of AFB_1 and AFM_1 are similar. It appears that AFB_1 and AFM_1 exert their carcinogenic effects by different mechanisms since human liver microsomes have very limited capacity to catalyze epoxidation of AFM_1. Moreover, the AFM_1-epoxide exhibits very low electrophilic activity compared to AFB_1. Indeed, toxic effects that are independent of metabolism of AFM_1 have been demonstrated in cultured cells.

Aflatoxicol (AFL) is the major metabolite of AFB_1 in many species, accounting for up to 25% of the total metabolites from AFB_1 in rats, a sensitive species. Little AFL is produced in mice and rhesus monkeys, which are considered resistant species. The alpha and beta isomers of AFL are produced as reduction products of AFB_1 and exhibit roughly 5% of the acute toxicity of AFB_1 in the duckling bioassay. However, AFL exhibits about 20% of the mutagenic potency of AFB_1, and nearly equivalent carcinogenic potency in trout. AFL is readily oxidized to AFB_1, in vivo. The AFB_1 reductase and AFL dehydrogenase enzymes that mediate the interconversion of AFB_1 and AFL are not well studied, but this mechanism is considered to be very important because AFL may play a role as a reservoir of AFB_1 in some organisms. Although several studies have reported that aflatoxicol was found in the serum, liver, urine, and stools of some human groups with compromised health, this metabolite has not been reported as a common metabolite in AFB_1-exposed healthy humans. Thus, the importance of AFL in AFB_1-induced human carcinogenesis is not fully understood.

Two additional major metabolites are aflatoxin P_1 (AFP_1) and Q_1 (AFQ_1), both of which are considered to be detoxification products of AFB_1. AFP_1 is the major urinary product of AFB_1 for rhesus monkeys but its excretion in humans is variable and not directly associated with the level of dietary intake. AFQ_1 is a major product of AFB_1 oxidation mediated by CYP3A4 in humans and is excreted in both urine and feces.

Decontamination

Because *A. flavus* and other aflatoxin-producing fungi are widespread in the world and their rapid growth on major food and feed crops can be difficult for control, especially in warm and moist climates, many methods have been developed for the removal of aflatoxins from contaminated products. Physical separation of obviously contaminated materials has proved to be successful in controlling aflatoxin contamination in peanuts. *A. flavus* and many other fungi emit a bright yellow-green fluorescence under ultraviolet light. This indicator of fungal contamination is used in the physical separation of contaminated peanuts and corn and some other crops. This method is used as an initial test for fungal contamination and cannot be relied on as an indicator of aflatoxin contamination since the fluorescence is due to other fungal products such as kojic acid. Heat treatment of contaminated crops also is used to detoxify feed materials. Although aflatoxins are quite stable under mild heat treatments, roasting reduces the AFB_1 content in peanuts by 80% after 30 minutes.

The most useful method for large-scale detoxification of contaminated feed is treatment with ammonia. The use of ammonia to detoxify corn and cottonseed meal increases the nutritional value of the feed. The detoxified feeds support the growth of trout, cows, and other animals without ill effects. The ammoniation process involves addition of aqueous ammonia and mild heating of the contaminated product in a sealed container, which in some cases can be very large (Figure 7.12). The process can reduce the levels of AFB_1 contamination in cottonseed meal and corn by 80%, and produces a product that meets regulatory standards for use as animal feed. Ammoniation is used around the world to inactivate aflatoxin in several contaminated commodities, including peanut meal, cottonseed, and corn. Ammoniation, as a decontamination process for aflatoxin reduction, is approved by major safety and regulatory agencies, including FAO, FDA, and USDA.

Other Mycotoxicoses

Other mycotoxicoses, such as vomitoxin, ochratoxin A, patulin and zearalenone, have also been shown to affect human health significantly

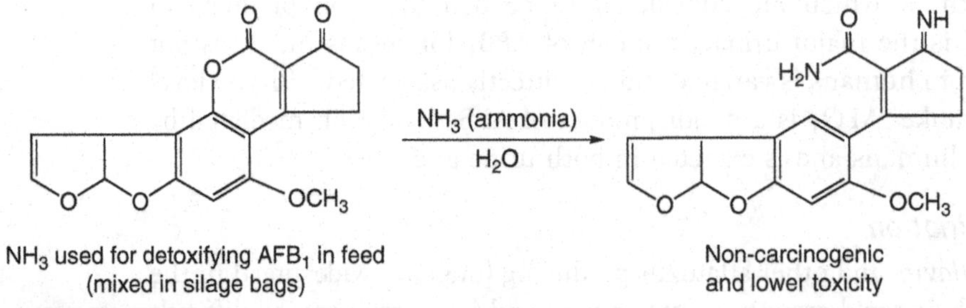

NH_3 (ammonia)
H_2O

NH_3 used for detoxifying AFB_1 in feed (mixed in silage bags)

Non-carcinogenic and lower toxicity

FIGURE 7.12 *Chemical deactivation of AFB1.*

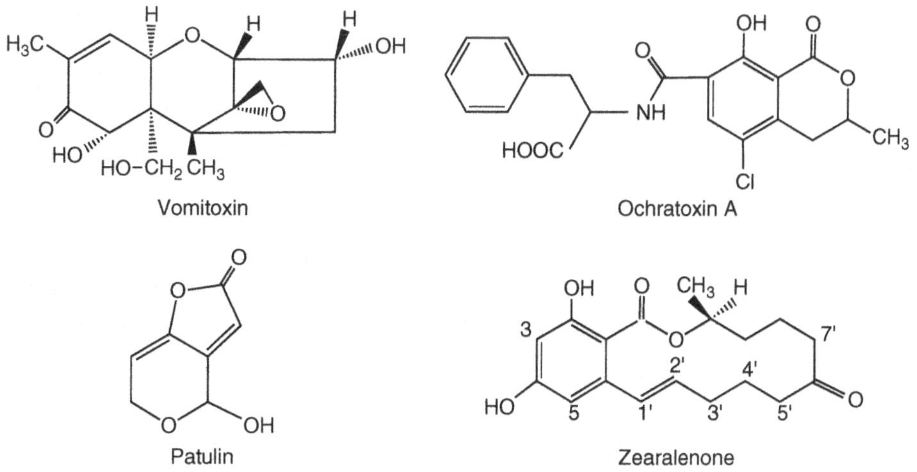

FIGURE 7.13 *Mycotoxins with possible toxicity in humans.*

(Figure 7.13). Because the etiologies of these diseases are poorly understood, these mycotoxicoses will be briefly mentioned.

Kashin-Beck disease is a disorder of the bones and joints of the hands and fingers, elbows, knees, and ankles of children and adolescents who slowly develop stiff deformed joints, shortened limb length, and short stature due to necrosis of the growth plates of bones and of joint cartilage. This disorder is endemic in some areas of eastern Siberia, Korea, China, and Tibet and is probably of environmental origin. Some studies have implicated the *Fusarium* metabolite, vomitoxin, as a causative agent.

Balkan endemic nephropathy (BEN) is a noninflammatory, chronic kidney disease that leads to kidney failure. It has been described in several rural regions of Bulgaria, Romania, and the former Yugoslavia (Serbia, Croatia, and Bosnia). At least 25,000 individuals may suffer from BEN or are suspected of having the disease, and the total number of people at risk in the three countries may exceed 100,000. One theory for the cause of this kidney disease poses that the mycotoxin, ochratoxin A, which can often heavily contaminate the food supplies in the endemic areas, is the cause of the disease possibly in combination with another environmental factor.

Patulin is a fungal metabolite that is secreted by many *Penicillium* species. Exposure generally results from consumption of moldy fruits and fruit products. Although patulin has been reported to produce toxic effects in cattle and poultry and has been shown to be a mutagen, a weak carcinogen, and a teratogen in experimental systems, a negative impact on human health has not been established. Nevertheless, an action limit of 50 ppb for patulin in juices has been established by the EU and most other countries, including the United States, with a provisional maximum daily intake of only 0.4 μg/kg body weight. This figure was obtained by the accepted method for ADI determination based, apparently, on the dose-response data for toxicity in rats, and with the assumption that children would be major consumers of the fruit drinks.

Zearalenone is a potent estrogenic substance that is produced by the fungus *Fusarium roseum* and its sexual stage, *Gibberella zeae*, and some related fungi. Consumption of corn infected with these fungi can cause a loss of reproductive capacity and other estrogenic effects in pigs and other domestic animals. Symptoms of zearalenone poisoning in young pigs include swelling and eversion of the vagina until, in some cases, the cervix is visible. Recent studies have implicated this mycotoxin in the cause of precocious puberty in a group of girls.

MUSHROOMS

Mushrooms are a delicacy to many people the world over. A few species are grown commercially in the United States and are consumed in vast quantities with considerable enjoyment. Health problems can arise, however, for collectors of wild mushrooms who are not expert in the identification of relatively safe varieties. Only about 50 of over 800 identified species in the United States are known to produce some toxic effects in people. In most cases, an unwary collector can consume a mildly toxic species of mushroom and will most likely suffer only simple gastrointestinal upset that will soon pass. For some of the potentially toxic mushroom species, special cooking processes have been developed to render them edible. In a few cases, however, the toxicity is not removed by cooking and the effects persist long after the offending mushroom has been digested or otherwise has been removed from the gastrointestinal tract.

Amanita phalloides

Amanita phalloides, also known as the death cap mushroom, is the most well-known and notorious of the toxic mushroom species. This elegant looking mushroom was known by the ancient Greeks and Romans to be a deadly poison, and it has been a favorite of poisoners down through the centuries (Figure 7.14). This species is thought to be responsible for the vast majority of the mushroom poisoning deaths worldwide. Although *A. phalloides* has been known in Europe for centuries, this species was comparatively rare in the United States until the early 1970s. At that time, the mushroom apparently entered the western United States as a stowaway in a nursery specimen from Europe. The mushrooms, or fruiting bodies, of *A. phalloides* usually appear in late summer or fall and can stand 8 inches tall. It is interesting to note, however, that the first report of death cap poisoning from freshly picked mushroom harvested in winter in Europe appeared in 2006, which may provide further evidence of the impact of global warming on potential environmental hazards. The color of the relatively large cap can range from greenish-brown to yellow. Because of its stunning appearance, which is similar to that of other highly desirable species of *Amanita*, *A. phalloides* is often a prize find of avid amateur mushroom hunters.

FIGURE 7.14 *Amanita phalloides.*

Amanita phalloides contains several complex cyclic peptides that account for the toxicity of the mushroom. The substances are of two types, the phalloidins and the amanitins. The phalloidins are potent hepatotoxins in both cultured hepatocytes and in vivo, following intravenous injection; however, they are poorly absorbed from the gastrointestinal tract and exhibit only weak toxicity when administered orally. Phalloidin binds to and stabilizes actin filaments and disrupts normal functioning of the cytoskeleton. The initial gastrointestinal effects of the mushroom are thought to result from the action of phalloidin.

The amanitins are considerably more toxic than the phalloidins. α-Amanitin exhibits a LD_{50} in rodents in the range of only 0.1 mg/kg when administered by either the intravenous or oral route. The target of this toxin is RNA polymerase II, the enzyme that mediates the synthesis of mRNA in cells. The cellular effects of α-amanitin are dramatic and include the disintegration of nucleoli, which prevents the synthesis of ribosomes that are the sites of protein synthesis. Metabolically active tissues dependent on high rates of protein synthesis such as the cells lining the gastrointestinal tract, hepatocytes, and the proximal convoluted tubules of kidney are disproportionately affected. The result is the destruction of the liver with a severity that resembles high doses of ionizing radiation. α-Amanitin also

PS: polysaccharide

cyclic peptides

MW: 10 kDa

amatoxin (9 different structures)
amanitin is just the cyclic peptide portion
(cyclic peptide has to be hydrolyzed off
of core to activate amanitin)

FIGURE 7.15 *Schematic structure of amatoxins.*

causes the destruction of the convoluted tubule of the kidney, which diminishes the kidney's effectiveness in filtering toxic nonelectrolytes from the blood.

The effects of *A. phalloides* poisoning may occur many hours following consumption of the mushroom. The initial signs of toxicity are abdominal pain, diarrhea, and vomiting, which can be lethal due to dehydration if the individual is not treated to restore proper electrolyte balance. Within three to five days following exposure to the toxins, long after the mushroom has been digested and the victim is beginning to think that recovery is likely, the individual will experience rapid, extensive, and lethal, so-called fulminant, failure of the liver and kidneys. Although many treatments have been tried in the effort to save the lives of the afflicted patients, including N-acetylcysteine and the plant flavonoid mixture called silymarin. The most successful are reported to be plasmapheresis to remove toxin from the blood in the early stages of toxicity, and liver transplantation to replace the damaged organ in the latter stages. The mushroom contains 0.2 to 0.3 mg α-amanitin/g fresh mushroom, and the ingestion of one 50-g mushroom provides a lethal dose for an adult.

The chemical structures of the amanitins and the phalloidins are complex indeed (Figure 7.15). Both substances are complexes of cyclic peptides that contain seven or eight amino acids. Although these cyclic peptides are responsible for the toxic effects of the mushroom, they exist in combination with polysaccharide components as large molecular weight complexes called myriamanins. Molecular masses as high as 60,000 Da have been assigned to some of these active complexes. The smaller cyclic peptides are obtained from the larger complexes by treatments with strong acids or bases.

FIGURE 7.16 *Amanita muscaria.*

Amanita muscaria

Several species of mushrooms, including *Inocybe* spp., *Conocybe* spp., *Paneolus* spp., *Pluteus* spp., and other *Amanita* spp. are not significant sources of food but are sought after for their psychotropic activities. *Amanita muscaria* is a classic example of such a mushroom (Figure 7.16).

This fleshy fungus grows throughout temperate areas of the world and is sought after as a hallucinogen. The reported pleasant effects of this mushroom and the slow degradation of the active principles combine to make *A. muscaria* a prized component of tribal and religious rituals in many parts of the world. The substances that are primarily responsible for the narcotic-intoxicant effects are a series of isoxazoles, such as muscimol (Figure 7.17), that comprise approximately 0.2% of the dry weight of the mushroom. The neurological symptoms produced by *A. muscaria* can vary but generally begin to occur within 30 to 90 minutes following ingestion. Following consumption of about 15 mg of muscimol, a state resembling alcoholic intoxication is generally produced. Confusion, restlessness, visual disturbance, muscle spasms, and delirium may follow. Patients are reported to pass into a deep sleep following the excited period, and upon waking, they may have little or no memory of the experience.

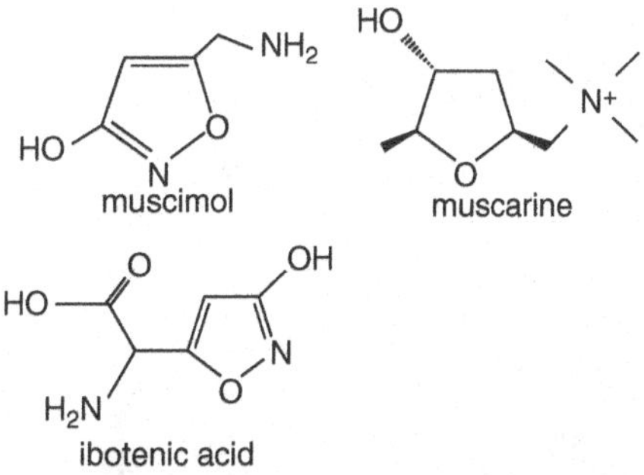

FIGURE 7.17 *Structures of active substances from A. muscaria.*

Ibotenic acid, another isoxazole present in *A. muscaria,* induces lassitude and sleep that is followed by a migraine and a lesser and localized headache that may persist for weeks.

A further important bioactive compound produced by *A. muscaria,* as well as some species of *Clitocybe* and *Inocybe*, is muscarine. This furan derivative is largely responsible for the insecticidal properties of *A. muscaria,* which also is known as fly agaric. The effects of muscarine poisoning are the result of its agonist activity on a subset of acetylcholine receptors that are designated as muscarinic receptors. Muscarine was the first parasympathomimetic substance studied and causes profound activation of the peripheral parasympathetic nervous system. Muscarine has no effects on the central nervous system because it does not cross the blood–brain barrier due to the high polarity of its positively charged nitrogen atom. The symptoms appear within 30 minutes of ingestion. The effects include increased salivation, lachrymation, and perspiration followed by vomiting and diarrhea. The pulse is slow and irregular and breathing is asthmatic. Death is uncommon, and patients generally respond well to atropine sulfate, which is an antagonist of the muscarinic receptors. Since severe cases of muscarine poisoning are rare following consumption of *A. muscaria,* it appears that the levels of muscarine are low in the mushroom relative to the levels of the other psychoactive compounds.

FIGURE 7.18 *Psilocybe spp.*

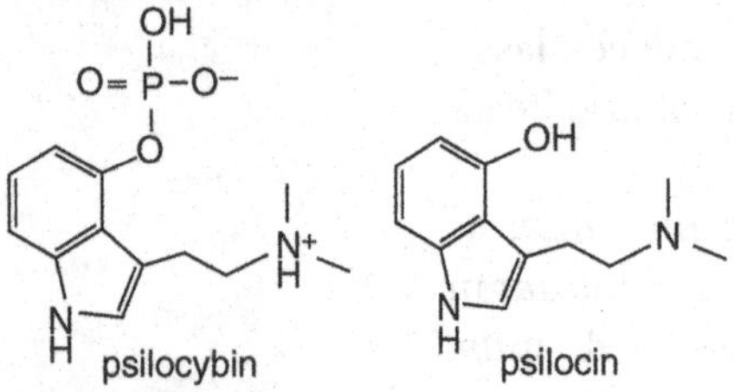

psilocybin psilocin

FIGURE 7.19 *Neuroactive products of Psilocybe spp.*

Psilocybe

The final group of mushrooms that are noted for their production of bioactive substances include species of *Psilocybe*, *Gymnopilus*, and *Conocybe*, and a few others that produce the hallucinogen psilocybin (Figure 7.18). Several of these mushroom species are eaten for their psychotropic effects in religious ceremonies of certain Native American tribes, a practice that dates to the pre-Columbian era. Some references suggest that *Psilocybe* mushrooms were even collected by our prehistoric ancestors. The neurological effects are caused by psilocin and psilocybin (Figure 7.19). Onset of symptoms is usually rapid and the effects generally subside within two hours. Poisonings by these mushrooms are rarely fatal. The most severe cases of psilocybin poisoning occur in small children, where large doses may cause the hallucinations accompanied by fever, convulsions, coma, and death. Psilocybin is rapidly dephosphorylated in the body to psilocin, which then acts as an agonist at the 5-HT2A serotonin receptor in the brain where it mimics the effects of serotonin (5-HT). The hallucinogenic effects of psilocybin are said to be comparable to a low dose of LSD.

ADDITIONAL READING

Berger, K.J., Guss, D.A. (2005). Mycotoxins revisited: Part II. *J. Emergency Med.* 28:175-183.

Broussard, C.N., Aggarwal, A., Lacey, S.R., Post, A.B., Gramlich, T., Henderson, J.M., Younossi, Z.M. (2001). Mushroom poisoning—From diarrhea to liver transplantation. *Am. J. Gastroenterol.* 96:3195-3198.

Groopman, J.D., Kensler, T.W. (2005). Role of metabolism and viruses in aflatoxin-induced liver cancer. *Toxicol. Appl. Pharmacol.* 206:131-137.

Hussein, H.S., Brasel, J.M. (2001). Toxicity, metabolism, and impact of mycotoxins on humans and animals. *Toxicol.* 167:101-134.

Michelot, D., Melendez-Howell, L.M. (2003). Amanita muscaria: Chemistry, biology, toxicology, and ethnomycology. *Mycol. Res.* 107:131-146.

Sudakin, D.L. (2003). Trichothecenes in the environment: Relevance to human health. *Toxicol. Lett.* 143:97-107.

CHAPTER 8

Food Contaminants from Industrial Wastes

CHAPTER CONTENTS

Through industrial activities in the modern era, many potentially hazardous substances have been deposited in the environment via ignorance, accident, and irresponsibility. Foods contaminated by these hazardous substances may cause adverse effect to humans and ecosystems. As our knowledge of toxicology has grown and analytical capabilities have improved, both a clearer understanding of the potential toxicity and an awareness of the health hazards caused by these contaminants have begun to receive much attention. The most widely studied of these substances produced by modern industrial activities are polychlorinated biphenyls, dioxins (tetrachlorodibenzo-*p*-dioxin), and heavy metals (lead, mercury, and cadmium).

CHLORINATED HYDROCARBONS

Polychlorinated Biphenyls (PCBs)

Polychlorinated biphenyls (PCBs) are a generic term used to describe a number of chlorinated derivatives of biphenyl (Figure 8.1). There are a total of 10 combinations possible, but there may be only 102 actually present.

Food Science and Technology

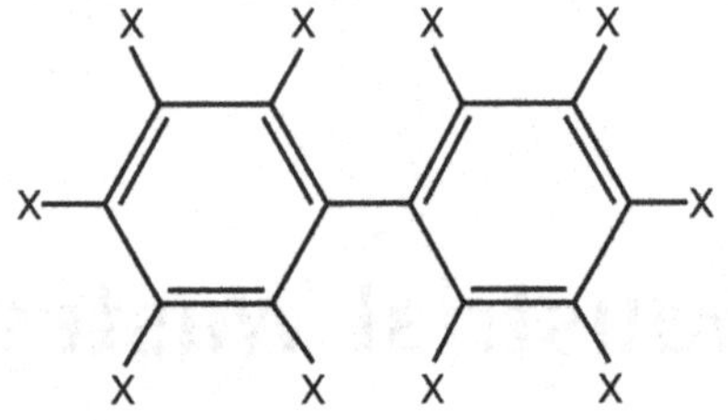

X = Cl or H

FIGURE 8.1 *Structure of polychlorinated biphenyls.*

PCBs that are generally available commercially are a mixture of different substances that are sold in formulations varying the percentage of chlorine in the substance. For example, in the United States, PCBs are sold by the trade name Aroclor, followed by different designations such as 1242, 1248, and 1260. The last two digits of these designations indicate the percentage of chlorine in the preparation. PCBs are heat stable, nonflammable, electrically nonconductive, and viscose liquid at room temperature. In addition, PCBs have unique natures, including high thermal conductivity, high electric resistance and water insolubility, and thus are an ideal material for electric appliances.

Not long after the first synthesis of PCBs, it became obvious that these compounds were unusual organic molecules. They were found to have considerable resistance to acids, bases, water, high temperatures, and electrical currents. Since the 1930s, with the discovery of these properties, PCBs have come to be used widely throughout the world for many purposes including insulating fluids in electrical transformers and capacitors, as well as in hydraulic fluids where resistance to high temperatures is required. Table 8.1 shows the use of PCBs in various products.

Environmental Impact

Production of PCBs began in 1930 and reached its maximum in 1970. During these years, vast quantities of PCBs were produced. It is estimated that by 1977, when production of PCBs in the United States was almost completely discontinued, over a billion pounds of these substances had been produced in the United States alone. Due to the clear, oily, and highly stable nature of these substances, PCBs have been used extensively in plasticizers, paints, lubricants, and insulating tapes (Table 8.1). PCBs also have found their way into fireproofing materials, inks, carbonless copy paper, and

Table 8.1 Use of PCB in Various Products

Usage	Products
Electric resistance	Transformer, condenser, fluorescent lamp, washing machine, refrigerator, etc.
Lubricant	Cutting oils, air compressor, vacuum pump oils, etc.
Adhesive	Electric cables, electric instruments, synthetic rubber, glue, etc.
Paint	Heat resistance paint, vinyl chloride paint, printing ink, etc.
Others	Carbon paper, copy paper, paper coating, color TVs, pesticides, water proof materials, etc.

Table 8.2 Worldwide PCB Sources and Amounts Entering the Environment up to 1970

Source	Amount (tons)
Disposal in dumps and landfills	18,000
Leaks and disposal of industrial fluids	4,500
Vaporization of plasticizers	1,500
Vaporization during open burning	400
Total	24,400

pesticide preparations. And, as a result of their high degree of stability, only about 600 million pounds of these substances are estimated to have degraded or been destroyed, which leaves about 450 million pounds of discarded PCBs still present in some form in the environment. For example, since 1966, General Electric purchased approximately 41,000 tons of PCBs from Monsanto Chemical Company, who discharged an average of 30 lbs PCB/day into the river. These figures represent only a part of the total quantity of PCBs in the world environment, since PCBs were produced by manufacturers in several other countries as well. Table 8.2 shows worldwide sources and estimated amounts of PCBs entering the environment up until 1970.

It wasn't until the 1960s when environmental contamination by PCBs received attention. At that time, the scientific community became concerned about the levels of chlorinated hydrocarbon pesticides such as DDT in food. PCBs, being close chemical relatives of DDT, began to show up as an unidentified contaminant in analytical samples. In many cases, PCBs were present in much higher quantities than was DDT. Subsequent concerted efforts to determine the levels of PCBs in foods indicated that the levels were highly dependent on the type of food analyzed.

Occurrence

As mentioned earlier, the high stability and lipid solubility of PCBs made it a highly persistent chemical in the environment. Consequently, PCBs entered the food chain in significant amounts. Market basket food analyses in 1971 and 1972 indicated PCB levels in foods available for general consumption to be always less than about 15 ppb, with the exception of certain food oils, which in some cases reached levels as high as 150 ppb. The highest levels of PCB contamination were found in fish from the Great Lakes. Chinook salmon from Lake Michigan contained levels of 10 to 24 ppm PCBs. Lake trout from Lake Michigan contained levels of 8 to 21 ppm PCBs.

As expected, other wildlife dependent on fish as part of their food also contained high levels of PCBs. Analyses of dolphin tissue from locations near the coast of California indicated levels on the order of 150 ppm based on fat. Analyses of blubber from porpoises and whales from East and West coasts of the United States indicated levels of 74 and 46 ppm. The accumulation of PCBs in birds was evident from analyses of eggs. Heron eggs from the Netherlands contained PCBs in the range of 34 to 74 ppm.

Massive surveys of PCB levels in human adipose tissue in the United States and surveys of PCBs in human milk showed levels generally in the range of about 0.1 to 3.0 ppm based on fat. These PCB levels in human tissues were dependent primarily on consumption of contaminated fish. Thus, consumption of fish from contaminated waters such as some of the Great Lakes was shown to be highly correlated with the level of PCBs found in human blood.

The revised permissible levels of PCBs in food established by the FDA are 2 ppm in fish and shellfish, 1.5 ppm in mild fats and dairy products, 3 ppm in poultry fat, and 0.3 ppm in eggs. Although the FDA has authority only over products shipped by interstate commerce, it has advised individual states where high levels of PCBs have been detected to consider establishing similar tolerances for PCB levels in their food products, especially fish. Several states have established PCB tolerances and in some cases this has meant the prohibition of commercial fishing in certain areas.

In Vivo Absorption and Metabolism

The absorption and excretion characteristics of PCBs are expected to be the same as those of stable fat-soluble substances. Thus, dietary PCBs are extensively absorbed from the gastrointestinal tract, with greater than 90% absorption common for most of these substances. The absorbed PCBs are stored primarily in adipose tissue with intermediate concentrations in the skin, adrenal glands, and aorta, and the lowest concentrations in the blood. Concentrations decrease over time more rapidly from the blood and more slowly from the adipose tissue. The biologic half-life of PCBs is estimated to be 8 weeks in male rats and 12 weeks in female rats.

The metabolism of PCBs is also strongly influenced by the extent of chlorination. The biphenyl derivatives that contain less chlorine are metabolized and excreted more rapidly than the biphenyl derivatives with higher percentages of chlorine. The route of metabolism of PCBs is primarily by converting to the corresponding phenols with loss of chlorine. The principal route of excretion of PCBs is through the feces with a relatively small percentage (less than 10%) showing up in the urine. This extensive elimination via the gastrointestinal tract, in spite of efficient absorption, suggests a major role for biliary excretion of PCBs. Excretion of PCBs via the milk in humans is generally minor in comparison to fecal and urinary excretion. However, the primary mode of excretion of PCBs in lactating cattle is via milk. Thus, cattle fed PCB-contaminated feed produce contaminated milk.

Mode of Toxic Action and Toxicities

The acute and chronic effects of PCBs have been studied extensively in several animal species including rabbits, mice, pigs, sheep, and monkeys. Chicks fed PCBs at levels of 50 ppm in the diet exhibit symptoms of depressed weight gain, edema, gasping for breath, hyperparacardia fluid, internal hemorrhaging, depression of secondary sexual characteristics, and increased liver size. Pigs and sheep appear to be less sensitive to the toxic effects of PCBs than monkeys. In pigs, dietary PCBs at levels of 20 ppm during the normal period of growth before slaughter resulted only in reduced feed efficiency and reduced rate of weight gain. The pigs had gastric lesions along with an increased frequency of pneumonia. Sheep appeared to be unaffected by dietary PCBs.

Acute toxic effects produced in adult rhesus monkeys at PCB levels of 250 to 400 mg/kg body weight include the development of gastric mucosal hypertrophy and hyperplasia, generalized hair loss, edema, and acne. In one study, the acne, edema, and hair loss were still present eight months after cessation of PCB administration, indicating a slow recovery from its toxic effects. Table 8.3 shows typical acute toxicities of PCBs.

Adverse reproductive effects of PCBs have been observed in several other animal species. Chickens fed 50 ppm PCBs produced chicks with beak, toe, and neck deformities. Also eggs produced from the exposed hens showed a significant increase in the percentage of nonviable embryos. In rabbits, 25 mg/kg PCBs in the diet for 21 days caused abortion in one out of four animals. Mice weaned on milk containing PCBs produced litters of fewer pups. Prolonged ingestion of dietary PCBs at 150 ppm reduced the reproductive capacity of female rats in addition to reducing plasma hormone (progresterone) level.

Table 8.3 Acute Toxicities of PCBs: LD_{50} (mg/kg)

Animal	Arcolor 1221	Arcolor 1254	Arcolor 1260
Rat (oral)	4,000	—	10,000
Weaning rats (oral)	—	1,200	1,300
Rabbit (skin)	approx. 2,000 to 3,000	—	approx. 1,300 to 2,000

In a study with female rhesus monkeys, relatively small doses (2.5–5.0 ppm) of PCBs in the diet resulted in symptoms of chloracne (severe acne-like skin eruptions) and edema. Menstrual cycles were changed, and problems of maintaining pregnancies occurred. Successful pregnancies produced infants of relatively low body weight. Fatty tissues of these infants contained PCBs at levels of approximately 25 ppm. Male rhesus monkeys receiving high doses of PCBs (5 ppm) showed only minor symptoms of toxicity with no evidence of reproductive effects.

The liver appears to be uniquely sensitive to the effects of PCBs. Increased liver weight, liver hypertrophy, proliferation of smooth endoplasmic reticulum, and an increase in microsomal enzyme activity characterize PCB effects in this organ. The activities and levels of certain cytochrome P450 enzymes increase dramatically following administration of PCBs, while the activity of glucose 6-phosphatase and the levels of vitamin A in the liver decrease dramatically.

PCBs may be involved in carcinogenesis in at least two ways. First, since many carcinogens require metabolic activation mediated by the cytochrome P450 enzyme system, PCBs may increase or decrease the carcinogenic potency of these compounds, depending on the role of specific metabolic processes in their conversion to active and inactive forms. Second, results of experiments with rats indicate that PCBs are themselves carcinogenic. In one experiment, long-term administration of PCBs (Aroclor 1260) at the level of 100 ppm to female rats resulted in liver tumors in 26 of 184 experimental animals administered PCBs, in comparison to only 1 in 73 controls. In addition, 146 of 184 test animals developed preneoplastic lesions in their livers.

Although much work remains to define the carcinogenic effects of PCBs and to establish the likelihood of human susceptibility, it appears that PCBs are relatively weak carcinogens. Even though PCB contamination in the environment in the United States has been widespread and in some cases quite significant, liver cancer is relatively rare in the United States. PCBs appear to play a relatively minor direct role in human carcinogenesis. However, the possible effects of these substances on the metabolism of other more potent environmental carcinogens should not be overlooked.

Outbreak

As is the case with many other environmental pollutants, documentation of the acute toxicity of PCBs in people has resulted from the widespread contamination of human food resulting from industrial accidents. In 1978 in Japan, the disease known as Yusho occurred as a result of the contamination of rice oil with 2000 to 3000 ppm of PCBs. The contamination resulted from a discharge of PCB heat transfer fluid into water that was subsequently used to irrigate rice fields. Over 1000 people showed symptoms of chloracne, pigmentation of the skin and nails, eye discharge, generalized swelling weakness, vomiting, diarrhea, and weight loss. Growth rate retardation occurred in young children and in the fetuses of mothers who were exposed to the PCBs. Again—as was the case with rhesus monkeys—these effects were slow to dissipate, lasting in many cases from several months to a year following the initial exposure. The patients had ingested about 2 g of PCBs from contaminated rice over 2 to 6 months.

Polychlorinated Dibenzo-*p*-dioxins (PCDDs)

Polychlorinated dibenzo-*p*-dioxins (PCDDs) are a large class of natural and synthetic substances that contain oxygen atoms held in ether linkages to carbon atoms generally within a six-membered ring (Figure 8.2). One comparatively small group of these substances, the dibenzodioxins, is of toxicological interest. The most notorious of dioxins is tetrachlorodibenzo-*p*-dioxin (TCDD). Although TCDD has 22 different isomers that generally occur in mixtures in the environment, the 2,3,7,8-tetrachloroisomer (2,3,7,8-TCDD) has received the greatest attention because of its unusual degree of toxicity. 2,3,7,8-TCDD is three times as potent a carcinogen as aflatoxin B_1 in female rats.

At room temperature TCDD is a colorless crystalline solid; it melts at 305 °C and is chemically quite stable, requiring temperatures of over 700 °C before chemical decomposition occurs. It is lipophilic and binds strongly to solids and other particulate matter that occurs in soils. The compound is sparingly soluble in water and most organic liquids.

The present concern about these compounds and the potential human health hazard they pose is a result of a growing awareness of their extraordinary toxicity and their inadvertent dispersion in the environment as trace contaminants in important commercial chemicals.

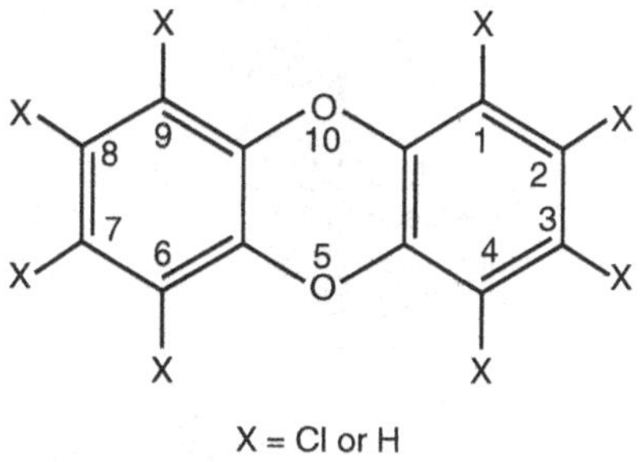

X = Cl or H

FIGURE 8.2 *Structure of polychlorinated dibenzo-p-dioxins (PCDDs).*

Occurrence

Dioxin is formed by burning chlorine-based chemical compounds with hydrocarbons. The major source of dioxin in the environment comes from waste-burning incinerators of various sorts and also from backyard burn-barrels. Dioxin pollution is also affiliated with paper mills, which use chlorine bleaching in their process, and with the production of polyvinyl chloride (PVC) plastics and certain

Table 8.4 Amounts of PCDDs Formed from Combustion of Plastics

Plastics	Concentration of PCDDs (ng/g sample) Total T4CDDs (22 Isomers)	Total PCDDs (77 Isomers)
Polyethylene (PE)	0.294	1.05
Polystyrene (PS)	0.123	0.214
Polyethylene terephthalate (PET)	0.050	0.566
Polyvinyl chloride (PVC)	7.45	53.5
PE + PVC	11.6	61.9
PS + PVC	5.38	18.4
PET + PVC	0.724	26.0

chlorinated chemicals such as pesticides. Table 8.4 shows the formation of dioxins from various plastics upon incineration.

Dioxins in the Diet

The major sources of dioxins are in our diet. Since dioxin is fat-soluble, it bioaccumulates, climbing up the food chain. Figure 8.3 shows levels of dioxin in the United States food supply reported in 1995.

Toxic equivalent (TEQ) of dioxin is calculated by looking at all toxic dioxins measured in terms of the most toxic form of dioxin, 2,3,7,8-TCDD. TEQ therefore takes into account all the relative strengths of dioxins and enables comparison with other forms. Therefore, some dioxins may count as only half a TEQ if it is half as toxic as 2,3,7,8-TCDD. A North American resident eating a typical Western diet will receive 93% of their dioxin exposure from meat and dairy products (23% is from milk and dairy alone; the other large sources of exposure are beef, fish, pork, poultry, and eggs). In fish, these toxins bioaccumulated up the food chain so that dioxin levels in fish are 100,000 times that of the surrounding environment. Figure 8.4 shows the daily intake of dioxins if one is eating a typical North American

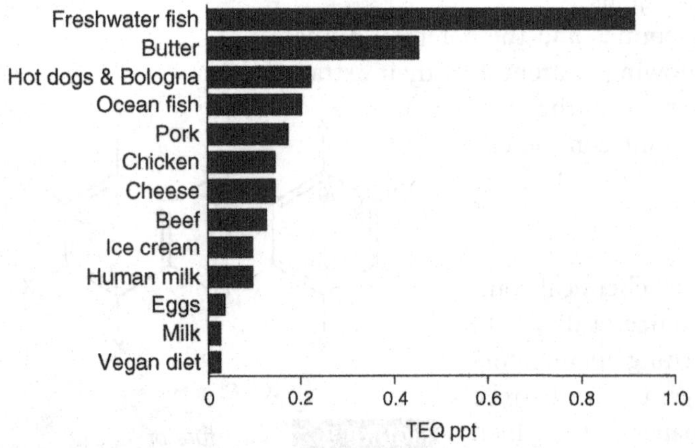

FIGURE 8.3 *Levels of dioxins found in various foods in the United States reported in 995.*

diet. The estimated daily intake of dioxins by an average North American is 116 pg/day.

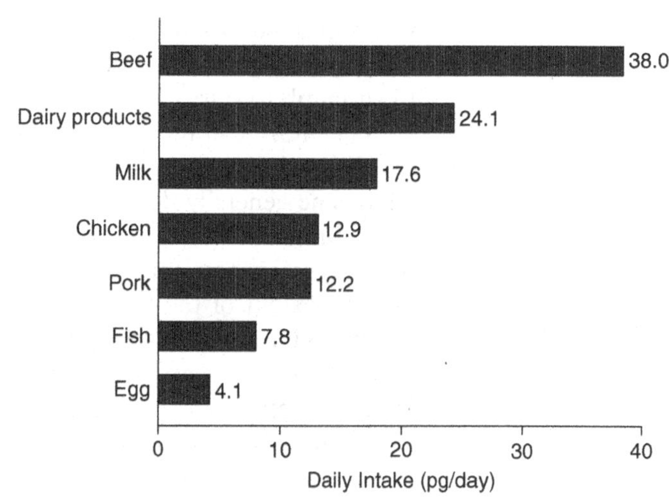

FIGURE 8.4 *Average daily intake of dioxins by a typical North American.*

Formation of TCDD from Herbicide

The chlorinated dibenzo-*p*-dioxin forms from the condensation of two orthochlorophenates. This particular dibenzo-*p*-dioxin formed depends on the chlorophenols present. Figure 8.5 shows the formation mechanism of 2,3,7,8-TCDD in the commercial synthesis of the herbicide, 2,4,5-trichlorophenoxy acetic acid (2,4,5-T). In this synthesis, the first step is the conversion of 1,2,4,5-tetrachlorobenzene (TCB) to sodium 2,4,5-trichlorophenate (TCP). At high temperatures, an unwanted contaminant, 2,3,7,8-tetrachlorodibenzo-*p*-dioxin, is formed by the condensation of two TCP molecules.

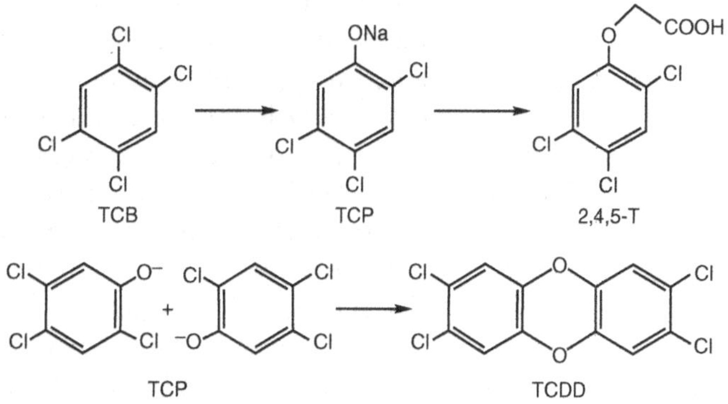

FIGURE 8.5 *Formation pathways of 2,4,5-T and TCDD.*

Herbicide synthesis is apparently not the only cause of environmental TCDD contamination. Significant levels of TCDD have been found in the incinerator smokestacks of chemical companies, even those that are not involved in the production of herbicides. Many reactions occur whenever organic and chlorine-containing substances are burned together. Some of these reactions apparently produce trace amounts of TCDD and various other dioxins.

Mode of Toxic Action and Toxicities

TCDD produces a wide range of physiological effects in humans and experimental animals. The guinea pig is the most sensitive species with an oral LD_{50} under 1 mg/kg. The hamster, the least sensitive species tested, has an LD_{50} some 10,000 times greater than that of the guinea pig. TCDD causes many effects in treated animals and the severity of these effects varies from species to species. For example, rats die from liver damage, but in the guinea pig the liver lesions seem less serious and the animals appear to succumb to starvation-like wasting of the adrenals.

The most common manifestation of TCDD poisoning is chloracne, which has been observed in rabbits, mice, monkeys, horses, and humans. Chloracne generally develops within 2 to 3 weeks following exposure and, depending on the severity of the exposure, may take from several months to many years to clear up. Kidney abnormalities appear in many species, and the extent of the lesions has prompted some researchers to conclude that the kidney is a specific target organ for TCDD exposure. Other physiological changes that have been observed in animal tests include hypothalamic atrophy, loss of fingernails and eyelashes, edema, bone marrow infection, polyneuropathy, and dry, scaly skin.

The carcinogenic effects of TCDD have been studied in some detail. TCDD enhances the potency of several carcinogens when administered, and is thus a promoter of carcinogenesis. TCDD alone is also an extremely potent carcinogen. Results of an experiment with rats treated with 10 or 100 ng of TCDD/kg body weight showed greatly increased numbers of liver tumors. At the higher dose level, both male and female rats developed increased numbers of tumors of the mouth, nose, lungs, and liver. In female rats, TCDD is three times as carcinogenic as aflatoxin B_1, which is considered one of the most potent hepatocarcinogens known. The no effect level or NOEL for TCDD in rats appears to be 10 ng/kg. Table 8.5 shows the structure–activity relationship of chlorinated dibenzo-*p*-dioxins for carcinogenicity.

TCDD is a highly potent teratogen in mice and rabbits. Rabbits that received a range of doses below 1 mg/kg during the sixth to fifteenth day of pregnancy showed increased resorption and higher rates of postimplantation loss. These effects were dose dependent. Severe effects on the bones and internal organs of surviving offspring were also apparent.

Table 8.5 Toxicity Equivalency Factor of Dioxins

PCDD	Toxicity Equivalency Factor
2,3,7,7-T_4CDD	1
1,2,3,7,8-P_5CDD	1
1,2,3,4,7,8-H_6CDD	0.1
1,2,3,6,7,8-H_6CDD	0.1
1,2,3,7,8,9-H_6CDD	0.1
1,2,3,4,6,7,8-H_7CDD	0.01
1,2,3,4,6,7,8,9-O_8CDD	0.0001

Results of small-scale studies with rhesus monkeys show that TCDD is fetotoxic to this species as well. Breeding monkeys fed TCDD at 1.7 ng/kg body weight per day for two years aborted four of seven pregnancies. Higher doses of TCDD produced higher levels of abortion and death in the pregnant females.

Although the results of experiments with laboratory animals clearly indicate the high toxicity, teratogenicity, and probable genotoxicity of TCDD, the outcomes of accidental human exposures suggest that people may be less sensitive to the toxic effects of TCDD than animal species. Many hundreds of cases of industrial accidents have been reported in which workers have been exposed to relatively high single doses of TCDD. In Seveso, Italy, approximately 37,000 residents may have been exposed to TCDD as result of an accident in trichlorophenol manufacture that occurred in 1976. By far the most significant finding from the Seveso incident and other cases of industrial exposure is that humans are much less sensitive to the immediate toxic effects of TCDD than are guinea pigs. To date, there has been no case of human fatality caused unequivocally by TCDD exposure. There are, however, many well-documented toxic effects in humans. Again, chloracne is the most obvious and initial effect in humans.

Other symptoms that develop with increased exposure to TCDD include a general sense of fatigue, disturbances in the responses of the peripheral nervous system (such as a reduction in the speed at which nerve impulses travel through the limbs), and liver toxicity (including enlargement of the organ as well as changes in the levels of many enzymes). Data from industrial exposures indicate that these conditions generally disappear after a few years, and the experience at Seveso seems largely to confirm these findings in the general population.

Further evidence to suggest that humans are much less sensitive to the toxic effects of TCDD than are animals, comes from an incident that occurred in Missouri in 1982 where waste chemicals containing high levels of TCDD were combined with oils and used as dust retardants in horse arenas and stables. Although many animals died, including rodents such as rats and mice, and large animals, such as cats, dogs, and horses, apparently as a result of TCDD exposure, there were no incidences of human deaths. One child who used the soil of the arenas as a sandbox became ill, but the symptoms of the illness subsided with no obvious continued effects.

The most important question in regard to TCDD toxicity may be related to the long-term human effects of TCDD exposure and its possible teratogenic effects. Unfortunately, teratogenic effects of TCDD in people have not been established, although numerous claims have been made about the teratogenic effects of the substance from Vietnam veterans who have been exposed to TCDD and from a group of people living in an area of Oregon where 2,4,5-T was routinely used to control undergrowth during forestry operations. Results of a few well-controlled studies have suggested that TCDD exposure may play a role in the formation of cancers of soft tissue in humans. The work of a Swedish group linked the use of TCDD-contaminated phenoxy herbicides with increased sarcomas of muscles, nerves, and fat tissue in people. In two studies, five- to six-fold increases in the incidences of this type of tumor were observed in people exposed to phenoxy herbicides compared to an unexposed control group. Although the validity of these studies has been criticized on several grounds, other studies of mortality rates in chemical workers exposed to TCDD through industrial accidents also show an increased level of soft tissue sarcoma.

In summary, TCDD has been shown to be one of the most toxic and carcinogenic substances in various animal species, occurring at low levels in apparently widespread areas of the environment. Humans appear to be quite resistant to the acute affects of TCDD. Much of this apparent resistance may be because humans ingest very small levels of TCDD in comparison to the levels that are administered to experimental animals. Several analyses of fish, milk, water, and beef fat used for human consumption were unable to detect TCDD at the detection limit of 1 ppt. Human exposures have been mainly through the skin. Evidence implicating TCDD exposure in the occurrence of soft tissue sarcomas in people is cause for concern and justifies efforts to minimize human exposure to this substance.

HEAVY METALS

Heavy metals are elements having atomic weight between 63.546 and 200.590 and a specific gravity greater than 4.0. Living organisms require trace amounts of some heavy metals, including cobalt (Co), iron (Fe), copper (Cu), manganese (Mn), molybdenum (Mo), vanadium (V), strontium (Sr), and zinc (Zn). They are essential to maintain the metabolism of the human body. Nonessential metals of particular concern to surface water systems are cadmium (Cd), chromium (Cr), mercury (Hg), lead (Pb), arsenic (As), and antimony (Sb). Table 8.6 shows the *in vivo* balance of typical metals in humans.

Excessive levels of heavy metals, including essential metals, can be detrimental to organisms. Heavy metal poisoning could result, for instance, from drinking-water contamination (e.g., lead pipes), high ambient air concentrations near emission sources, or intake via the food chain. Heavy metals can enter a water supply by industrial and consumer waste, or even from acidic rain breaking down soils and releasing heavy metals into

Table 8.6 *In vivo* Balance of Typical Metals in an Average American

	Source	As	Cd	Cr	Hg	Pb
Essential for human		No	No	Maybe	No	No
Body pool in human (mg)		17	330	6	?	80
Daily input (μg)	food	810	200	50	7.5	310
	water	10	15	10	NF	20
	air	NF	1	0.3	NF	20
Daily output (μg)	feces	675	163	59.4	7.0	314
	urine	225	50	0.6	0.6	30
	others	NF	0	0	NF	NF
Retained in (μg)		Liver	Kidney	Lung	Nerve	Bone
		0	2	0.3	?	6

NF: None found.

streams, lakes, rivers, and groundwater. Therefore, once heavy metals are deposited in the ecosystem, they readily contaminate various foods and subsequently are ingested by human and animal.

Arsenic

Occurrence

Arsenic was first written about by Albertus Magnus (Germany) in 1250. In the Victorian era, arsenic was mixed with vinegar and chalk and eaten by women to improve the complexion of their faces, making their skin paler to show they did not work in the fields. In 2005, China was the top producer of white arsenic with almost 50% world share followed by Chile and Peru.

The most important compounds of arsenic are white arsenic (III) oxide (As_2O_3), yellow sulfide orpiment (As_2S_3), red realgar (As_4S_4), Paris Green or copper (II) acetoarsenite [$Cu(C_2H_3O_2)_2 \bullet 3Cu(AsO_2)_2$], calcium arsenate [$Ca_3(AsO_4)_2$], and lead hydrogen arsenate ($PbHAsO_4$). The later three have been used as agricultural insecticides and poisons. Orpiment and realgar formerly were used as painting pigments, though they have fallen out of use due to their toxicity and reactivity. In addition to the inorganic forms, arsenic also occurs in various organic forms in the environment. Inorganic arsenic and its compounds, upon entering the food chain, are progressively metabolized to a less toxic form of arsenic through a process of methylation. For example, certain molds produce significant amounts of trimethylarsenic if inorganic arsenic is present. Also, arsenic containing pesticides such as copper acetoarsenite had been widely used and found in marine animals such as oysters, mussels, and shrimp.

Arsenic Contamination in Drinking Water

The most serious concern about arsenic contamination is its presence in drinking water. For example, arsenic contamination of groundwater has caused a massive epidemic of arsenic poisoning in Bangladesh and neighboring countries. It is estimated that approximately 57 million people are drinking groundwater with arsenic concentrations elevated above the World Health Organization's standard of 10 ppb. The arsenic in the groundwater is of natural origin, and is released from the sediment into the groundwater due to the anoxic conditions of the subsurface. The northern United States, including parts of Michigan, Wisconsin, Minnesota, and the Dakotas are known to have significant concentrations of arsenic in their groundwater. Increased levels of skin cancer have been associated with arsenic exposure in Wisconsin, even at levels below the 10 ppb in drinking water standard.

Arsenic Contamination in Foods

Most foods (except fish) contain less than 0.25 μg/g arsenic. Many species of fish contain between 1 and 10 μg/g. Arsenic levels at or above 100 μg/g, however, have been found in bottom feeders and shellfish. Both lipid- and water-soluble organo-arsenic compounds have been found but the water-soluble forms constitute the larger portion of the total arsenic content. Eventually arsenic is accumulated in marine animals: 3 to 10 ppm in oysters, 42 to 174 ppm in mussels, and 42 to 174 ppm in shrimp. Therefore, the amount of arsenic ingested in human diets is greatly influenced by the proportion of seafood consumed. It is estimated that ingestion of 200 g of seafood containing 1 mg total arsenic/100 g by a person may result in an absorption of 2 mg arsenic. Assuming that up to 10% of arsenic is ingested as in organic arsenic the consumption of 200 g seafood may eventually lead to absorption of 200 μg inorganic arsenic. Table 8.7 shows the arsenic content in typical seafood in Belgium.

Table 8.7 Total Arsenic Content in Seafood Reported in Belgium

Seafood	Content (μg/100 g)
Tuna	46.6
Salmon	40.3
Crab	56.6
Anchovy	86.2
Mackerel	172.0

Dietary arsenic intake estimates from various countries range from less than 10 μg/day to 200 μg/day. Although there are other routes of arsenic intake, including air and drinking water, dietary arsenic represents the major source of arsenic exposure for most of the general population. Persons who are high consumers of fish may ingest significant amounts of arsenic from seafood.

Mode of Action and Toxicity

Arsenic absorption occurs predominantly from ingestion from the small intestine, though minimal absorption occurs from skin contact and inhalation. Arsenic exerts its toxicity by inactivating up to 200 enzymes, especially those involved in cellular energy pathways and DNA synthesis and repair. Arsenic disrupts ATP production through several mechanisms. At the level of the citric acid cycle, arsenic inhibits pyruvate dehydrogenase, and by competing with phosphate it uncouples oxidative phosphorylation, thus inhibiting energy-linked reduction of NAD^+, mitochondrial respiration, and ATP synthesis. Hydrogen peroxide production is also increased, which might form reactive oxygen species and oxidative stress. These metabolic interferences lead to death from multisystem organ failure probably from necrotic cell death, not apoptosis. A postmortem reveals brick-red colored mucosa, due to severe hemorrhage.

It was found that the sulfhydryl groups of various enzymes, such as pyruvate dehydrogenase and α-ketoglutarate dehydrogenase, were the targets for arsenic poisoning. Meantime, an antidote, the so-called British Anti-Lewisite (BAL), was developed for the arsenic poisoning. The mechanism of BAL antiarsenic activity is shown in Figure 8.6. However, use of BAL is limited because it causes some adverse effects such as fever, conjunctivitis, lacrimation, headache, nausea, and pain at the injection site.

Acute arsenic poisoning is associated initially with nausea, muscular weakness, vomiting, brown pigmentation, localized edema, and severe diarrhea. Acute toxicities by experimental animals are LD_{50} (oral) of As^{5+} and As^{3+} are 1000 mg/kg and 5 mg/kg, respectively. The decreasing order of toxicities among arsenic compounds is $AsH_3 > As^{3+} > As^{5+} > R - As - X$.

Chronic arsenic toxicity results in multisystem disease. Chronic symptoms of arsenic poisoning are liver enlargement, anemia, and reduction of white blood cells. Also, arsenic is a well-documented human carcinogen affecting numerous organs. Epidemiological studies on inorganic arsenic

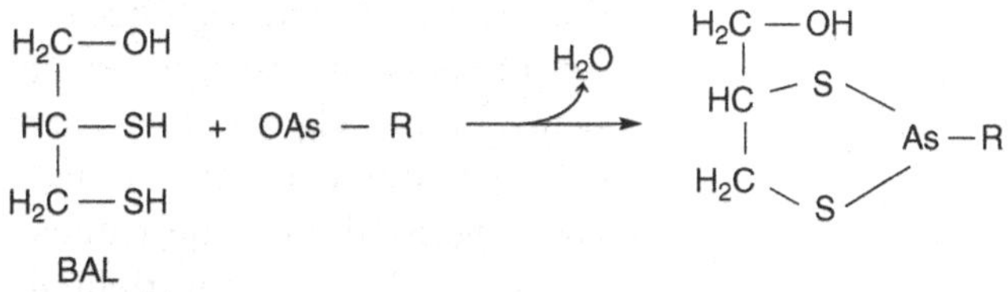

FIGURE 8.6 *Reaction mechanisms of BAL with arsenates.*

exposure suggest that a small but measurable risk increase for bladder cancer would occur when 10 ppb of arsenic were consumed for a prolonged time period.

Outbreak

In the early summer 1955, physicians in the western part of Japan worried about outbreaks characterized by anorexia, skin pigmentation, diarrhea, vomiting, fever, and abdominal distention among infants, mostly less than 12 months of age. It was reported that 130 infants died after exhibiting those symptoms. As an apparent link, most of the patients were bottle-fed, often using the popular Morinaga dried milk brand. It was found that the symptoms shown by these patients, including fever, skin pigmentation, hepatomegaly, and anemia were all consistent with clinical symptoms of arsenic poisoning. Subsequently, arsenic was identified from the Morinaga dried milk that the patients had ingested. Later, arsenic was found in a disodium phosphate product added to cow's milk as a stabilizer to preserve constant acidity. This sodium phosphate consisted of approximately 45% crystal water, 14% P_2O_3, 28% Na_2O, 2% V_2O_5, and 6% As_2O_5. The final product of Morinaga dried milk contained approximately 21 to 34 μg arsenic per gram. It was estimated that a total of 2.5 mg arsenic was ingested by one-month-old infants, 3.2 mg by two-month-old infants, and 4.6 mg by six-month-old infants. The poisoned infants were treated by BAL and other therapy. The symptoms seemed to disappear in the infants who survived. However in 2006, more than 600 surviving victims, now in their 50s, have been reported to suffer from severe sequelae, such as mental retardation, neurological diseases, and other disabilities.

Lead

Occurrence

Lead occurs widely throughout the environment and has been found in all bodies of water and soils tested. Lead, occurring chiefly as the sulfide galena, is the most abundant of the heavy metals found in the earth's crust. The heating process required to convert the ore to metal was used by the early Phoenicians, Egyptians, Greeks, Chinese, and East Indians in the preparation of eating utensils, water ducts, and vessels for various liquids, ornaments, and weights. The early Romans used lead for their extensive aqueduct system and for storage vessels for wine and food. Some historians have suggested that the extensive use of lead-containing vessels by the Romans may have contributed to mental and emotional instability in Roman leaders and thus to the downfall of the Roman Empire. Reports by ancient Roman physicians show that roughly two-thirds of the emperors who reigned between 30 BCE and 220 CE drank a great deal of lead-tainted wine. Additionally the majority of the Roman emperors also had gout, a condition occasionally reported in lead poisoning due to moonshine whiskey and indicative of lead-induced kidney deterioration. The most likely source of lead was the lead-lined vessels used to boil down grape syrup to prepare Roman wines. Contemporary efforts to prepare the syrup according to the ancient recipes yielded a mixture with 240 to 1,000 mg of lead/liter. Thus, very high levels of lead could occur in the wine from the combined practices of using lead-lined vessels for wine storage and consumption, and by the use of lead-tainted syrup in wine preparation.

Environmental Contamination

The principal causes of environmental lead contamination in recent times stem from its use in lead storage batteries and gasoline antiknock additives (tetra-ethyl lead). Lead-containing pesticides also have been used, but this use has decreased in recent years.

In the early 1970s, typical premium gasoline contained between 2 and 4 g of lead per gallon with an average of about 2.8 g. Regular gasoline averaged about 2.3 g of lead per gallon. Thus, in California, where over 13 million motor vehicles were in use, the total lead consumed in gasoline was approximately 26,000 tons per year. On average 79 to 80% of the lead in gasoline is exhausted out the tailpipe as particulate. Less is exhausted at low-speed driving and more is exhausted under freeway conditions. Figure 8.7 shows

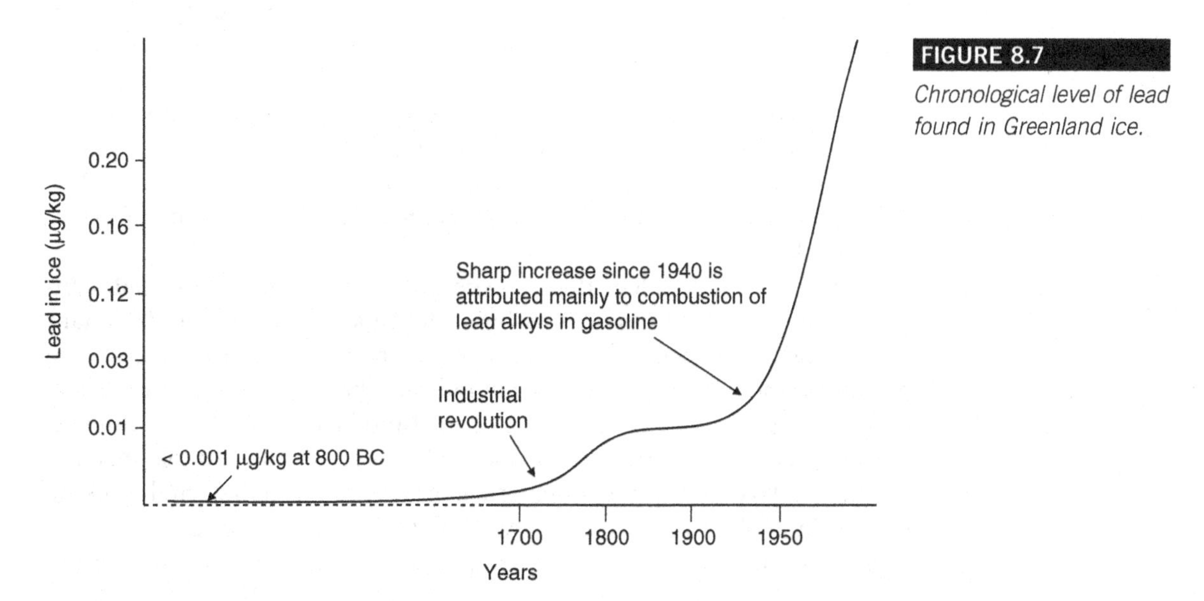

FIGURE 8.7

Chronological level of lead found in Greenland ice.

the chronological level of lead found in Greenland ice. The sudden increase of lead levels in the ice in Greenland, as shown, indicates that the use of lead-containing antiknocking agents in gasoline contributed to the high level of lead contamination in the environment.

During the early 1970s there was intense public debate over the question of whether or not lead should be banned from gasoline. The issue was resolved when the automobile industry met lower automobile emissions standards, resulting in a reduction of airborne lead in busy traffic cities such as New York.

Food Contamination

The major nonindustrial sources of lead contamination for people are food and water. Supplies of drinking water for most cities in the United States contain level of lead lower than the 15 μg/liter maximum allowed by the US Environmental Protection Agency. However, water that has been allowed to stand in lead pipes or in lead-containing vessels for a period of only a few days may contain lead in concentrations of 1 mg/liter.

Lead has been detected in all foods examined, even those grown far from industrialized areas. Recent analyses suggest a natural level of lead to be on the order of 0.3 ppb in wide-ranging marine fish. These fish are comparatively free of geographically localized contamination and are considered good indicators of general environmental contamination. Levels considerably in excess of this, however, have been detected in commonly consumed foods. In general, plant foods grown in industrialized areas exhibit higher lead levels compared to foods grown in remote areas. For example, in bean pods and corn husks, although these same plant components obtained from urban gardens exhibit relatively low levels, those obtained near highways had lead levels at least 10 times higher, respectively. For animal products, the highest levels of lead are found in bone. However, as a food group, seafood tends to contain the highest levels of lead, with a range of 0.2 to 2.5 ppm. On average, the total lead intake from food is estimated at roughly 0.01 mg/day.

In the past, lead-soldered cans were a well-known source of lead, contaminating such food products as infant formulas, evaporated milk, infant juices, and baby foods. Lead levels of 0.5 ppm in these products were not uncommon. Analyses of albacore indicate that conventional processing of food increases lead levels in the canned product considerably. Butchering and packing the albacore in unsoldered cans results in a 20-fold increase in the levels of lead, and butchering, grinding, air-drying, and packing results in a 400-fold increase. A 4000-fold increase in lead levels of albacore results from butchering and packing in lead-soldered cans. In the

United States, the FDA and the canning industry have combined efforts to institute the use of nonsoldered cans. The result has been a drop in lead levels in products used primarily by infants to one-fifth to one-tenth the levels observed when soldered cans were used.

Mode of Toxic Action

The extent of lead absorption in the gastrointestinal tract depends on several factors. One factor is the chemical form of the lead. Organic lead compounds, such as tetraethyl lead, are readily absorbed from the gastrointestinal tract (> 90%), eventually concentrating primarily in the bone, and to lesser extent in the liver, kidney, muscle, and central nervous system. Under normal circumstances in the adult, inorganic lead compounds are poorly absorbed from the alimentary tract (5–10%). Absorption of inorganic lead in infants and children is considerably higher, however, with estimates in the range of 40–50%. Absorbed lead is excreted primarily in the urine (16%).

Several dietary factors affect the level of lead absorption in people and experimental animals. For example, approximately three times as much lead is absorbed following a 16-hour fast than when lead is administered following normal periods of food or feed consumption. In animal studies, increasing the corn oil content in the diet from 5 to 40% resulted in a 7- to 14-fold increase in lead content in many tissues.

A low level of dietary calcium also was shown to produce increased levels of lead absorption and resultant lead toxicity. Lead exposure of rats consuming diets low but not deficient in calcium produced four-fold higher blood lead levels than in lead-exposed rats consuming a diet with adequate calcium levels. This is because calcium appears to compete with lead in the gastrointestinal tract for a common absorption site. Iron deficiency also affects lead absorption from the gastrointestinal tract. As much as six-fold increases in tissue lead have been reported for rats in which iron stores had been reduced. Decreased zinc intakes also result in increased gastrointestinal absorption of lead and increased toxicity. Zinc also is reported to influence lead levels in the fetuses of animals that have been treated with lead.

Dietary factors and aging may also affect distribution of lead in the body. A very low calcium diet limits the amount of lead that can be stored in bone because bone formation is slower. Under normal circumstances, lead has a biological half-life of about 10,000 days in bone. At times of low calcium intake, however, a significant amount of lead may be released into the blood stream due to bone resorption. Under such conditions, higher lead content in the kidney and blood has been observed in experiments with animals.

This may prove especially significant for elderly people. Aging frequently is accompanied by bone demineralization, and previously immobilized lead may be released. The higher incidence of kidney disease and urinary problems in the elderly further increases susceptibility to lead poisoning since urinary excretion of lead is inhibited.

Three stages of lead poisoning have been recognized. The first stage, called the asymptomatic stage, generally is not associated with behavioral disorders or organ dysfunction, but is characterized by changes in the blood. Anemia is a well-established early symptom of relatively mild lead poisoning. Lead decreases the lifetime of erythrocytes and the synthesis of heme. Although the interaction of lead with the hematopoietic system is quite complex (Figure 8.8), most of the observed effects of lead on blood can be explained by its inhibitory influences of δ-aminolevulinic acid synthetase or ALA synthetase, ALA dehydrase, and ferrochelatase. Stage 1 of lead poisoning is characterized by an increase of uroporphyrinogen III level in the blood resulting from decreased insertion of iron into uroporphyrinogen III mediated by ferrochelatase. Also at this stage, urinary ALA is increased since the ALA dehydrase-mediated conversion of ALA to porphobilinogen (PBG) is decreased. In the later phases of this asymptomatic period, urinary ALA is increased further and hematocrit and hemoglobin values decrease.

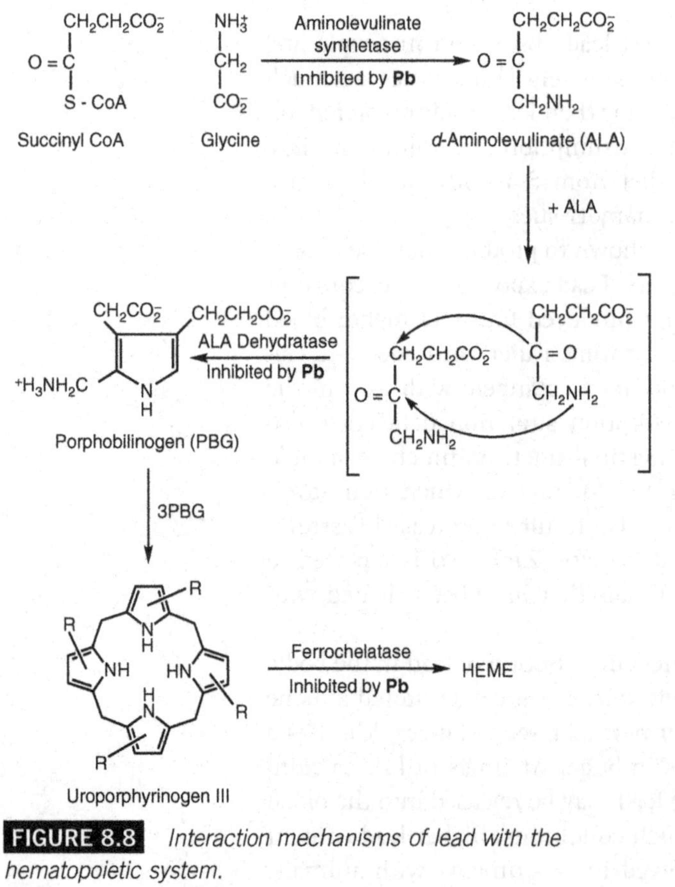

FIGURE 8.8 *Interaction mechanisms of lead with the hematopoietic system.*

In Stage II lead poisoning, the symptomatic stage or the symptomatic period, anemia may be quite obvious, and disorders of the central nervous system, including hyperactivity, impulsive behavior, perceptual disorders, and slowed learning ability, appear. In more severe cases, the symptoms include restlessness, irritability, headaches, muscular tremor, atoxia, and loss of memory. With continued exposure to lead, Stage III ensues, with symptoms that may culminate in kidney failure, convulsion, coma, and death. These symptoms have been reported in industrial accidents or following consumption of house paint or moonshine whiskey.

Of major concern are the neurological effects produced in children by continued exposure to relatively low levels of lead. In one study designed to assess the magnitude of this problem, groups of schoolchildren were placed into low-lead and high-lead exposure groups according to the lead level found in deciduous teeth. Although none of the children exhibited the symptoms of clinical lead poisoning, children in the high-lead group showed increased distractibility, inability to follow directions, and increased impulsivity compared to children in the low-lead group. Children in the high-lead group also scored lower on standard IQ tests and verbal tests.

The levels of lead in the blood associated with various symptoms are indicated in Table 8.8. A dose within the range of 0.2 to 2.0 mg lead/day in the adult human generally is required to produce measurable effects in the blood and to produce blood lead levels in the range of 20 to 30 μg/dl. However, in a child 1 to 3 years of age, consumption of approximately 135 μg/day with 50% absorption from the gastrointestinal tract results in approximately 20 μg/dl in the blood. The actual levels of lead consumption from the diet, estimated to be in the order of 15–20 μg/day for adults and 5 μg/day for infants, are thought to be below the range of doses that produce measurable effects in blood and may be in excess of the doses required to adversely affect neurological functions.

These dangerously high levels of lead consumption, especially for children, have been pointed out by many scientists, and the problem has been described clearly by a panel of the National Academy of Sciences. The consensus of scientific opinion is that human exposure to lead must be reduced wherever possible.

Table 8.8 Symptoms of Lead in Blood

Level of Pb in Blood (μg/100 ml)	Symptoms
25–30	Uroporphyrinogen increase in blood, ALA increase in urine
40–50	Hematocrit decrease, hemoglobin decrease, ALA increase
50–60	Anemia
>60	Hyperkinesis, short attention span, aggressive conduct (minimal brain dysfunction)
>120	Mental retardation, blindness, death

Mercury

Occurrence

Mercury occurs primarily in geographical belts in the earth's crust as red sulfide, cinnabar. The heating process required to prepare free metallic mercury from its sulfide was discovered in very early times. Because of its shiny, fluid appearance and its great density, magical powers were attributed to metallic mercury. As a result, mercury was put to many questionable uses in alchemy and medicine. Although metallic mercury was used until relatively recent times to remedy bowel obstructions, it is no longer used in remedies in modern medicine.

Due to economic reasons, and especially to the well-established toxicity of mercury and many of its derivatives, uses of mercury today are quite restricted. There is also an increased effort to reprocess and reclaim mercury used in many products. Metallic mercury and mercury salts are used primarily by the electrical and chemical industries in switches, coatings, and catalysts. Organo-mercurial compounds are used to some extent as diuretics, but are being increasingly replaced by nonmercurial compounds. Also, organomercurials in various forms are very popular antiseptics and are used, in some cases, in sterilizing solutions for medical instruments. Mercury continues to be used in the dental preparations for fillings. Projections for world mercury use by the year 2000 are in the range of 10^5/year.

Until relatively recent times mercury was thought to be environmentally stable, so discarded metallic mercury was simply buried in the ground or deposited in waterways. In addition, salts of mercury were thought to readily form complexes with various components in water and to form inert precipitates. However, as indicated in Figure 8.9, mercury can be converted from one form to another in the environment.

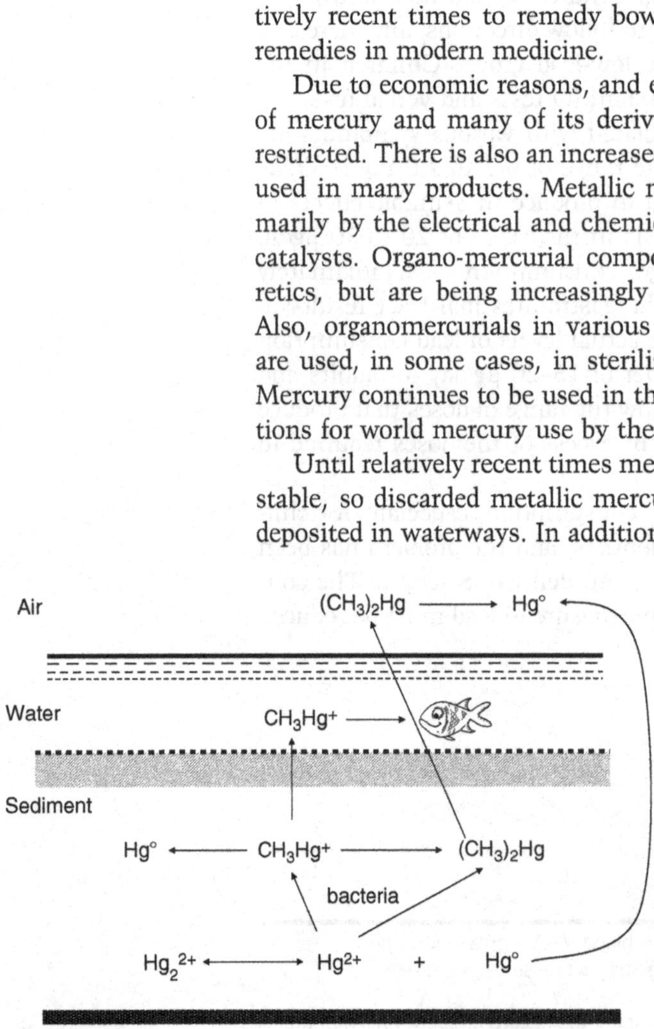

FIGURE 8.9 *Conversion pathways of mercury in the environment.*

Metallic mercury deposited in a lake settles into the sediment where bacteria may carry out oxidations and alkylations producing both water-soluble substances (salts) and lipid-soluble substances (alkyl compounds). The dialkyl mercury compounds are relatively volatile and can be transported in the air. These metabolic conversions, therefore, provide a means whereby mercury deposited in one form in the environment will eventually be converted to another form and may be found at sites distant from the original site or disposal location. For example, a highly toxic organo-mercurial compound used primarily as an algaecide will eventually be converted, at least in part, to the less toxic inorganic forms and to metallic mercury. In addition, such organo-mercurial compounds are eventually converted to methyl mercury, the most toxic of the mercurial compounds.

Contamination in Foods

Mercury contamination of food depends, to a great extent, on the chemical form that is placed in the environment or the extent of environmental interconversion of mercury compounds. Mercury levels of most plant foods and meats are generally considered to be quite low, with recent estimates for meat products in the United States indicating a range of 1 to 7 ppb. Mercury levels of other foods, including potatoes, legumes, and cereals, are generally less than 50 ppb. In the few cases that have been examined, methyl mercury accounts for the greater percentage of mercury compounds in meat products and fish. Inorganic mercury is thought to be the major form of mercury in plant foods.

Fish is the primary source of dietary mercury. Fish and other creatures, such as predatory birds, which occupy advanced levels in the food chain, are known to concentrate mercury as much as 1000-fold above the levels that occur in their immediate environment. Larger marine fish have higher concentrations of mercury than smaller fish. For example, analyses by the FDA indicate that large tuna contain an average of 0.25 ppm of mercury, and smaller tuna have an average of 0.13 ppm. In addition, the amount of methyl mercury relative to total mercury increases with the age of the fish. Shellfish accumulate mercury from the aquatic environment at levels nearly 3000 times greater than the levels to which they are exposed in the water. Analyses of fish in many countries indicate that 99% of the world's fish catch has a total mercury content not exceeding 0.5 mg/kg.

The FDA has set this level of 0.5 ppm as a maximum acceptable level of mercury in fish and shellfish, and the World Health Organization (WHO) recommended a maximum level of mercury of 0.05 ppm mercury in food items other than fish. Based on analyses of mercury levels associated with human poisoning, the WHO committee suggested a provisional tolerable intake of mercury of 0.3 mg total/week, which is not to include more than 0.2 mg of methyl mercury. Thus, weekly consumption of about 600 g (or approximately 86 g/day) of fish containing 0.5 mg/kg of mercury would not exceed the tolerable intake. However, according to a 1972 report by the WHO, per capita fish consumption in the United States was 18 g/day, 56 g/day in Sweden, and 88 g/day in Japan. These data indicate that potentially large numbers of individuals are consuming enough fish to put them at risk of mercury poisoning.

Due to the dramatic and tragic demonstrations of the acute effects of mercury poisoning in humans, regulatory agencies throughout the world were concerned about the natural levels of mercury in the diet. Extensive analyses of fish in the United States and elsewhere revealed that mercury levels far in excess of the 0.5 ppm guidelines occurred in commonly consumed species. For example, in Lake Erie, levels as high as 10 ppm were found in some fish. Larger marine fish also were consistently found to have mercury levels in excess of the 0.5 ppm. Swordfish in particular were incriminated when analyses revealed that only a small percentage of the total samples of this fish had levels of mercury less than 0.5 ppm. As a result, in 1971 the FDA seized over 800,000 pounds of commercial swordfish, which led to the near collapse of the swordfish industry in the United States and elsewhere. These apparently high levels of mercury in marine fish were blamed on contemporary pollution of the marine environment until analyses of museum specimens of fish revealed that the levels of mercury had not changed significantly over the past 100 years. In addition, mercury in fish is less toxic than mercury by itself. The toxicity is reduced at an approximate 1:1 ratio by selenium that occurs along with the mercury in fish. These as yet unexplained phenomena are the subjects of continuing research.

In Vivo Absorption and Toxicity

The extent of absorption of mercury compounds is dependent on the site of absorption and the chemical form of the element. Less than 0.01% of ingested metallic mercury is absorbed from the gastrointestinal tract. Thus, the use of metallic mercury in the treatment of bowel obstruction was probably of little hazard. However, approximately 80% of inhaled metallic mercury in the vapor state is absorbed in the respiratory tract. Estimates of absorption of mercury salts range from approximately 2% of the daily intake of mercuric chloride in mice to approximately 20% of mercuric acetate in rats.

Absorption and distribution of alkyl mercury compounds, because of their greater lipid solubility, are much more extensive than for either of the inorganic forms. Results of experiments with human volunteers and with several animal species indicate that methyl mercury is absorbed almost completely from the gastrointestinal tract. Following absorption, methyl mercury moves to the plasma, where it is bound in the red blood cells. The compound then moves primarily to the kidneys, as well as to the colon, muscles, and other tissues, including those of the fetus. Concentrations of methyl mercury in fetal blood are increased over concentrations in maternal blood. This apparently is due to the increased concentration of hemoglobin in the fetal blood. Although total transport across the blood–brain barrier is relatively slow, removal of methyl mercury from the brain appears to lag behind removal from other tissues; the concentration of methyl mercury in the brain is roughly 10 times higher than that in the blood for humans.

Most mercury compounds are metabolized to the mercuric state and excreted in the urine and feces. In contrast, methyl mercury is excreted mainly in the feces by processes of biliary excretion and exfoliation of intestinal epithelial cells. Extensive reabsorption of methyl mercury from the intestines is one of the factors increasing the biological half-life (80 days) of this compound.

Toxic effects of mercury compounds in humans have been known for many years. Inorganic mercury affects primarily the kidneys, affected by inorganic mercury leading to uremia and anuria. The early symptoms of acute inorganic mercury poisoning are gastrointestinal upset, abdominal pain, nausea, vomiting, and bloody diarrhea. Doses required for acute kidney effects appear to be in excess of 175 ppm of mercuric salts in the diet.

Outbreaks

Perhaps the first well-established incidences of poisoning from organomercurial compounds involving relatively large numbers of people began in 1954 in Japan. Severe neuromuscular and neurological defects began to appear in people living near Minamata Bay and later in people living near Niigata, Japan. As of 1970, poisoning of over 120 people in these two locations was officially documented with 50 deaths reported. Early symptoms of methyl mercury poisoning, also dubbed Minamata disease, included loss of sensation at the extremities of the fingers, toes and areas around the mouth, loss of coordination, slurred speech, diminution of vision called tunnel vision, and loss of hearing. Pregnant women exposed to methyl mercury gave birth to infants with mental retardation and cerebral palsy. Later symptoms include progressive blindness, deafness, lack of coordination, and mental deterioration. Neuromuscular development was abnormal in individuals exposed in utero. The cause of the disease was traced to consumption of fish from the local waters that were heavily contaminated with methyl mercury. Chemical factories that were located upstream from these towns were the source of the methyl mercury that eventually accumulated in the fish.

During the 1950s to the 1970s, several outbreaks of human poisoning due to consumption of wheat products treated with mercury pesticides were reported with many thousands of individuals affected. In an incident of methyl mercury poisoning that occurred in 1972 in Iraq, 6530 people were officially admitted to hospitals and 459 of these patients died. The outbreak occurred as a result of consumption of bread made from seed wheat. The seed, treated with methyl mercury fungicide and colored with a brownish-red dye, was distributed in bags that were labeled in English and marked with appropriate warnings. Farmers were apparently not familiar with the warnings, and the brownish-red dye could be removed by washing, giving the impression that the poison was removed. The seed was ground into flour and subsequently used to make bread.

Cadmium

Occurrence

Cadmium has chemical characteristics similar to those of zinc. Cadmium occurs in nature wherever zinc is found and is produced as a by-product in the mining of zinc and lead. It is used in the galvanizing of other metals to prevent rusting and in the manufacture of storage batteries and plastics.

The major sources of cadmium exposure for the general population are water, food, and tobacco. The ultimate source of contamination of these products is difficult to discover. Cadmium in water often results from the cadmium alloys used to galvanize water pipes. However, no single source of cadmium contamination in the environment has been identified.

Food generally contains less than 0.05 ppm of cadmium, providing an estimated 0.5 mg cadmium/week. Analyses of the world food supply conducted on a periodic basis by WHO indicate that the foods with the highest levels of cadmium contamination are consistently shellfish and the kidneys of various animals including cattle, chicken, pigs, sheep, and turkeys. Cadmium levels in kidneys, for example, are often in excess of 10 ppm, and cadmium levels in shellfish, such as oysters, reach levels as high as 200 to 300 ppm. Cadmium levels are about 0.03 ppm in most meats in the United States and 0.075 ppm in hamburger. The higher levels of cadmium in the hamburger are probably due to processing methods. Soybeans are reported to contain 0.09 ppm cadmium, and most other plant foods are very low in cadmium (e.g., 0.003 ppm in apples).

When considered as part of the normal diet, these various dietary components provide an estimated 73 μg cadmium/day per person. This value is not significantly different from the provisional tolerable intake set by the WHO (57–71 μg/day). The provisional allowable intake has been developed based on the results of experiments with animals and on analyses of human exposures in industrial accidents resulting in cadmium poisoning. The value includes a safety margin of a factor of approximately four from a level that can cause minimal kidney damage with long-term consumption. Thus, the levels of cadmium present in the diet do not appear to present an imminent health hazard to people in a general population. However, for individuals who consume unusually large amounts of shellfish or kidneys, the daily intake of cadmium may be in excess of the provisional tolerable intake.

In Vivo Absorption and Toxicity

Data on the biological effects of cadmium in people are incomplete compared to the data for mercury and lead. Apparently only about 5% of orally administered cadmium is absorbed in the gastrointestinal tract. The various salts of cadmium differ in their water solubility, and therefore, may be absorbed to somewhat differing extents. Cadmium does not occur as stable alkyl derivatives that would be expected to have increased lipid solubility. The extent of absorption of cadmium from the gastrointestinal tract of rats is considerably greater in the newborn animal than in the older animal. Rats dosed with cadmium fluoride at 2 and 24 hours of age absorbed over 20 and 10 times, respectively, more than the amount of cadmium absorbed by animals dosed at 6 weeks of age. In addition, the extent of absorption was dependent on other factors. For example, the absorption of cadmium taken with milk was roughly 20 times greater than absorption of cadmium taken without milk. For humans the extent of cadmium absorption can double under the influence of calcium, protein, or zinc deficiency.

The distribution of cadmium has been studied in laboratory animals. In rats, absorbed cadmium is distributed to the liver, spleen, adrenals, and duodenum within 48 hours following administration. Accumulation is slower in the kidneys, where peak levels are reached by the sixth day. Unless the dose of cadmium is unusually high, levels in the kidneys are roughly 10 times the levels in the liver. With very high doses, the levels of cadmium in these two organs become similar. With continued low doses, concentrations of cadmium in the other organs remain small, with about 50% of the total cadmium in the body occurring in the kidneys and liver.

Cadmium is highly stable in the body with estimates of biological half-life ranging from 20 to 40 years. It is thought that a metal-binding protein known as metallothionein, present in the kidneys, is responsible for cadmium's long biological half-life. Metallothionein synthesis is increased in response to cadmium and zinc exposure. As the levels of cadmium increase in the body, so do the levels of metallothionein. It is not clear, however, why the kidneys are the major site of cadmium concentration since metallothionein is present in several other organs as well. Also, metallothionein in itself does not reduce the toxicity of cadmium. The cadmium-metallothionein complex is actually more toxic than cadmium alone.

Biological effects of cadmium have been studied extensively in rats. In these studies the most sensitive organ to cadmium exposure is the kidney, with symptoms of toxicity arising in rats at a dose of 0.25 mg/kg. Symptoms include increased excretion of glucose, protein, amino acids, calcium, and uric acid. The liver also is affected as indicated by increased gluconeogenesis, leading to hyperglycemia, and pancreatic effects are indicated by a decreased circulation insulin level. At higher doses (2 mg/kg), the testes and prostate glands atrophy and the adrenals hypertrophy with increased levels of circulating epinephrine and norepinephrine. Also, there is evidence of increased levels of dopamine in the brain. There are some indications that cadmium is a hypertensive agent, a teratogen, and a carcinogen, although these results have not been confirmed in humans.

Several components in the diet reduce or eliminate some of the toxic effects of cadmium. Zinc, selenium, copper, iron, and ascorbic acid are protective agents. The mechanism by which these protective effects occur is by no means clear.

Outbreaks

Cadmium has been implicated as the cause of at least one epidemic of human disease resulting from contaminated food. This epidemic occurred in the Jintsu River Valley of Japan following World War II. Local health practitioners observed many cases of kidney damage and skeletal disorders accompanied by great pain. The disease was called in Japanese, *Itai-Itai Byo*, which is loosely translated as "ouch-ouch" disease. It soon was realized that certain other host factors in addition to consumption of contaminated food were required to initiate the disease. These other host factors include pregnancy, lactation, aging, and calcium deficiency. The dietary source of human cadmium poisoning was thought to be contaminated rice that had cadmium levels of 120 to 350 ppm of ash, compared to 20 ppm of rice ash in the control area. The source of this cadmium was thought to be a lead-zinc-cadmium mine upstream from the rice fields. Although the weight of scientific opinion is consistent with the idea that cadmium is involved in the etiology of Itai-Itai disease, reservations about this hypothesis have been raised by several investigators. One piece of evidence that certainly suggests that cadmium is not the only causative agent of the disease is that high levels of cadmium are found in rice from many areas of Japan where no cases of the disease have been reported.

SUGGESTIONS FOR FURTHER READINGS

Fowler, B. A. (Ed.) (1983). Biological and environmental effects of arsenic, topics in environmental health, Vol. 6. Elsevier Science Publisher, New York.

Grandjean, P., Landrigan, P. J. (2006). Developmental neurotoxicity of industrial chemicals. A silent pandemic. *Lancet* 368:2167-2178.

Masuda, Y., Yoshimura, H. (1982). Chemical analysis and toxicity of polychlorinated biphenyls and dibenzofurans in relation to Yusho. *J. Toxicol. Sci.* 7: 161-175.

Matschullat, J. (2000). Arsenic in the geosphere—A review. *Sci. Total Environ.* 249: 297-312.

National Academy of Science. (1999). Arsenic in drinking water. National Academy Press, Washington, DC.

Tsuchiya, K. (1977). Various effects of arsenic in Japan depending on type of exposure. *Environ. Health Persp.* 19:35-42.

CHAPTER 9

Pesticide Residues in Foods

CHAPTER CONTENTS

WHAT IS A PESTICIDE?

A pesticide is a substance or mixture of substances used for preventing, controlling, or lessening the damage caused by a pest. A pesticide may be a chemical substance (synthetic or naturally occurring), biological agent (such as a virus or bacteria), antimicrobial, disinfectant, or any other device used against any pest, including insects, plant pathogens, weeds, mollusks, birds, mammals, fish, nematodes (roundworms), and microbes. Therefore, pesticides are divided into several groups depending on the purpose for which it is used:

- Insecticides for the control of insects; these can be ovicides (substances that kill eggs), larvicides (substances that kill larvae), or adulticides (substances that kill adult insects)
- Herbicides for the control of weeds

Food Science and Technology

- Fungicides for control of fungi and oomycetes
- Rodenticides for control of rodents
- Bactericides for the control of bacteria
- Miticides or acaricides for the control of mites
- Molluscicides for the control of slugs and snails
- Virucides for the control of viruses
- Nematicides for the control of nematodes

Insecticides are used to kill noxious insects such as mosquitoes, bees, wasps, and ants, which cause diseases in animals and humans. Herbicides are used to prevent growth of weeds in many commodities. They also are applied in parks and wilderness areas to kill invasive weeds and to protect the environment. Fungicides are used to protect agricultural crops from various fungi.

HISTORY

There are records indicating that humans have utilized pesticides to protect their crops before 2,500 BCE. In Sumeria, about 4,500 years ago, elemental sulfur dusting was used to protect crops from pests for the first time. By the fifteenth century, some heavy metals such as arsenic, mercury, and lead were being applied to crops to kill pests. In the seventh century, nicotine sulfate was extracted from tobacco leaves for use as an insecticide. In the nineteenth century, two naturally occurring chemicals (pyrethrum from chrysanthemums and rotenone from tropical vegetables) were used as a pesticide. Pesticides have played an important role in securing the food supply in the world for over 60 years. Pesticides, either synthetic or naturally occurring, are a diverse group of chemicals used to control the undesirable effects of target organisms. Pesticides may be classified according to their target organisms as mentioned earlier. In regards to food toxicology, significant categories of pesticides are insecticides, fungicides, and herbicides. Acaricides (which affect mites), molluscicides, and rodenticides are less important.

Commercial production is a relatively recent innovation in the development of synthetic pesticides. As an example of this, the value of the insecticidal effects of DDT, which was synthesized in 1874, was not reported until 1939, after commercial production had begun. The Swiss chemist,

Paul Hermann Müller, received a Nobel Prize in 1948 for his discovery of the insecticidal effects of DDT. Its value to public health in controlling insect vectors of disease was demonstrated following World War II. In response to a shortage of nicotine during World War II in Germany, the first organophosphate insecticide, tetraethyl pyrophosphate (TEPP), was developed. Parathion, synthesized by the German scientist Gerhard Schrader in 1944, was among the first insecticides of commercial importance, and it continues to be widely used today.

The introduction of these and other synthetic pesticides has made an enormous contribution not only to agriculture but also to human health. For example, pre- and post-World War II, thousands of lives were saved by the use of DDT to control the mosquito vector of malaria transmission in Europe and Asia. In recent decades, a dramatic increase in agricultural output, nicknamed the "Green Revolution," is a result of the use of synthetic pesticides to control weeds and insects that would otherwise limit crop production. Insecticides and fungicides also are used to reduce post harvest losses of valuable crops and to maintain the nutritional value and freshness of food until it is consumed. On the other hand, much attention has turned to the possible hazards of pesticide residues in foods, which have become one of today's most important food safety issues.

Until quite recently, far too little attention was paid to the possible toxic effects of pesticides on humans and other nontarget organisms. This lack of concern has resulted in disastrous ecological consequences. For example, contamination of both groundwater and soil is now a serious problem in many agricultural areas of the United States. In some cases, contamination persists for decades after use of the offending product is discontinued. In the past, poor management practices were the cause of an extreme overuse of such pesticides as DDT, which in turn led to the development of resistance in target organisms.

Since the 1962 publication of Rachel Carson's book, *Silent Spring*, the role of pesticides in modern society has become an emotional and contentious issue. More recently, controversy has broadened to include questions about the safety of pesticide residues in foods, which can now be analyzed in great detail as a result of dramatic increases in chemists' analytical capabilities. There have been increasing questions about the use of pesticides because of their possible adverse effects on humans. Some people insist that all food commodities should be grown and harvested organically (pesticide free). However, damage caused by pests on food crops is significant. An example of damage on typical crops without pesticides is shown in Table 9.1. As shown, world food production would decrease significantly

Table 9.1 Damage of Crops Caused by Pests Without Pesticides

Commodity	% Lost
Avocado	43
Banana	33
Cabbage	37
Carrot	44
Cauliflower	49
Grain	25
Lettuce	62
Mango	30
Orange	26
Pineapple	70
Sweet potato	95
Tomato	30

without pesticides. Among 6 billion people in the world, nearly 1 billion are starving today. Therefore, it is almost impossible to abandon the use of pesticide unless some dramatic method to increase food production without pesticides is discovered.

PESTICIDES IN THE FOOD CHAIN

Pesticides have been used in many different ways during the production of food pre- and post-harvest. Pesticides used on animals may also be found in foods in trace levels. Unfortunately, some pesticides, such as DDT, persist and remain in the environment and are consequently found in various foods grown on contaminated soil, or in the fish that live in contaminated waters.

In recent years, contamination of surface and groundwater by pesticides has been recognized as a serious and growing problem in agricultural regions. Whereas many pesticides degrade rapidly in the environment, bind tightly to soil, or are simply too insoluble or nonvolatile to move throughout the environment, others are both persistent and mobile. Certain pesticide application methods, particularly aerial spraying, are notoriously inefficient in delivering the pesticide to the target. Large amounts enter the environment directly, where runoff from agricultural fields may contaminate both surface and groundwater. Livestock that drink the contaminated water may have detectable pesticide residues in their meat or milk.

In some areas, nonagricultural applications of pesticides may also be a source of environmental and water contamination. Home use of pesticides, which constitutes a significant percentage of the total use of pesticides, is subject to the same laws and regulations as agricultural uses. However, although home use involves the least hazardous pesticides, the possibility of misuse by homeowners is significant. Not only can contamination of home-grown produce occur, but accidental poisoning due to improper storage and disposal occurs frequently. Another example of nonagricultural application of pesticides is forest management, which often involves large quantities of herbicides and insecticides. Maintenance of golf courses and other large expanses of turf is pesticide-intensive and often involves fungicides that pose a significant risk to mammals. Also, in many areas, commercial and recreational fishing is restricted because of environmental contamination by persistent pesticides.

To the extent that contaminated water is used in the processing or preparation of food, pesticides may enter the food supply through this medium. Human exposure to contaminated water through drinking or washing also has an indirect effect on issues of pesticide residues in foods since it may comprise a significant portion of pesticide contamination within the exposed population. Thousands of samples of food are examined by the FDA each year to determine compliance with established pesticide tolerances on raw agricultural products. Residues of pesticide chemicals are found in about half of the samples, and generally about 3% of the samples contain residues in excess of, or not authorized by, legal tolerances. Table 9.2 shows the frequency of occurrence of typical pesticide residues found in total diet study foods by FDA in 2003.

Table 9.2 Frequency of Occurrence of Pesticide Residues Found in Total Diet Study Foods by FDA in 2003*

Pesticide	Total No. of Findings	Occurrence (%)	Range (ppm)
DDT	123	12	0.0001–0.171
Malathion	71	7	0.0006–0.121
Endosulfan	70	7	0.0001–0.116
Dieldrin	64	6	0.0001–0.141
Chlorpyrifos	50	5	0.0002–0.110
Carbaryl	20	2	0.0003–0.190
Methamidophos	19	2	0.0002–0.123
Diazinon	19	2	0.0002–0.043
Indane	16	2	0.0001–0.007

**Based on four market baskets analyzed consisting of 1039 total items; only those found in more than 2% of the samples are shown.*

REGULATIONS

In the United States, before a new pesticide can be sold for use on a farm, it had to be registered for agricultural use by the Environmental Protection Agency (EPA). Regulation of pesticides is under the jurisdiction of the EPA, the Food and Drug Administration (FDA), and the United States Department of Agriculture (USDA). The EPA handles pesticide registration and the establishment of tolerances, and the FDA and the USDA monitor pesticide residue levels in the food supply. Additionally, a number of states, most notably California and Texas, have implemented their own more extensive regulations.

Under Section 408 of the Food, Drug, and Cosmetics Act (FDCA), residues of pesticides in processed foods are regulated as food additives. Congress has specifically exempted residues in raw agricultural products from the scope of the FDCA. By law, no pesticide that will leave a detectable residue—whether of parent compound, breakdown product, or metabolite—in a processed food can be registered until a tolerance for such residue levels has been set by the EPA. For new products or uses, tolerances are set based on results of extensive toxicological testing. In general, a tolerance is set no higher than the "no observable effect level" (NOEL) found in such testing, divided by a safety factor of 100. However, if good agricultural practice can result in an even lower level of residues, the legal tolerance will be reduced accordingly.

Compounds that are considered possible or probable human carcinogens as a result of animal testing are regulated more stringently. Under Section 409 of the FDCA, specifically the Delaney Clause, no tolerance can be set for any carcinogen in processed foods. Therefore Section 409/Delaney Clause does not permit food additives that have been established as animal carcinogens, and thus the compound then becomes an illegal food additive. The EPA's current practice is to deny any tolerance even for raw agricultural products since the Delaney Clause prohibits setting a Section 409 Tolerance. Unfortunately, at the time, these regulations did not apply to the pesticides that were in use before these laws were adopted. The EPA has required reregistration of such products, along with any additional toxicological testing required to meet current standards. Many compounds or uses will be eliminated when this program of reregistration is completed. However, in the meantime, these products remain in use, giving rise to much controversy.

The Department of Agriculture controls approval of pesticides for production under the Federal Insecticide, Fungicide, and Rodenticide Act

(FIFRA) of 1964. This law provides that any "economic poison" or chemical pesticide must be registered before being marketed in interstate commerce.

The majority of components in most formulated pesticides are listed on the label only as inert ingredients. However, these components are listed inert only so far as they exert no pesticidal action. These components include solvents, surfactants, carriers, antioxidants, and other compounds. Current pesticide regulation does not deal directly with inert ingredients nor their residues or possible toxic effects. Future amendments to FIFRA are likely to deal with this aspect of pesticide manufacture and use.

In 1996, Congress unanimously passed a landmark pesticide food safety legislation supported by the administration and a broad coalition of environmental, public health, agricultural, and industry groups. President Clinton promptly signed the bill on August 3, 1996, and the Food Quality Protection Act (FQPA) of 1996 became law. One of the most significant issues of this act is the suspension of the Delaney Clause. The major contents of this law are:

1. Mandating a single, health-based standard for all pesticides in all foods.
2. Providing special protections for infants and children.
3. Expediting approval of safer pesticides.
4. Creating incentives for the development and maintenance of effective crop protection tools for American farmers.
5. Requiring periodic reevaluation of pesticide registrations and tolerances to ensure that the scientific data supporting pesticide registrations will remain up-to-date in the future.

By successfully implementing the FQPA, the EPA dramatically amended the FIFRA and FFDCA. FQPA also increased protection standards specifically for infants and children, in which the use of an extra ten-fold margin of safety was applied. This FQPA safety factor provision takes into account the potential for pre- and postnatal toxicity and the completeness of the toxicology and exposure, unless reliable data support a different factor. Table 9.3 shows the frequency of occurrence of pesticide residues found in selected baby foods by FDA in 2003, after the FQPA was implemented (based on four market baskets consisting of a total of 93 items).

Table 9.3 Frequency of Occurrence of Pesticide Residues Found in Selected Baby Foods by FDA in 2003*

Pesticide	Total No. of Findings	Occurrence (%)	Range (ppm)
Thiabendazole	9	10	0.001–0.102
Endosulfan	8	9	0.0001–0.0054
DDT	6	6	0.0001–0.0007
Carbaryl	6	6	0.001~ 0.030
Malathion	3	3	0.001–0.033
Dieldrin	3	3	0.0001
Chlorpyrifos	2	2	0.0002–0.0011
Benomyl	2	2	0.032–0.052

**Based on four market baskets analyzed, consisting of 93 total items.*

INSECTICIDES

DDT

Although DDT [1,1-(2,2,2-trichloroethylidene)bis(4-chlorobenzene)] (Figure 9.1) has been banned in the United States since 1972, it remains one of the best-known synthetic pesticides.

The incidence of malaria before and after the use of DDT is shown in Table 9.4. Because DDT is a very nonpolar molecule, it has high lipid solubility. Since DDT is also extremely stable, it accumulates in animal tissues and in the food chain. As illustrated earlier in Table 9.2, DDT is still one of the most abundant pesticide residues in food.

During the 40 years following DDT's commercial introduction in the 1940s, more than 4 billion pounds were used to control insect-borne diseases. Until 1972, DDT was widely used in the United States, mostly on cotton, peanuts, and soybeans. As a result of its use, DDT residues are now ubiquitous in the environment, and at the present time, some level can be detected in almost all biological and environmental samples.

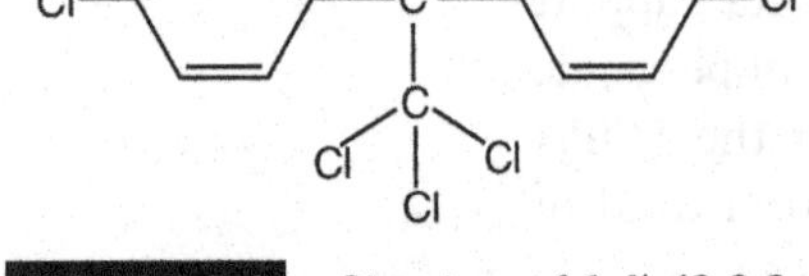

In addition, due to its high lipid solubility, DDT concentrates in milk. When DDT was widely used, levels in human milk and adipose tissue were found to be higher than concentrations permitted in meat and dairy products. However, since its use has been prohibited, storage levels of DDT in human tissue have declined significantly. DDT is, however, still in use in other countries, largely to control insect-borne diseases that pose a substantial threat to public health.

H
Cl C Cl
C
Cl Cl
Cl

FIGURE 9.1 *Structure of 1,1′-(2,2,2-trichloroethylidene)bis(4-chlorobenzene) (DDT).*

Table 9.4 Incidence of Malaria Before and After the Use of DDT

Country	Year	No. of Cases
Cuba	1962	3519
	1969	3
Jamaica	1954	4417
	1969	0
Venezuela	1947	8,171,115
	1958	800
India	1935	>100 million
	1969	285,962
Yugoslavia	1937	169,545
	1969	15
Taiwan	1945	>1 million
	1969	0

Toxicity

The possible clinical effects of many repeated doses of DDT were first explored in 1945 when a scientist conducted a test, lasting a total of 11.5 months, where he daily inhaled 100 mg of pure DDT and drank water dusted at the rate of 3240 mg/m^2 (see Table 9.5 for acute toxicities). Much of the inhaled dust must have been deposited in the upper respiratory tract and swallowed. Later, for one month he consumed food that had been sprayed with DDT at a rate of 2160 mg/m^2. No ill effect of any kind was observed in either case.

Later studies of DDT in volunteers were designed to explore the details of storage and excretion of the compounds in people and to search for possible effects of doses considered to be safe. In initial studies, each man was given 0, 3.5, and 35 mg DDT/day. These administered dosages, plus DDT measured in the men's food, resulted in dosage levels of 2.1, 3.4, 38, 63, and 610 μg DDT/kg body weight/day.

Table 9.5 Acute Oral and Acute Dermal LD_{50} of DDT in Various Laboratory Animals

Species	Oral (mg/kg)	Dermal (mg/kg)
Rat	500–2,500	1,000
Mouse	300–1,600	375
Guinea pig	250–560	1,000
Rabbit	300–1,770	300–2,820
Dog	>300	—
Cat	100–410	—

In Vivo Metabolism

In a study using Swiss mice and Syrian golden hamsters, DDT was metabolized to base-labile glucuronide of 4-chloro-α-(4-chlorophenyl)benzene acetic acid (DDA) and excreted in urine. The more stable glycine and alanine conjugates of DDA also were found. DDT metabolic intermediate 1,1′-(2-chloroethenylidene)bis(4-chlorobenzene) (DDMU) is partially metabolized in vivo by mice to 4-chloro-α-(4-chlorophenyl)-α-hydroxylbenzene acetic acid (OHDDA) and other metabolites that are excreted in urine. As illustrated in Figure 9.2, metabolic detoxification sequence of DDT conversion in rats was shown to be:

DDT — DDD — DDMU — DDMS — DDNU — DDOH — DDA

Oral doses of DDT (5, 10, 20 mg/day) administered to human volunteers, in part, were excreted as DDA. Ingested DDA is promptly and efficiently excreted in the urine, undergoing virtually no tissue storage during ingestion. These results suggest that measurement of urinary DDA excretion offers a useful method of monitoring DDT exposure.

Methoxychlor is a DDT analogue that has replaced DDT in many applications. The structure of methoxychlor is shown in Figure 9.3. Enzymes in both mammals and soil organisms are able to catalyze the demethylation of the methoxy oxygen atoms, producing a more polar degradation product that may be conjugated and excreted. Thus, methoxychlor does not accumulate in animal tissues and does not persist in the environment. Mammalian LD_{50} values for methoxychlor range from 5,000 to 6,000 mg/kg, 40 to 60 times higher than for DDT. However, methoxychlor also shows less toxicity to its target organisms than does DDT.

Chlorinated Cyclodiene Insecticides

Cyclodiene insecticides are an important group of chlorohydrocarbons, most of which are synthesized by the principle of the Diels–Alder reaction. This reaction is named in honor of Otto Paul Hermann Diels and Kurt Alder, who first documented this novel reaction in 1928 and thusly were awarded the Nobel Prize in Chemistry in 1950. The structures of typical cyclodiene insecticides are shown in Figure 9.4. Table 9.6 lists LD_{50} of cyclodiene insecticides, with the addition of endrin, in rats.

Mode of Toxic Action

Like DDT, cyclodiene compounds are neurotoxic. However, as a class they are much more toxic to mammals than DDT and tend to produce more severe symptoms, for example, convulsions. The mechanism of neurotoxic

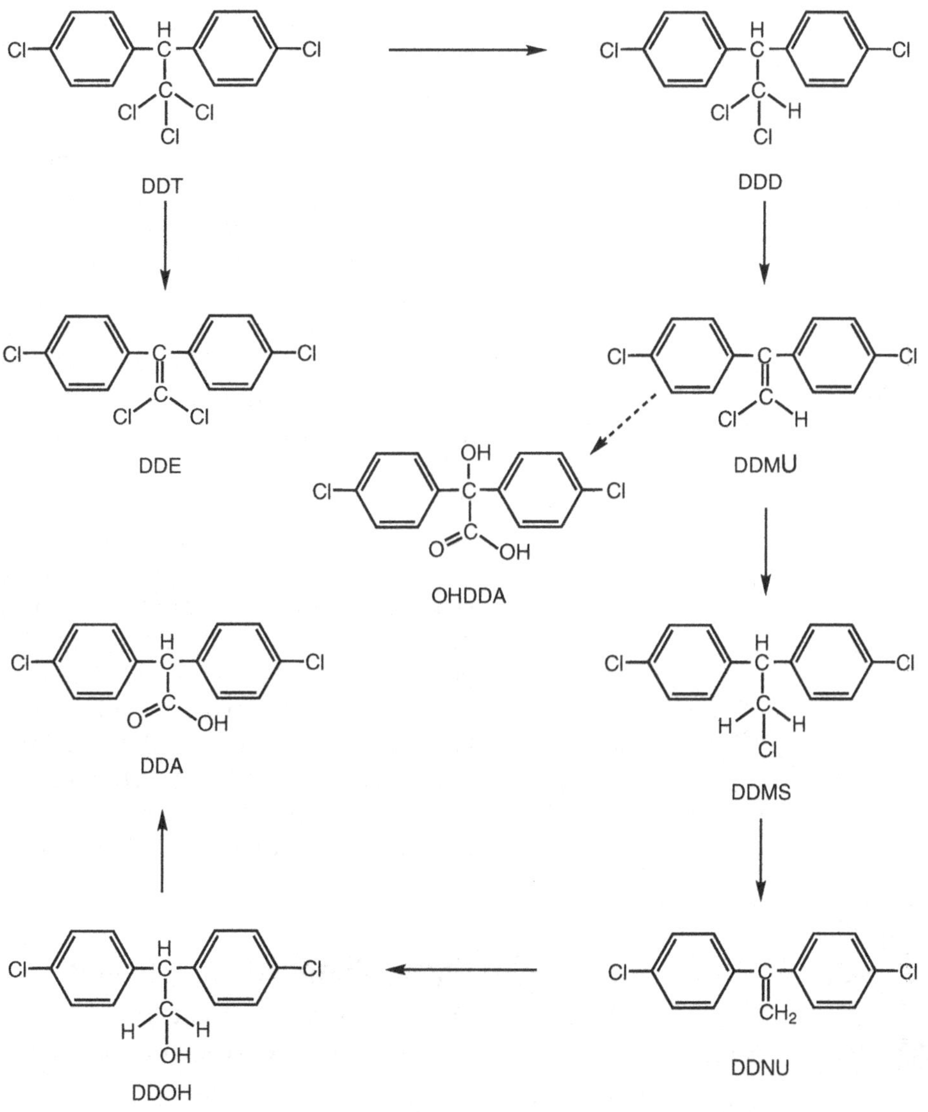

DDT: 1,1'-(2,2,2-trichloroethylidene)bis(4-chlorobenzene)

DDD: 1,1'-(2,2-dichloroethylidene)bis(4-chlorobenzene)

DDE: 1,1'-(2,2-dichloroethenylidene)bis(4-chlorobenzene)

DDMU: 1,1'-(2-chloroethenylidene)bis(4-chlorobenzene)

DDA: 4-chloro-α-(4-chlorophenyl)-benzene acetic acid

DDMS: 1,1'-(2-chloroethylidene)bis(4-chlorobenzene)

DDOH: 4-choro-β-(4-chlorophenyl)-benzene ethanol

DDNU: 1,1'-ethenylidenebis(4-chlorobenzene)

OHDDA: 4-chloro-α-(4-chlorophenyl)-α-hydroxylbenzene acetic acid

FIGURE 9.2 *Metabolic conversion pathways of DDT.*

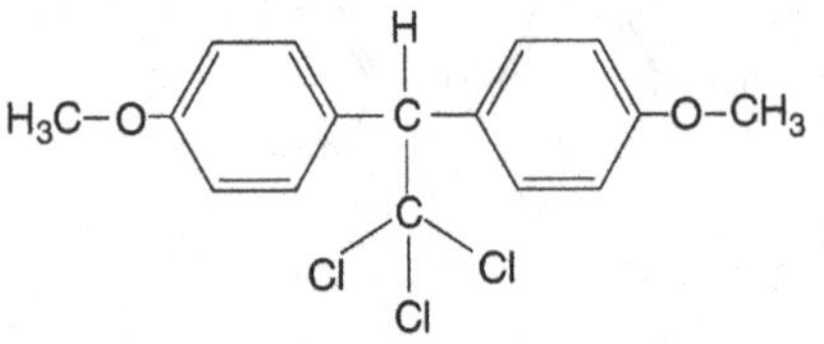

FIGURE 9.3 *Structure of methoxychlor.*

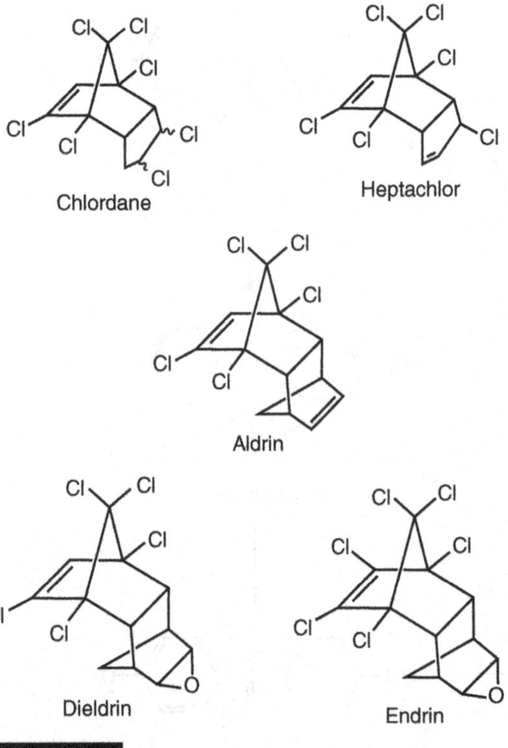

FIGURE 9.4 *Structures of typical cyclodiene insecticides.*

action is not understood but is thought to involve disruption of nerve impulse transmission by interfering with control of Ca^{2+} and Cl^{-} concentrations. A number of human poisonings and fatalities have resulted from accidental exposure to endrin and dieldrin. Chronic feeding studies in a variety of mammalian species have shown that increased liver weight and histological changes in the liver similar to those caused by DDT are produced by doses of endrin ranging from 5 to 150 ppm, depending on the species.

Some reproductive toxicity has been reported, but only at doses high enough to cause histological changes in the maternal liver. Many studies of the carcinogenicity of these compounds have been made. Most have been inconclusive; however, there is sufficient evidence overall to consider many of these compounds as probable animal carcinogens. Like DDT, cyclodiene compounds are highly lipid soluble and quite stable. Hence, they accumulate in animal tissues and bioconcentrate in the blood chain. As a result, the production and use of cyclodiene compounds have been sharply reduced, and many have been banned entirely, including chlordane and dieldrin.

Organophosphate Insecticides

Organophosphate insecticides (OPs) are among the oldest of the synthetic pesticides and are

Table 9.6 LD_{50} of Chlorinated Cyclodiene Insecticides in Rats

Insecticide	LD_{50} (mg/kg)
Chlordane	150–700
Heptachlor	100–163
Aldrin	25–98
Dieldrin	24–98
Endrin	5–43

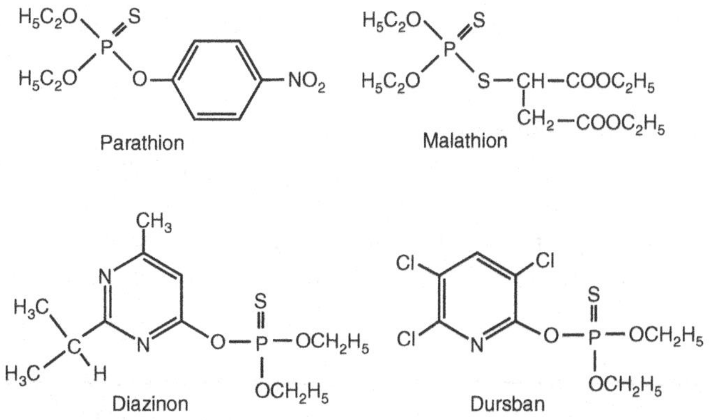

FIGURE 9.5 Structures of typical organophosphate insecticides.

currently the most widely used class of insecticides. Although French chemist, Jean Louis Lassaigne, first synthesized OPs from the reaction of phosphoric acid and alcohol in 1820, it was not until the 1930s that Gerhard Schrader, a German chemist, discovered their insecticidal properties. At this time, the agricultural industry was rapidly expanding and eagerly used synthetic insecticides in addition to such natural insecticides as nicotine, rotenone, and pyrethrum. There are many OPs whose structures are chemically modified. Figure 9.5 shows the structures of some typical OPs. Table 9.7 lists the LD_{50}s of typical OPs for mice.

In Vivo Metabolism

The OPs do not accumulate in the body because they are rapidly metabolized and excreted. They also undergo a number of metabolic reactions in mammals. Malathion, for instance, is quite susceptible to hydrolysis by esterases and so has very low mammalian toxicity. Parathion, on the other

Table 9.7 LD_{50} of Typical Organophosphate Insecticides in Mice

Insecticide	LD_{50} (mg/kg)
Parathion	10–12
Methyl parathion	30
Methyl paraoxon	1.4
Paraoxon	0.6–0.8

hand, contains an aromatic phosphate ester group that is more resistant to enzymatic hydrolysis. Activation to parathion's toxic analogue, paraoxon, can thus proceed to a greater extent, resulting in a much higher mammalian toxicity. Thus, malathion is registered for use by home gardeners, whereas the use of parathion is restricted to trained applicators. It must be noted that oxidation of malathion can also occur upon exposure to air. In addition, improper or extended storage can give rise to contamination with the quite toxic malaoxon. Metabolic pathways of OPs are summarized as:

1. Oxidation	Thiono → oxo-form Oxidative dealkylation Thioether → sulfoxide → sulfone Oxidation of aliphatic substituents Hydroxylation of an aromatic ring
2. Reduction	Nitro → amino group
3. Isomerization	
4. Hydrolysis	Enzymatic Nonenzymatic
5. Dealkylation at the carboxy group	Ester → acid Saponification
6. Conjugation	Hydroxy compounds with glucuronic acid Hydroxy compounds with sulfate

Mode of Toxic Action

OPs inhibit the activity of acetylcholinesterase (AChE), which is a neurotransmitter in mammals. Normally, acetylcholine (ACh) is rapidly broken down following its release by a group of enzymes known as cholinesterase. OPs, or their metabolites, can compete with acetylcholine for its receptor site on these enzymes, thus blocking the breakdown of ACh. The extent of inhibition of the enzyme depends strongly on steric factors; that is, on how well the inhibitor "fits" on the enzyme, as well as on the nature of the organic groups present. Aromatic groups with electron-withdrawing substituents, as are present in parathion and related compounds, enhance binding to AChE and thus increase toxicity. The resulting accumulation of ACh at smooth muscular junctions causes continued stimulation of the parasympathetic nervous system, producing such symptoms as tightness of the chest, increased salvation, lacrimation, increased sweating, peristalsis (which may lead to nausea, vomiting, cramps, and diarrhea), bradycardia, and a characteristic constriction of the pupils of the eye.

Although OPs are a significant occupational hazard to agricultural workers, residues on food products normally do not result in exposures sufficient to lead to toxic symptoms in humans.

Carbamate Insecticides

The chemical structures of several important carbamates are shown in Figure 9.6. These compounds are synthetic analogues of the toxic alkaloid physostigmine found in calabar beans. This compound is the toxic principle upon which the "trial-by-poison" of certain West African tribes was based. Related compounds have clinical use in the treatment of glaucoma and other diseases.

The carbamate insecticides are active against a relatively narrower range of target organisms than the organophosphates, but they are highly toxic to such beneficial insects as honeybees. In general, these compounds are quite toxic to mammals via oral exposure, although in most cases their dermal toxicity is low. Table 9.8 gives the LD_{50} in rats for representative carbamate insecticides.

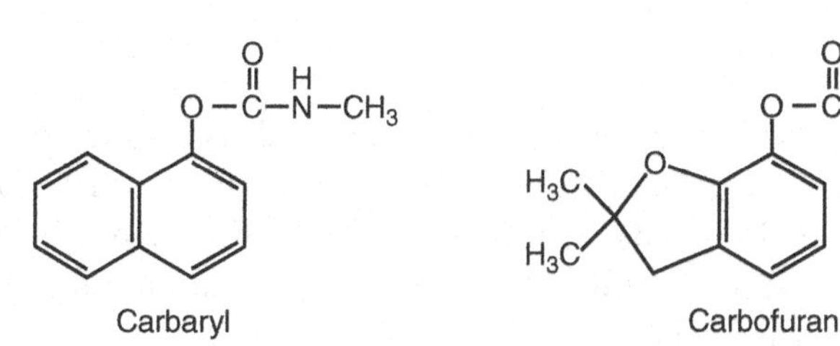

FIGURE 9.6

Structure of typical carbamate insecticides.

Table 9.8 LD_{50} of Typical Carbamate Insecticides in Rats

Insecticide	LD_{50} (mg/kg)
Carbaryl	850
Carbofuran	8–14
Aldicarb	0.93
Methomyl	17–24
Aldoxycarb	27

Carbamate insecticides have been involved in a large number of human poisoning incidents, both as a result of occupational exposure and as contamination of food products. For example, the carbamate aldicarb was the cause of 281 people in California becoming ill in 1985 as a result of contaminated watermelons. Because aldicarb is quite water soluble, it can accumulate to dangerous levels in foods possessing high water content. Accordingly, aldicarb is not registered for such applications. However, because it is widely used on other crops, the possibility of contamination exists, as shown by the watermelon incident.

Mode of Toxic Action

Like the organophosphates, the carbamate insecticides are AChE inhibitors in mammals. Carbamates are direct-acting inhibitors of AChE; however, they are not able to "age" the neurotoxic esterase. Therefore, they are not associated with the delayed neuropathy syndrome. The symptoms of poisoning are typically cholinergic with lacrimation, salivation, miosis, convulsions, and death.

HERBICIDES

The value of harvest losses by pests, diseases, and weeds is estimated worldwide to be about 35% of the potential total harvest. About 9–10% of the reduced yield is caused by weeds. Numerous chemicals have been used as herbicides to prevent weed growth. Consequently, trace amounts of herbicides are present in final food products.

Chlorophenoxy Acid Esters

Chlorophenoxy acid esters and their salts are widely used as herbicides. They mimic the plant hormone indole acetic acid and are able to disrupt the growth of broad-leaf weeds and woody plants. Familiar compounds in this class include 2,4-D and 2,4,5-T (Figure 9.7). These compounds gained considerable notoriety because they are active ingredients in the defoliant

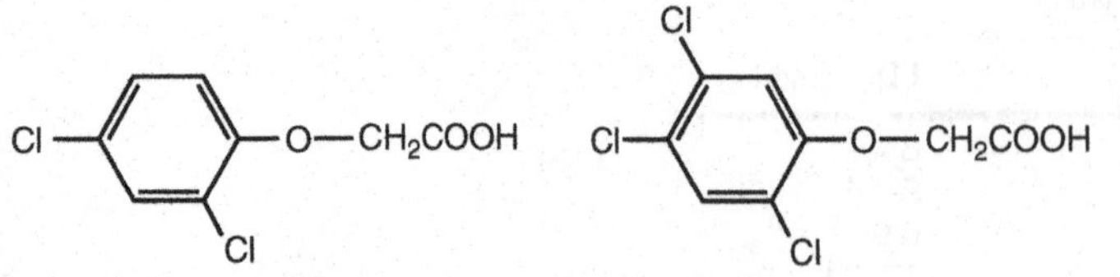

2,4-D (2,4-dichlorophenoxy acetic acid) 2,4,5-T (2,4,5-trichlorophenoxy acetic acid)

FIGURE 9.7

Structure of typical chlorophenoxy herbicides.

Agent Orange used during the Vietnam War. However, this class of compound has relatively low acute toxicity toward mammals. The acute oral toxicities (LD_{50}) of 2,4-D and 2,4,5-T in rats are 375 and 500 mg/kg, respectively.

Mode of Toxic Action and Toxicity

The mechanisms of mammalian toxicity of chlorophenoxy herbicides are not clear. Sublethal doses cause nonspecific muscle weakness. Higher doses lead progressively to stiffness of the limbs, ataxia, paralysis, and coma. The chlorophenoxy esters are readily hydrolyzed to the acid form. The acids are in some cases sufficiently soluble in water to be excreted directly in urine. In other cases, easily excreted conjugates are formed. Because of the rapid elimination of the acids and conjugates, accumulation in mammalian systems does not occur and chronic effects resulting from low-level exposures are not generally seen.

Formulated chlorophenoxy herbicides have been found to be teratogenic in many animal species. This effect is now thought to be due to a contaminant, TCDD, often referred to as "dioxin" in the popular press. Details of TCDD are described in Chapter 7.

NATURALLY OCCURRING PESTICIDES

Naturally occurring pesticides have been used in agriculture for a long time. It was recognized early in the nineteenth century that crushed flowers (pyrethrum powders) from plants in the chrysanthemum family could control insects. By 1851, pyrethrum powders were used worldwide. It is now known that at least six active esters are present in pyrethrum, and various synthetic pyrethroids modeled on these natural esters are currently in widespread use. Natural as well as synthetic pyrethroids have very low toxicity to mammals. Nicotine, another natural insecticide, is produced by plants in the tobacco family and was used as an insecticide at least as early as 1763. It is a potent insecticide, with an LD_{50} between 10 and 60 mg/kg for various target species; it also has very high mammalian toxicity by both oral and dermal exposure. Many other plants (walnut trees, for example) secrete chemicals that prevent the growth of competitive plants within their root zone, and thus provide their own pesticide. Finally, the use of various herbs to control particular pests is a recognized part of gardening history, indicating that farmers have accumulated much knowledge regarding the use of chemicals in agriculture.

SUGGESTIONS FOR FURTHER READING

Aardema, H., Meertens, J.H., Ligtenberg, J.J., Peters-Polman, O.M., Tulleken, J.E., Zijlstra, J.G. (2008). Organophosphorus pesticide poisoning: Cases and developments. *Neth. J. Med.* 66:149-153.

Accomplishments under the Food Quality Protection Act (FQPA), Pesticides, US EPA (2007). http://www.epa.gov/pesticides/regulating/laws/fqpa/fqpa_accomplishments.htm.

Boobis, A.R., Ossendorp, B.C., Banasiak, U., Hamey, P.Y., Sebestyen, I., Moretto, A. (2008). Cumulative risk assessment of pesticide residues in food. *Toxicol Lett.* 180:137-150.

Eskenazi, B., Rosas, L.G., Marks, A.R., Bradman, A., Harley, K., Holland, N., Johnson, C., Fenster, L., Barr, D.B. (2008). Pesticide toxicity and the developing brain. *Basic Clin. Pharmacol Toxicol.* 102:228-236.

Gold, L.S., Slone, T.H., Ames, B.N., Manley, N.B. (2001). Pesticide residues in food and cancer risk: A critical analysis. In "Handbook of Pesticide Toxicology," 2nd Edition, R. Krieger (Ed.), pp. 799-843. Academic Press, San Diego.

Ohkawa, H., Miyagawa, H., Lee, P.W. (Eds.) Pesticide Chemistry: Crop protection, public health, environmental safety. Wiley-VCH, Verlag GmbH & Co. KGaA, Weinheim.

Zweig, G., Sherma, J. (Eds.) (1964). Analytical Methods for Pesticides, Plant Growth Regulators, and Food Additives. Academic Press, New York.

Food Additives

CHAPTER CONTENTS

A food additive is a substance or mixture of substances, other than basic food components, added to food in a scientifically controlled amount. They are added to food to preserve flavor or improve its taste and appearance. Some additives have been used for centuries; for example, preserving food by pickling—with vinegar or salting, as with bacon, preserving sweets; or

Food Science and Technology

using sulfur dioxide, as in some wines. With the advent of processed foods in the second half of the twentieth century, many more additives have been introduced, of both natural and artificial origin. These additions can be made during production, processing, storage, and packaging. It is natural for people to desire better foods, not only from the perspective of health but also for taste, color, or texture. Hence, a tremendous number of substances have been used since the beginning of the twentieth century to enhance food acceptance.

There are two categories of food additives. The first category, intentional additives, are purposely added to perform specific functions. They include preservatives, antibacterial agents, bleaching agents, antioxidants, sweeteners, coloring agents, flavoring agents, and nutrient supplements. The second type of additives are incidental, and may be present in finished food in trace quantities as a result of some phase of production, processing, storage, or packaging. An incidental additive could be a substance present in food due to migration or transfer from the package or processing equipment. Since most food additives are intentionally added substances, only intentional additives are discussed here.

Substances intentionally added to foods vary from preservatives to flavoring materials. Table 10.1 indicates the approximate number of substances used for each purpose to date. Approximately 300 substances are recognized as food additives, and 60 to 70 food additives are ingested daily by every person in the United States. Table 10.2 lists the most common food additives used for various purposes.

Table 10.1 Approximate Number of Different Types of Food Additives

Purpose of Additive	Number of Different Additives
Preservatives	30
Antioxidants	28
Sequestrants	44
Surfactants	85
Stabilizers	31
Bleaching, maturing agents	24
Buffers, acids, alkalies	60
Coloring agents	35
Special sweeteners	9
Nutrient supplements	>100
Flavoring agents	>700
Natural flavoring materials	>350

Table 10.2 Most Common Chemicals Developed as Food Additives	
Purposes	**Chemicals**
Preservatives	Benzoic acid, sorbic acid, *p*-oxybenzoic acid, hydrogen peroxide, AF-2[a]
Antioxidants	Ascorbic acid, DL-α-tocopherol, BHA, propyl gallate
Sweeteners	Saccharine, dulcin, sodium cyclamate[a]
Coloring agents	Food Red No. 2,[a] Food Yellow No. 4, Scarlet Red, Indigo carmine
Flavoring agents	Safrole, methyl anthranilate, maltol, carbon
Bleaching agents	$CaOCl_2$, NaOCl, $NaClO_2$, SO_2
Nutrient supplements	Vitamins

[a]*Banned for use in food.*

As defined by the FDA, the five main reasons to use additives in foods are as follows.

- *To maintain product consistency.* Emulsifiers give products a consistent texture and prevent them from separating. Stabilizers and thickeners give smooth uniform texture. Anticaking agents help substances such as salt to flow freely.
- *To improve or maintain nutritional value.* Vitamins and minerals are added to many common foods such as milk, flour, cereal, and margarine to make up for those likely to be lacking in a person's diet or lost in processing. Such fortification and enrichment has helped reduce malnutrition among the US population. All products containing added nutrients must be appropriately labeled.
- *To maintain palatability and wholesomeness.* Preservatives retard product spoilage caused by mold, air, bacteria, fungi, or yeast. Bacterial contamination can cause food-borne illness, including life-threatening botulism. Antioxidants are preservatives that prevent fats and oils in baked goods and other foods from becoming rancid or developing an off-flavor. They also prevent cut fresh fruits such as apples from turning brown when exposed to air.
- *To provide leavening or control acidity/alkalinity.* Leavening agents that release acids when heated can react with baking soda to help cakes, biscuits, and other baked goods to rise during baking. Other additives help modify the acidity and alkalinity of foods for proper flavor, taste, and color.

- *To enhance flavor or impart desired color.* Many spices and natural and synthetic flavors enhance the taste of foods. Colors, likewise, enhance the appearance of certain foods to meet consumer expectations. Examples of substances that perform each of these functions are provided in Table 10.2.

Despite the fact that food additives undergo extensive laboratory testing before they are put into commercial food products, the use of additives in foods has engendered great controversy and widespread public concern. There are two basic positions concerning the use of food additives. One is that all additives are potential health threats and should not, on that basis, be used. The other is that unless an additive is proved to be hazardous, using it to protect food from spoilage or to increase its nutritional completeness, palatability, texture, or appearance is well justified. The former opinion has been voiced by some consumers. Their concern is that basic food materials already are contaminated by many toxic substances such as pesticides and microorganisms. Nonetheless, once additives are approved for use in a food product, people will be ingesting them continuously. Therefore, even when an acceptable daily intake (ADI) has been officially established and each product remains within those limits, total ingestion of certain additives from various sources may exceed the ADI. This position holds that the chronic toxicities, such as carcinogenicity and teratogenicity, of food additives have not yet been sufficiently studied. In fact, most food additives are used without any information being made available to consumers about their chronic toxicities. Due to the high costs of testing and other factors, progress in research on the chronic toxicities of food additives is very slow.

The second basic position regarding food additives points out their many benefits: Were it not for food additives, baked goods would go stale or mold overnight, salad oils and dressings would separate and turn rancid, table salt would turn hard and lumpy, canned fruits and vegetables would become discolored or mushy, vitamin potencies would deteriorate, beverages and frozen desserts would lack flavor, and wrappings would stick to the contents.

Within the current structure of the food processing industry in the United States, it would be virtually impossible to abandon food additives entirely. However, in order to provide the maximum protection to the consumers, it is wise to study any potential toxicity of the food additives in current use.

REGULATIONS

The quality of the food supply in the United States is regulated by numerous state and federal laws. Before the turn of the last century, most states had laws that protected consumers from hazardous or improperly processed foods, but there was no corresponding federal regulatory framework. Public sentiment became focused on the wholesomeness of the food supply following the publication of Upton Sinclair's novel, *The Jungle*, in which the deplorable conditions in slaughter houses were described. In addition Dr. Harvey W. Wiley, who was a chemist in the US Department of Agriculture from 1883 to 1930, began conducting chemical and biological analyses of substances in food and discovered instances of misbranded and adulterated food. Since colonies of experimental animals had not yet been developed, Wiley conducted biological testing on himself and a group of young men, who became known as the Poison Squad. Because of the work of Wiley and his group, reliable information on the adulteration, toxicity, and misbranding of food was obtained. Thus, the Pure Food and Drug Act finally passed in 1906 after years of effort to enact such a law. However, it has been suggested that it passed only because it was presented along with the Meat Inspection Act, a response to the public outcry generated by Sinclair's novel.

The 1906 Pure Food and Drug Act forbade the production of misbranded or adulterated food products in the District of Columbia, and prohibited interstate distribution of fraudulent or unhealthy products. It banished such chemical preservatives as boric acid, salicylic acid, and formaldehyde; additionally, it defined food adulteration as the addition of poisons or deleterious materials, the extraction of valuable constituents and the concealment of the resulting inferiority, substitution of other constituents, and the mixture of substances that would adversely affect health.

The next major piece of legislation was the Federal Food, Drug, and Cosmetic Act of 1938 (referred to as the 1938 Act), which added further provisions to the 1906 legislation. It defined food as:

- Substances used as food or drink by people or animals
- Chewing gum
- Substances used for components of any such food materials

The law also established standards of product identity and fill levels, prohibited adulteration, mandated truthful labeling, and restricted the use of chemicals to those required in the manufacturing of the food, with specific tolerance levels set for chemicals with appreciable toxicity.

The Food Additive Amendment that was added to the law in 1958 had a great impact on the food manufacturing industry. Although this amendment to the 1938 Act gave official recognition of the US government's tolerance regarding the use of food additives and acceptance of the necessity of additives for a wholesome and abundant food supply, it also took the government out of the business of toxicity testing. The amendment stipulated that any food additive was to be proven safe by the manufacturer, who must also prove toxicity. Thus, the additive manufacturer was requested to bear the burden of lengthy delays in production and the cost of the millions of dollars required for such testing.

The amendment, however, did not apply to additives in use prior to 1958. The potential hazards of these substances initially were assessed based on the opinions of specialists working in the general field of toxicology. Those substances that were considered unsafe for general use in foods were prohibited from further use or strict limits were placed on the levels that could be added to foods. Those substances for which no concerns were expressed were considered "generally recognized as safe" (GRAS). Substances on the GRAS list can be used by manufacturers within the general tenets of good manufacturing practices.

Also included in the Food Additive Amendment of 1958 was the Delaney Clause, which was proposed in response to the increasing concern over the possible role of food additives in human cancer. It states that no substance shall be added to food if it is found by proper testing to cause cancer in people or animals. Although the Delaney Clause seems to be a very straightforward and simple way of handling a potentially dangerous class of substances, it provides little room for scientific interpretation of data and remains controversial. The Delaney Clause gives no consideration to the specific dose of a substance likely to be encountered in foods and there is no provision for interpretation of animal results in terms of the site of carcinogenesis, known sensitivity of the experimental model, or the specific dose of carcinogens required to produce the cancer. Although several substances have been banned from food use because of apparent carcinogenicity, in no case has the Delaney Clause been invoked. All substances banned from use have been banned under the general safety provisions of the 1938 Act.

The Color Additive Amendments of 1960 took special notice of color additives and their potential for misuse and toxicity. The amendments differentiated between natural and synthetic color additives and required:

- Listing of all color additives on labels
- Certification of batches of listed colors where this was deemed necessary to protect public health
- Retesting for safety of previously certified colors using modern techniques and procedures where any questions of safety had arisen since the original listing

- Testing of all color additives for food in long-term dietary studies that included carcinogenesis and teratogenesis studies in two or more animal species

There are currently two classifications considered suitable for use in food colors, GRAS colors and certified colors. GRAS colors are generally pigments that occur naturally in plants and animals or that are simple organic or inorganic compounds. GRAS colors are exempt from the certification procedure required for most of the synthetic colors. In order to be certified, a sample of each batch of a color must be submitted to the FDA for analyses. The batch receives approval for use if the sample meets previously established standards of quality. Each batch must be analyzed to ensure that the chemical compositions in the new batches are the same as the composition of the batch that was subjected to biological testing.

The practice of periodic review is now also used for all food additives. Since 1970, the FDA has been reviewing GRAS substances and ingredients with prior sanctions according to current standards of safety. Under the current regulatory framework, no substance that appears directly or indirectly in the food may be added to or used on or near food, unless:

- The substance is GRAS
- The substance has been granted prior sanction or approval
- A food additive regulation has been issued, establishing the conditions for its safe use

All nonflavor substances on the GRAS list are under review individually in accordance with present-day safety standards while all of the roughly 1000 flavoring agents used in food are reevaluated under a separate program that considers these substances by chemical class. After a GRAS review is completed, a substance with previous GRAS status will be:

- Reaffirmed as GRAS
- Classified as a food additive, and conditions of use, levels of use, and any other limitations will be specified
- Placed under an interim food additive regulation, which indicates that further toxicological information must be obtained for the substance
- Prohibited from use

Substances in the interim food additive category can be used in foods while the testing is going on if there is no undue risk to the public. A substance with GRAS status has no specific quantitative limitations on its use; however, its use is restricted by general definitions of adulterated food as specified in the 1938 Act and by what is designated good manufacturing practice (GMP). GMP includes the following conditions for substances added to food:

- The quantity of a substance added to food does not exceed the amount reasonably required to accomplish its intended physical, nutritional, or other technical effect in food.
- The quantity of a substance that becomes a component of food as a result of its use in the manufacturing, processing, or packaging of food in which it is not intended to accomplish any physical or other technical effect in the food itself shall be reduced to the extent reasonably possible.

GMP also specified that the substance is of appropriate food grade and is prepared and handled as a food ingredient.

Generally, food additives are regulated in amendments under Section 409 of the 1938 Act. The law requires not only that food additives be safe at the levels used, but also effective in accomplishing their intended effect. Ineffective additives cannot legally be used regardless of their safety. However, this law specifically excludes certain classes of compounds. Pesticide residues in raw agricultural products are not legally considered to be food additives in the United States, although they are subject to regulation by the EPA and the individual states. Colorants are regulated separately under the Color Additive Amendments in the 1938 Act passed in 1960.

Prior to the 1958 food additive amendments in the 1938 Act, most food additives were not regulated directly, although certain individual substances were banned by the FDA. Additives that were in common and widespread use in 1958 were exempted from requirements of safety testing on the basis of long-term experience with these compounds. They were termed GRAS under conditions of their intended use, usually at the lowest practical level and in accordance with good manufacturing practices. Although the FDA has specifically listed hundreds of additives as GRAS, the published list is not inclusive.

Food additives also have been targeted by the Food Quality Protection Act (FQPA), which took effect in 1996, even though this law was intended mainly to apply to pesticide residues in food. The major impact of FQPA on food additives is elimination of the Delaney Clause. The elimination of the Delaney Clause was part of the 1996 FQPA that substituted a more manageable "risk cup" approach to assessing total human exposure and risk to many chemicals including food additives.

PRESERVATIVES

One of the most important functions of food additives is to preserve food products from spoilage. Preservatives prevent the spoilage of foods caused by the action of microorganisms or oxidation. The development of methods of food preservation was essential to the prehistoric transition of humans from nomadic hunter-gatherer tribes, to settled agricultural communities. Since the dawn of history, people have struggled to preserve enough food to survive from one growing season to the next. Smoke was probably the first preservative agent to be discovered. Common salt also was used in prehistoric times. Ancient Egyptians made use of vinegar, oil, and honey, substances that still find application today. In the thirteenth century, Wilhelm Beukels discovered "pickling," a process of preserving food by anaerobic fermentation in brine, a solution of salt in water.

In Ancient Assyria, Greece, and China, sulfur dioxide, which normally was described as a fumigant, was also utilized as a preservative. By the late Middle Ages, it was widely used throughout Europe for preserving wine and possibly other applications as well. However, during the late fifteenth century, in what may have been the first legal actions directed against chemical food additives, a number of decrees were promulgated to regulate the use of sulfur in wine production. No other chemical preservatives were introduced until the late eighteenth century when Hofer suggested that borax (hydrated sodium borate) be used. Yet, even at the present time, it is estimated that up to one-third of the agricultural production of the United States is lost after harvest.

Synthetic chemicals began to be used as food preservatives at the beginning of the twentieth century, and the widespread use of these chemicals made a broad variety of food available to more people for longer periods of time. As mentioned earlier, there are many criticisms of the use of food preservatives. However, the considerable time between the production and the consumption of food today makes some use of preservative necessary in order to prevent spoilage and undesirable alterations in color and flavor. Many microorganisms, including yeasts, molds, and bacteria, can produce undesirable effects in the appearance, taste, or nutritional value of foods. A number of these organisms produce toxins that pose high risks to human health. Atmospheric oxygen can also adversely affect foods, as for example in the development of rancidity in fats.

The word *preservative* gradually has been given broader connotation, covering not only compounds that suppress microbes but also compounds that prevent chemical and biochemical deterioration. The action of preservatives is not to kill bacteria (bactericidal) but rather to delay their action by suppressing their activity (bacteriostatic).

Before being considered for use in food, a chemical preservative must fulfill certain conditions:

- It must be nontoxic and suitable for application.
- It must not impart off-flavors when used at levels effective in controlling microbial growth.
- It must be readily soluble.
- It must exhibit antimicrobial properties over the pH range of each particular food.
- It should be economical and practical to use.

Benzoic Acid

Benzoic acid, which usually is used in the form of its sodium salt, sodium benzoate (Figure 10.1), long has been used as an antimicrobial additive in foods. It is used in carbonated and still beverages, syrups, fruit salads, icings, jams, jellies, preserves, salted margarine, mincemeat, pickles and relishes, pie, pastry fillings, prepared salads, fruit cocktail, soy sauce, and caviar. The use level ranges from 0.05 to 0.1%.

Benzoic acid in the acid form is quite toxic but its sodium salt is much less toxic (Table 10.3). The sodium salt is preferred because of the low aqueous solubility of the free acid. In vivo, the salt is converted to acid, which is the more toxic form.

Subacute toxicity studies of benzoic acid in mice indicated that ingestion of benzoic acid or its sodium salt caused weight loss, diarrhea, irritation of internal membranes, internal bleeding, enlargement of liver and kidney, hypersensitivity, and paralysis followed by death. When benzoic acid (80 mg/kg body weight) and sodium bisulfate (160 mg/kg body weight) or their mixture (benzoic and/sodium bisulfate = 80 mg/160 mg) were fed to mice for 10 weeks, the death rate was 66% from the mixture and 32% from benzoic acid alone.

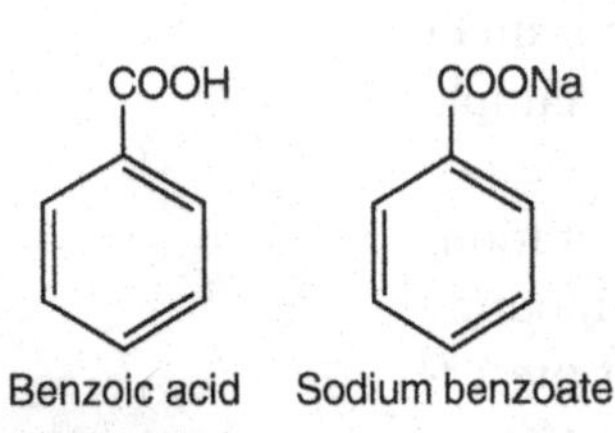

FIGURE 10.1 *Structures of benzoic acid and sodium benzoate.*

Four-generation reproductive and developmental toxicities of benzoic acid were examined using diets containing 0, 0.5, and 1% of benzoic acid fed to male and female rats housed together for eight weeks. The second generation was observed through its entire life cycle and the third and fourth generations were examined by autopsy. No changes in normal patterns of growth, reproduction, or lactation during life were recorded and no morphological abnormalities were observed from the autopsies.

Table 10.3 Acute Toxicity of Sodium Benzoate

Animal	Method	LD_{50} (mg/kg)
Rat	Oral	2,700
Rat	Intravenous injection	1714 ± 124
Rabbit	Oral	2,000
Rabbit	Subcutaneous injection	2,000
Dog	Oral	2,000

Degradation pathways for benzoic acid also have been studied in detail and the results have supported the harmlessness of this substance. The metabolic breakdown pathways of benzoic acid are shown in Figure 10.2. The total dose of benzoic acid is excreted within 10 to 14 hours and 75 to 80% is excreted within 6 hours. After conjugation with glycine, 90% of benzoic acid appears in the urine as hippuric acid. The rest forms a glucuronide,

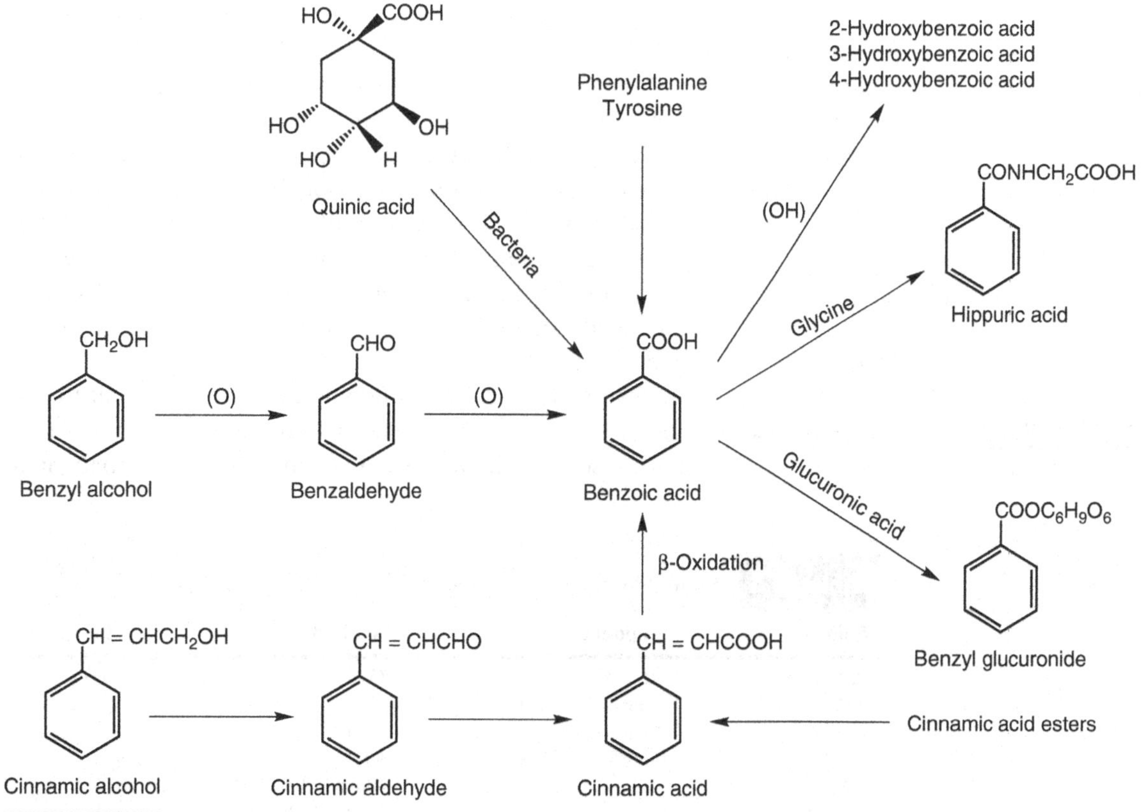

FIGURE 10.2 *The metabolic breakdown pathways of benzoic acid.*

1-benzoylglucuronic acid. The lower aliphatic esters of benzoic acid are first hydrolyzed by esterase, which abounds in the intestinal wall and liver. The resulting benzoic acid subsequently is degraded in the usual manner.

Sorbic Acid and Potassium Sorbate

Sorbic acid and its salts have broad-spectrum activity against yeast and molds, but are less active against bacteria. The antimicrobial action of sorbic acid was discovered independently in the United States and Germany in 1939, and since the mid-1950s sorbates have been increasingly used as preservatives. The structures of these compounds are shown in Figure 10.3. Sorbates generally have been found superior to benzoate for preservation of margarine, fish, cheese, bread, and cake. Sorbic acid and its potassium salts are used in low concentrations to control mold and yeast growth in cheese products, some fish and meat products, fresh fruits, vegetables, fruit beverages, baked foods, pickles, and wines.

Sorbic acid is practically nontoxic. Table 10.4 shows acute toxicity of sorbic acid and its potassium salt. Animal studies have not shown obvious problems in tests performed with large doses for longer time periods. When sorbic acid (40 mg/kg/day) was injected directly into the stomach of male and female mice for 20 months, no differences were observed in survival rates, growth rates, or appetite between the injected mice and the control. When the dose was increased to 80 mg/kg/day for three additional months, however, some growth inhibition was observed. When potassium sorbate (1 and 2% in feed) was fed to dogs for three months, no pathological abnormalities were observed. This evidence indicates that the subacute toxicity of sorbic acid is negligible.

$CH_3CH = CHCH = CHCOOH$
Sorbic acid

$CH_3CH = CHCH = CHCOOK$
Potassium sorbate

FIGURE 10.3 *Structures of sorbic acid and potassium sorbate.*

Two-generation reproduction and developmental toxicities of sorbic acid using various animals have shown that neither sorbic acid nor its potassium salt induces malignant

Table 10.4 Acute Toxicity of Sorbic Acid and Its Potassium Salt

Animal	Compound	Method	LD_{50} (g/kg)
Rat	Sorbic acid	Oral	10.5
Rat	Potassium sorbate	Oral	4.2
Mouse	Sorbic acid	Oral	>8
Mouse	Potassium sorbate	Oral	4.2
Mouse	Sorbic acid	Intraperitoneal	2.8
Mouse	Potassium sorbate	Intraperitoneal	1.3

growths in animals. For example, rats fed with 5% sorbic acid in feed for two generations (1000 days) showed no changes in growth rates, rates of reproduction, or any other behaviors.

As a relatively new food additive, sorbate has been subject to stringent toxicity-testing requirements. It may well be the most intensively studied of all chemical food preservatives. In 90-day feeding studies in rats and dogs and a lifetime feeding study in rats, a 5% dietary level of sorbates procured no observable adverse effects. However, at a 10% dietary level in a 120-day feeding study, rats showed increased growth and increased liver weight. This has been attributed to the caloric value of sorbate at these high dietary levels since it can act as a substrate for normal catabolic metabolism in mammals. Sorbates are not mutagenic or tumorigenic, and as noted previously, no reproductive toxicity has been observed.

Hydrogen Peroxide

Hydrogen peroxide is used as an agent to reduce the number of bacteria in dairy products or other foodstuffs. In the dairy industry, hydrogen peroxide also has been used as a substitute for heat pasteurization in the treatment of milk and as a direct preservative in keeping the quality of the milk. In Japan, it has been used as a preservative for fish-paste products. Hydrogen peroxide also has a bleaching effect. The use of highly pure hydrogen peroxide in manufactured cheese has been approved by the United States Food and Drug Administration (industrial grade hydrogen peroxide is usually a 3–35% aqueous solution; a commercial home product is a 3% aqueous solution).

Acute toxicities (LD_{50}) of hydrogen peroxide for rats are 700 mg/kg/b.w. and 21 mg/kg/b.w. by subcutaneous injection and intravenous injection, respectively. When large amounts of hydrogen peroxide were injected directly into the stomachs of rats, weight and blood protein concentrations were changed slightly. When hydrogen peroxide was mixed with feed, however, no abnormalities were observed. The use of bactericides has been limited due to their toxicity to humans, and only hydrogen peroxide currently is recognized for use.

AF-2 [2-(-furyl)-3-(5-nitro-2-furyl)acrylamide]

The antibacterial activity of nitrofuran derivatives was first recognized in 1944. The disclosed antibacterial properties of these compounds had created a new group of antimicrobial agents. Such activity is dependent on the presence of a nitro group in the 5-position of the furan ring.

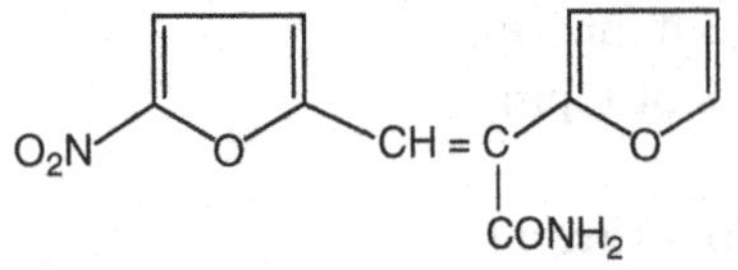

FIGURE 10.4 *Structure of 2-(2-furyl)-3-(5-nitro-2-furyl)acrylamide (AF-2).*

Numerous 5-nitrofuran derivatives have been synthesized and certain compounds have been used widely in clinical and veterinary medicine and as antiseptics for animal foods.

AF-2 (Figure 10.4) was legally approved for use in Japan in 1965 and was added to soybean curd (tofu), ham, sausage, fish ham, fish sausage, and fish paste (kamaboko). The safety-testing data on which the compound was approved were those obtained for acute and chronic toxicity for two years and reproductive potential for four generations using mice and rats. At the time, no attention was paid to mutagenicity. In 1973, AF-2 was proved to be mutagenic in various microbial test systems. The mutagenicity of this food additive strongly suggested its carcinogenicity and the risk of its use as a food additive. Within a year or two after the discovery of its mutagenicity, the actual carcinogenicity of this chemical was demonstrated in animal studies. Since that time, more emphasis has been placed on finding causative agents of cancer.

Short-term bioassays had received much attention because they were able to screen suspected carcinogens. The mechanism of carcinogenesis is not yet clear, but a close relation between carcinogens and mutagens has been demonstrated. AF-2 was the first example of compound that was shown to be a carcinogen. The fact that AF-2 was discovered first to be mutagenic proved the value of mutagenicity testing as a screening method for carcinogens.

ANTIOXIDANTS

One of the most common types of food deterioration is an undesirable change in color or flavor caused by oxygen in air (oxidative deterioration). Oxidation causes changes not only in color or flavor but also decreases the nutritional value of food and sometimes produces toxic materials. Since most foods consist mainly of carbohydrates, fats, proteins, and water, microbiological spoilage is one of the most important factors to be considered in preserving the carbohydrate and protein portions of food products. However, oxidation, particularly atmospheric oxidation, is the chief factor in the degradation of fats and fatty portions of foods. Oxidative deterioration of fat results not only in the destruction of vitamins A, D, E, K, and C, but also in the destruction of essential fatty acids and the development of a pungent and offensive off-flavor. In extreme cases, toxic by-products have resulted from oxidative reactions.

The most efficient method of preventing oxidative degradation is the use of antioxidative agents. The antioxidants may be classed as natural and synthetic. Naturally occurring antioxidants exhibit relatively weak antioxidant properties. As a consequence, synthetic antioxidants have been developed for use in foods. In order for these substances to be allowed in foods, they must have a low toxicity; should be effective in low concentrations in a wide variety of fats; should contribute no objectionable flavor, order, or color to the product; and should have approval by the Food and Drug Administration.

L-Ascorbic Acid (Vitamin C)

L-Ascorbic acid, or vitamin C, is widely present in plants. The structures of ascorbic acid and dehydroascorbic acid are shown in Figure 10.5. Vitamin C is not only an important nutrient but is also used as an antioxidant in various foods. However, it is not soluble in fat and is unstable under basic conditions. Vitamin C reduces cadmium toxicity and excess doses prolong the retention time of an organic mercury compound in a biological system. Overdoses of vitamin C (106 g) induce perspiration, nervous tension, and lowered pulse rate. WHO recommends that daily intake be less than 0.15 mg/kg. Toxicity due to ascorbic acid has not been reported. Although repeated intravenous injections of 80 mg dehydroascorbic acid was reported to be diabetogenic in rats, oral consumption of 1.5 g/day of ascorbic acid for six weeks had no effect on glucose tolerance or glycosuria in 12 normal adult males and produced no change in blood glucose concentrations in 80 diabetics after five days. The same report noted that a 100-mg intravenous dose of dehydroascorbic acid given daily for prolonged periods produced no signs of diabetes. Ascorbic acid is readily oxidized to dehydroascorbic acid, which is reduced by glutathione in blood.

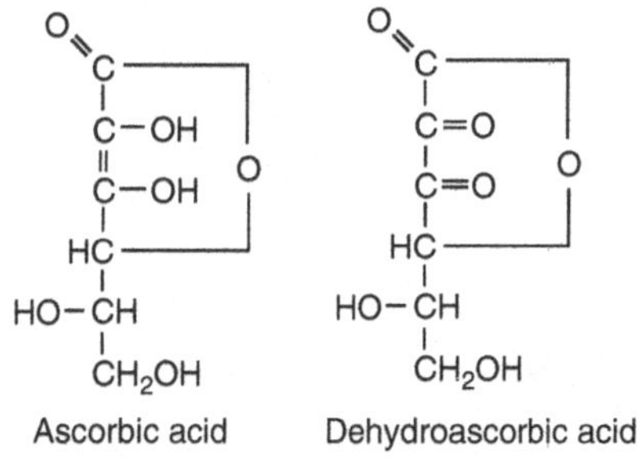

Ascorbic acid Dehydroascorbic acid

FIGURE 10.5 *Structures of ascorbic acid and dehydroascorbic acid.*

dl-α-Tocopherol (Vitamin E)

α-Tocopherol is known as vitamin E and exists in many kind of plants, especially in lettuce and alfalfa. Its color changes from yellow to dark brown when exposed to sunlight. The structure of α-tocopherol is shown in Figure 10.6.

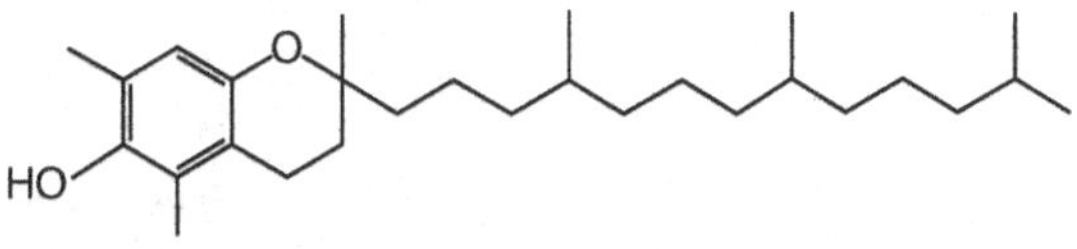

FIGURE 10.6 *Structure of α-tocopherol (vitamin E).*

Natural vegetable oils are not readily oxidized due to the presence of tocopherol. During refining processes, however, tocopherol may be removed from oils; consequently, refined vegetable oils can become unstable toward oxidation. In one experiment, vitamin E appeared to be relatively innocuous, having been given to patients for months both orally and parenterally at a dosage level of 300 mg/day without any observed ill effects. However, in another experiment, 6 out of 13 patients given similar doses complained of headache, nausea, fatigue, dizziness, and blurred vision.

Although the chronic toxicity of vitamin E has not been thoroughly studied, WHO recommends 2 mg/kg/day as the maximum daily dose.

Propyl Gallate

Propyl gallate (n-propyl-3,4,5-trihydroxybenzoate, Figure 10.7) is used in vegetable oils and butter. When 1.2 or 2.3% propyl gallate was added to feed for rats, loss of weight was observed. This may be due to the rats' reluctance to eat food that was contaminated with the bitter taste of propyl gallate. When it was given for 10 to 16 months at the 2 to 3% level, 40% of the rats died within the first month and the remainder showed severe growth inhibition. Autopsies of rats indicated kidney damage resulting from the ingestion of propyl gallate. However, no other animal studies show serious problems and further studies indicated that propyl gallate does not cause serious chronic toxicities.

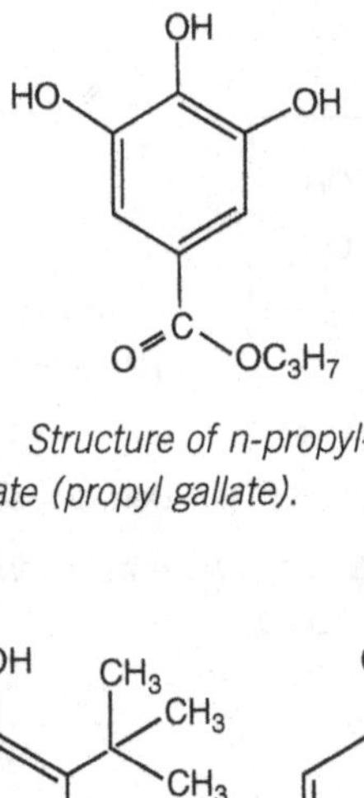

OH
HO
OH
C
O
OC_3H_7

FIGURE 10.7 *Structure of n-propyl-3,4,5-trihydroxybenzoate (propyl gallate).*

Butylated Hydroxyanisol and Butylated Hydroxytoluene

Butylated hydroxyanisole (BHA) and butylated hydroxytoluene (BHT) are the most commonly used antioxidants and present constant intractable problems to the food industry. The structures of BHA and BHT are shown in Figure 10.8.

BHA produces mild diarrhea in dogs when it is fed continuously for four weeks at the level of 1.4 to 4.7 g/kg. It also causes chronic allergic reactions, malformations, and damage to the metabolic system. When BHT was fed to rats at levels of 0.2, 0.5, and 0.8% mixed with feed for 24 months, no pathological changes were observed. The same results were obtained when the dose was increased to 1% of the feed.

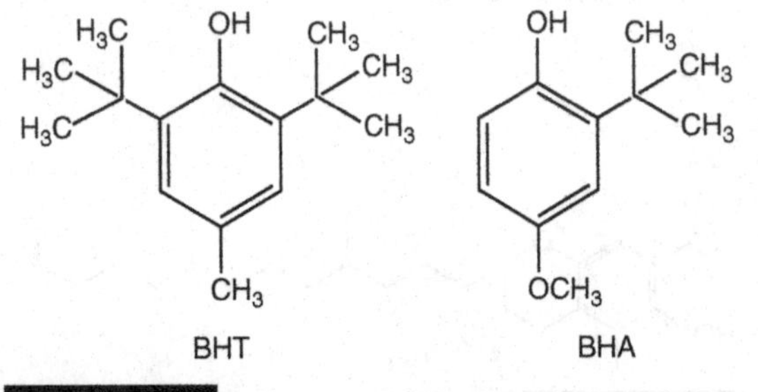

H_3C
H_3C
H_3C
OH
CH_3
CH_3
CH_3
CH_3
BHT
OH
CH_3
CH_3
CH_3
OCH_3
BHA

FIGURE 10.8 *Structure of butylated hydroxytoluene (BHT) and butylated hydroxyanisole (BHA).*

Antioxidants—including vitamin C, vitamin E, BHA, and BHT—have some anticarcinogenic activities. If BHT is mixed with a known carcinogen such as N-2-fluorenyl acetamide (FAA) or azoxymethane (AOM) in the feed, the rate of tumor induction in rats is diminished. The mechanisms of antioxidants in chemical carcinogenesis are not well understood yet, but the relationship between chemical carcinogens and BHT has been intensively investigated.

SWEETENERS

Naturally occurring sweeteners such as honey and sucrose were known to the ancient Romans. However, sweeteners obtained from natural sources have been limited. To supplement the demand, sweetening agents such as saccharin have been synthesized since the late nineteenth century. Recently, these nonnutritive sweeteners have begun to receive much attention as ingredients in low-calorie soft drinks. The synthetic or noncarbohydrate sweeteners provide sweetened foods for diabetics who must limit sugar intake, for those who wish to limit carbohydrate calorie intake, and for those who desire to reduce food-induced dental caries.

Saccharin and Sodium Saccharin

Saccharin, which is 300 to 500 times sweeter than sucrose, is one of the most commonly used artificial sweeteners. The name saccharin is a commercial name of the Fahlberg and List Company. The sodium salt is the form actually used in the formulation of foods and beverages (Figure 10.9). Its acute toxicity is shown in Table 10.5.

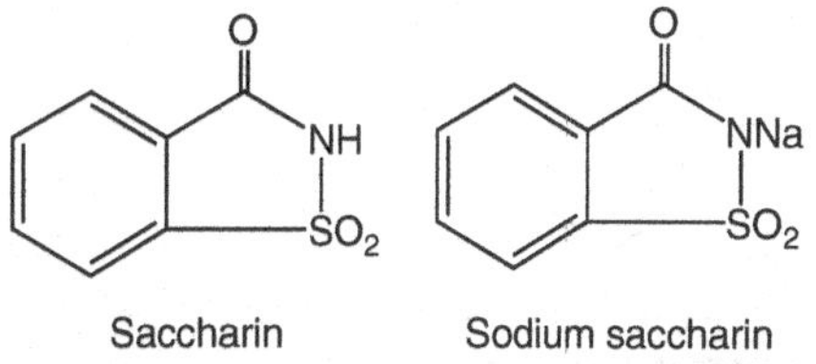

FIGURE 10.9 *Structures of saccharin and sodium saccharin.*

Table 10.5 Acute Toxicity of Sodium Saccharin

Animal	Method	LD_{50} (g/kg)
Mouse	Oral	17.5
Mouse	Intraperitoneal	6.3
Rat	Oral	17.0
Rat	Intraperitoneal	7.1
Rat	Oral	14.2 ± 1.3
Rabbit	Oral	5–8 (LD)

It has been questioned whether or not saccharin is a health hazard. In 1972, it was found that 7.5% of saccharin in feed produced bladder cancer in the second generation of rats. Some reports, however, showed contradictory results. Consequently, WHO has recommended that daily intake of saccharin be limited to 0 to 0.5 mg/kg. The carcinogenicity of saccharin is still under investigation. When pellets of saccharin and cholesterol (1:4) were placed in the bladders of mice, tumors developed after 40 to 52 weeks.

When 2.6 g/kg of a mixture of sodium cyclamate and saccharin (10:1) was given to rats for 80 days, eight rats developed bladder tumors after 105 weeks. When only sodium cyclamate was fed to rats for two years, bladder cancer also appeared. The main attention, therefore, has been given to the carcinogenicity of sodium cyclamate.

Sodium Cyclamate

Sodium cyclamate is an odorless powder. It is about 30 times as sweet as sucrose in dilute solution. The structure of sodium cyclamate is shown in Figure 10.10 and its acute toxicity is shown in Table 10.6.

Capillary transitional cell tumors were found in the urinary bladders of 8 out of 80 rats that received 2600 mg/kg body weight per day of a mixture of sodium cyclamate and sodium saccharin (10:1) for up to 105 weeks. When the test mixture was fed at dietary levels designed to furnish 500, 1120, and 2500 mg/kg body weight to groups of 35 and 45 female rats, the only significant finding was the occurrence of papillar carcinomas in the bladders of 12 of 70 rats fed the maximum dietary level of the mixture (equivalent of about 25 g/kg body weight) for periods ranging from 78 to 105 weeks (except for one earlier death). In vivo conversion from sodium cyclamate to cyclohexylamine was observed particularly in the higher dosage group. Cyclohexylamine is very toxic (LD_{50} rat oral = 157 mg/dg) compared to sodium cyclamate (LD_{50} oral = 12 g/kg).

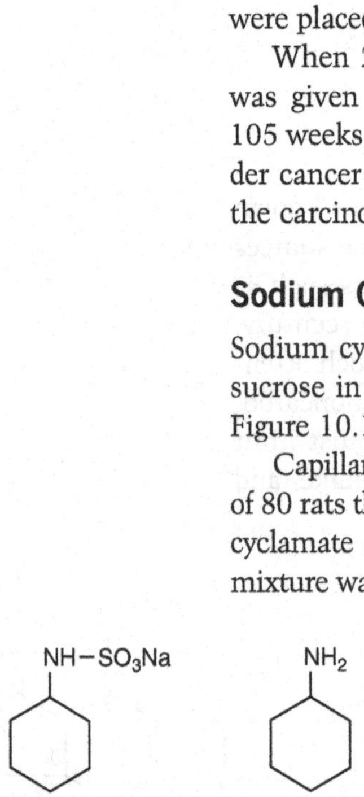

$NH-SO_3Na$ NH_2

Sodium cyclamate Cyclohexylamine

FIGURE 10.10 *Structures of sodium cyclamate and its metabolite, cyclohexylamine.*

Table 10.6 Acute Toxicity of Sodium Cyclamate

Animal	Method	LD_{50} (g/kg)
Mouse	Oral	10–15
Mouse	Intraperitoneal	7
Mouse	Intramuscular	4–5
Rat	Oral	2–17
Rat	Intraperitoneal	6
Rat	Intramuscular	3–4

In 1968, the FDA discovered teratogenicity of sodium cyclamate in rats and prohibited its use in food.

COLORING AGENTS

Coloring agents have been used to make food more attractive since ancient times. The perception and acceptability of food, in regards to taste and flavor, is strongly influenced by its color. Nutritionists have long known that without expected color cues, even experts have difficulty in identifying tastes. Certain varieties of commercial oranges have coloring applied to their peels because the natural appearance—green and blotchy—is rejected by consumers as unripe or defective. Consumers reject orange juice unless it is strongly colored, even if it is identical in taste and nutritional value. Congress has twice overridden the FDA (in 1956 and 1959) when it proposed a ban of the coloring agents used.

The natural pigments of many foods are unstable in heat or oxidation. Thus, storage or processing can lead to variations in color even when the nutritional value remains unchanged. Changes in the appearance of a product over time may cause consumers to fear that a "bad" or an adulterated product has been purchased, particularly in light of highly publicized incidents of product tampering. The use of food coloring can resolve this problem for retailers and manufacturers. Ripe olives, sweet potatoes, some sauces and syrups, as well as other foods, are colored mainly to ensure uniformity and consumer acceptability.

Candies, pastries, and other products such as pharmaceutical preparations are often brightly colored. Pet foods are colored for the benefit of human owners, not their color-blind pets. Such applications are criticized by some as unnecessary, or even frivolous, even when natural food dyes are involved.

Red color can be produced naturally from a dried sugar beet root (beetred) or from insects such as cochineal (Central America) and lac dye (Southeast Asia). Natural dyes, however, are usually not clear, and their variety is limited. Synthetic coloring agents began to replace them in the late nineteenth century. Twenty-one synthetic chemicals were recognized for use in 1909 at the Second International Red Cross Conference.

About 80 synthetic dyes were being used in the United States for coloring foods in 1900. At that time, there were no regulations regarding their safety. Many of the same dyes were used to color cloth and only the acute toxicity of these coloring agents was tested before they were used in foods. The chronology of the addition of new colors is shown in Table 10.7. The number of coloring agents currently permitted as food additives is much fewer than previously and still is being reduced.

Table 10.7 Chronology of Newly Developed Synthetic Coloring Agents

Year	Agent
1916	Tartrazine
1918	Yellow AB & OB
1922	Guinea Green
1927	Fast Green
1929	Ponceau SX
1929	Sunset Yellow
1929	Brilliant Blue
1950	Violet No. 1
1966	Orange B
1971	FD&C Red No. 40

A large class of synthetic organic dyes is the azo group. Many commercial dyes belonged to this class, particularly due to their vivid colors, especially reds, oranges, and yellows. Azo compounds contained the functional group (R—N = N—R′), in which R and R′ can be either aryl or alkyl. There are two kinds of azo dyes, those that are water-soluble and those that are not. In general, the water-soluble azo dyes are less toxic because they are more readily excreted from the body. Either type, however, can be reduced to form the toxic amino group, —NH_2, in the body in conjunction with the action of microorganisms such as *Streptococcus, Bacillus pyocyaneus*, and *Proteus* sp. One particular bioassay report on azo compounds showed that only 12 out of 102 azo compounds did not reduce into amine. In 1937, dimethyl amino azobenzene or butter yellow was found to induce malignant tumors in rats. The carcinogenicity of butter yellow was later confirmed.

Amaranth (FD&C Red No. 2)

Amaranth is 1-(4-sulfo-1-paphthylazo)-9-naphthol-3,6-disulfonic acid, trisodium salt (Figure 10.11) and is an azo dye. It is a reddish-brown powder with a water solubility of 12 g/100 ml at 30°C. Before it was prohibited by the FDA in 1976 following indications that it induced malignant tumors in rats, amaranth had been used in almost every processed food with a reddish or brownish color, including soft drinks, ice creams, salad dressings, cereals, cake mixes, wines, jams, chewing gums, chocolate, and coffee as well as a variety of drugs and cosmetics at the level of 0.01 to 0.0005%. In 1973, an estimated $2.9 million worth of amaranth was added to more than $10 billion worth of products.

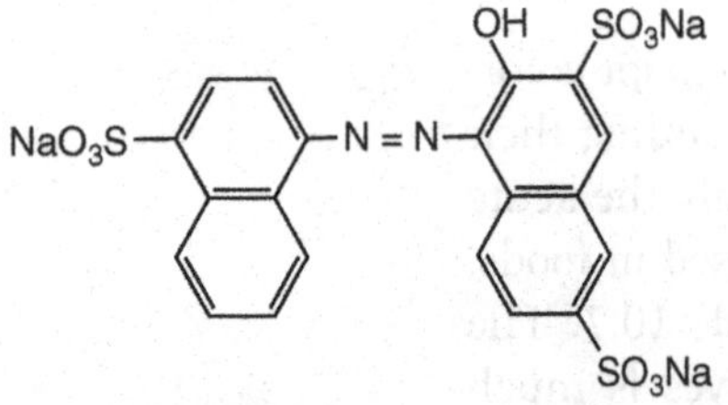

FIGURE 10.11 *Structure of amaranth.*

In toxicity studies, amaranth (0.5 ml of a 0.1% solution) was injected under the skin of rats twice a week for 365 days, in which no tumor growth was observed. When 0.2% of amaranth in feed (average 0.1 g/kg/day) was fed to rats for 417 days, no induction of tumors was observed. However, one case of intestinal cancer was observed when feeding was continued for an additional 830 days. The FAO/WHO special committee determined its ADI as 0 to 1.5 mg/kg.

It was determined that amaranth is metabolized into amine derivatives in vivo. Amaranth is reduced by aqueous D-fructose and D-glucose at elevated temperatures to form a mixture of hydrazo and amine species, which may have toxicological significance. The interactions between additives such as amaranth and other food components should be considered from the viewpoint of toxicology.

Tartrazine (FD&C Yellow No. 4)

This coloring agent is 5-hydroxy-1-p-sulfophenyl-4-(p-sulfophenylazo)-pyrazol-3-carboxylic acid, trisodium salt (Figure 10.12). It is a yellow powder and has been used as food coloring additive since 1916.

Tartrazine is known as the least toxic coloring agent among synthetic coloring chemicals. The median acute oral lethal dose of tartrazine in mice is 12.17 g/kg. Beagle dogs received tartrazine as 2% of the diet for two years without adverse effects, with the possible exception of pyloric gastritis in one dog. Tumor incidence was unchanged relative to controls, in rats receiving tartrazine at 1.5% of the diet for 64 weeks, and in rats administered this dye at 5.0% of the diet for two years.

Human sensitivity to tartrazine has been reported with some frequency and has been estimated to occur in 1/10,000 persons. Anaphylactic shock, potentially life-threatening, has been reported but symptoms more commonly cited are urticaria (hives), asthma, and purpura (blue or purple spots on the skin or mucous membrane). Of 97 persons with allergic symptoms in one trial, 32 had adverse reactions to challenge with 50 mg tartrazine. Physicians use 0.1 to 10 mg tartrazine to test for its sensitivity.

In the United States, tartrazine can be used in foods only as a coloring agent (FDA Regulations 8.275, 8.501). It is also permitted in Great Britain for use as a food-coloring agent. An ADI was set in 1966 at 0 to 7.5 mg/kg by the FDA. Although many synthetic coloring agents are toxic if used in large enough amounts, and many are suspected carcinogens as well, natural coloring agents are

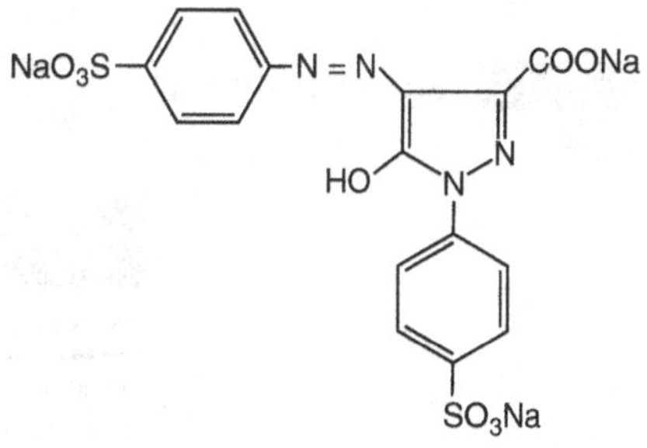

FIGURE 10.12 *Structure of tartrazine.*

not always safe. Caramel, which gives a light brown color, contains carcinogenic benzo[a]pyrene in small amounts; curcumin, which gives the yellow color to curry, is 15 times more toxic than tartrazine.

FLAVORING AGENTS

For many years, natural or synthetic flavor and fragrance chemicals have been used to enhance the palatability, and thus, increase the acceptability of foods. They also are used to produce imitation flavors for various food products including ice cream, jam, soft drinks, and cookies. Since the mid-nineteenth century, numerous flavor chemicals have been synthesized. Coumarin was synthesized in 1968; vanilla flavor, vanillin, was synthesized in 1874; cinnamon flavor was made in 1884 (cinnamic aldehyde). By the twentieth century, nearly 1000 flavoring chemicals had been developed. At the present time, over 3000 synthetic chemicals are used as flavor ingredients.

As was the case with coloring agents, the toxicity of flavoring agents began to receive attention in the 1960s. Most natural flavors used in the United States are generally recognized as safe (GRAS) on the basis of their long time occurrence in foods and wide consumption with no apparent ill effects. These chemicals have been used in large quantities in most food products with little regard for safety. Food industries even attempt to produce flavor ingredient chemicals that are found naturally in plants (so-called natural identical substances). Since general attention to food safety has focused on incidental additives, such as pesticides, it has been assumed that the natural flavoring chemicals are not health hazards. However, there are many toxic chemicals in natural products and their chronic toxicity should be carefully reviewed.

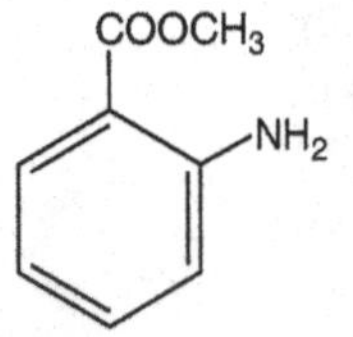

$COOCH_3$
NH_2

FIGURE 10.13 *Structure of methyl anthranilate.*

Methyl Anthranilate

Methyl anthranilate (Figure 10.13) is a colorless liquid that has a sweet, fruity, grape-like flavor. It is found in the essential oils of orange, lemon, and jasmine and has been widely used to create imitation Concord grape flavor. Table 10.8 shows the acute toxicity of methyl anthranilate. Methyl

Table 10.8 Acute Toxicity of Methyl Anthranilate

Animal	LD_{50} (oral, mg/kg)
Mouse	3,900
Rat	2,910
Guinea pig	2,780

anthranilate promotes some allergic reactions on human skin, which has led to it being prohibited for use in cosmetic products.

Safrole (1-Allyl-3,4-Methylenedioxybenzene)

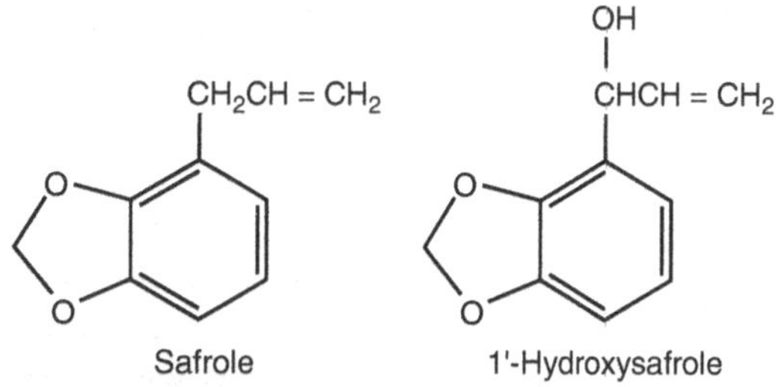

Safrole is a colorless oily liquid possessing a sweet, warm-spicy flavor. It has been used as a flavoring agent for more than 60 years. Oil of sassafras, which contains 80% safrole, also has been used as a spice. In the United States, the FDA banned the use of safrole in 1958 and many other countries followed this lead and also banned the use of safrole in flavors. Safrole, either naturally occurring in sassafras oil or the synthetic chemical, has been shown to induce liver tumors in rats. The continuous administration of safrole at 5,000 ppm in the total diet of rats caused liver tumors. Studies in dogs showed extensive liver damage at 80 and 40 mg/kg, lesser damage at lower levels, but no tumors. In vivo, safrole metabolites into 1′-hydroxysafrole. The structures of safrole and 1′-hydroxysafrole are shown in Figure 10.14.

$CH_2CH=CH_2$ OH $CHCH=CH_2$ O O O O

Safrole 1'-Hydroxysafrole

FIGURE 10.14 *Structure of safrole and its metabolite, 1′-hydroxysafrole.*

Diacetyl (2,3-butane dione)

Diacetyl is an intensely yellowish or greenish-yellow mobile liquid. It has a very powerful and diffusive, pungent, buttery odor and typically used in flavor compositions, including butter, milk, cream, and cheese. Diacetyl was found to be mutagenic in Ames test conducted under various different conditions with *Salmonella typhimurium* strains. For example, diacetyl was mutagenic by TA100 in the absence of S9 metabolic activation at doses up to 40 mM/plate. It was mutagenic in a modified Ames assay in *Salmonella typhimurium* strains TA100 with and without S9 activation.

The acute oral LD_{50} of diacetyl in guinea pigs was calculated to be 990 mg/kg. The acute oral LD_{50} of diacetyl in male rats was calculated to be 3400 mg/kg, and in female rats, the LD_{50} was calculated to be 3000 mg/kg. When male and female rats were administered via gavage a daily dose of 1, 30, 90, or 540 mg/kg/day of diacetyl in water for 90 days, the high-dose produced anemia, decreased weight gain, increased water consumption, increased leukocyte count, and an increase in the relative weights of liver, kidneys, and adrenal and pituitary glands. The data for teratogenicity and carcinogenicity are not available. Although the FDA has affirmed diacetyl GRAS as a flavoring agent, low molecular weight carbonyls, such as formaldehyde, acetaldehyde, and glyoxal have been reported to possess a certain chronic toxicity. Therefore, diacetyl should be reviewed carefully for its toxicity.

FLAVOR ENHANCERS

A small number of food additives are used to modify the taste of natural and synthetic flavors even though they do not directly contribute to flavor; such substances are known as flavor cnhancers. Most people are familiar with the use of table salt to enhance the flavor of a wide variety of foods. Salt can be an effective enhancer even at levels far below the threshold for salty taste and is widely used in processed foods such as canned vegetables and soups.

Another well-known enhancer is monosodium glutamate (MSG). Its use as an enhancer has periodically aroused concern about the potential toxicity of a quantity of free glutamate ingested at once. MSG became controversial because of its association with the so-called Chinese restaurant syndrome. Symptoms of the syndrome, which is usually self-diagnosed, include headache and drowsiness. Because of these concerns, both the acute and chronic toxicity of MSG have been widely studied. After a review of the available data, the FDA affirmed the GRAS status of MSG in the United States. Although some individuals are susceptible to transient discomfort following ingestion of MSG, it does not pose any risk of lasting injury.

SUGGESTIONS FOR FURTHER READING

Ayres, J.C., Kirschman, J.C. (Ed.) (1981). "Impact of Toxicology on Food Processing." AVI Pub. Co., Westport, Connecticut.

Cosmetic Ingredient Review Expert Panel. (2007). Final report on the safety assessment of HC Red No. 7. *Int. J. Toxicol*. 27:45-54.

Matkowski, A. (2008). Plant in vitro culture for the production of antioxidants—A review. *Biotechnol. Adv*. 2008, Jul 16. [Epub ahead of print]

Whitehouse, C.R., Boullata, J., McCauley, L.A. (2008). The potential toxicity of artificial sweeteners. *AAOHN J*. 56:251-259.

CHAPTER 11

Toxicants Formed During Food Processing

CHAPTER CONTENTS

Food processing practices have been used since the Ancient Era. Roasting and drying were applied to eggs and honey in the Old Stone Age (15,000 BCE). Smoking techniques were used to preserve milk and wine in the New Stone Age (9,000 BCE). Fermentation of rice and onions was recorded in the Bronze Age (35,000 BCE). Flavoring food started in the Iron Age (1,500 BCE) and adulteration methods are observed in the Roman era (~600 BCE). And, in the Modern Era, the development of food processing technology—which includes evaporation, smoking, sterilization, pasteurization, irradiation, pickling, freezing, and canning—greatly expanded the longevity of food storage. For example, smoke treatment made a year-round supply of fish possible; and canned foods could be sent anywhere in the world. Another important method of food processing is home cooking. Cooking techniques such as frying, toasting, roasting, baking, broiling, steaming, and boiling increases the palatability—flavor, appearance, and texture. Cooking also improves the stability as well as digestibility of foods. Moreover, it kills toxic microorganisms and deactivates such toxic substances as enzyme inhibitors. Since antiquity, people appreciated home-cooked food.

In the United States, commercial food processing is subject to regulation by the FDA and must meet specified standards of cleanliness and safety. Some particular methods of food processing are considered under the category of food additives, since they may intentionally alter the form or nature of food. Chemical changes in food components, including amino acids, proteins, sugars, carbohydrates, vitamins, and lipids, caused by high-heat treatment have raised questions about the consequence of reducing nutritive values and even the formation of some toxic chemicals such as polycyclic aromatic hydrocarbons (PAHs), amino acid or protein pyrolysates, and *N*-nitrosamines. Among the many reactions that occur in processed foods, the Maillard Reaction plays the most important role in the formation of various chemicals, including toxic ones.

During processing, undesirable foreign materials may accidentally be mixed into foods. Although most modern food factories are engineered to avoid any occurrence of food contamination during processing, low-level contamination is hard to eliminate entirely. Many instances of accidental contamination by toxic materials have been reported. In 1955 in Japan, sodium phosphate, a neutralizing agent, was contaminated with sodium arsenite and added to milk during a drying process. The final commercial dried milk contained 10 to 50 ppm of arsenic. Subsequently, many serious cases of arsenic poisoning were reported.

It is a common misunderstanding that gamma irradiation, which is most often used for food irradiation, produces radioactive materials in foods. In fact, although the electromagnetic energy used for irradiation is sufficient to penetrate deep into foods and can kill a wide range of microorganisms, it is far below the range required to produce radioactivity in the target material. However, there are still uncertainties about the toxicity of chemicals that may be produced during irradiation. The energies used are sufficient to produce free radicals, which may in turn produce toxic chemicals.

POLYCYCLIC AROMATIC HYDROCARBONS (PAHs)

Polycyclic aromatic hydrocarbons (PAHs) occur widely in the environment and in everyday products such as water, soil, dust, cigarette smoke, rubber tires, gasoline, roasted coffee, baking bread, charred meat, and in many other foods. Typical PAHs are shown in Figure 11.1. For over 200 years, carcinogenic effects have been ascribed to PAHs. In 1775, Percival Pott, an English physician, made the association between the high incidence of

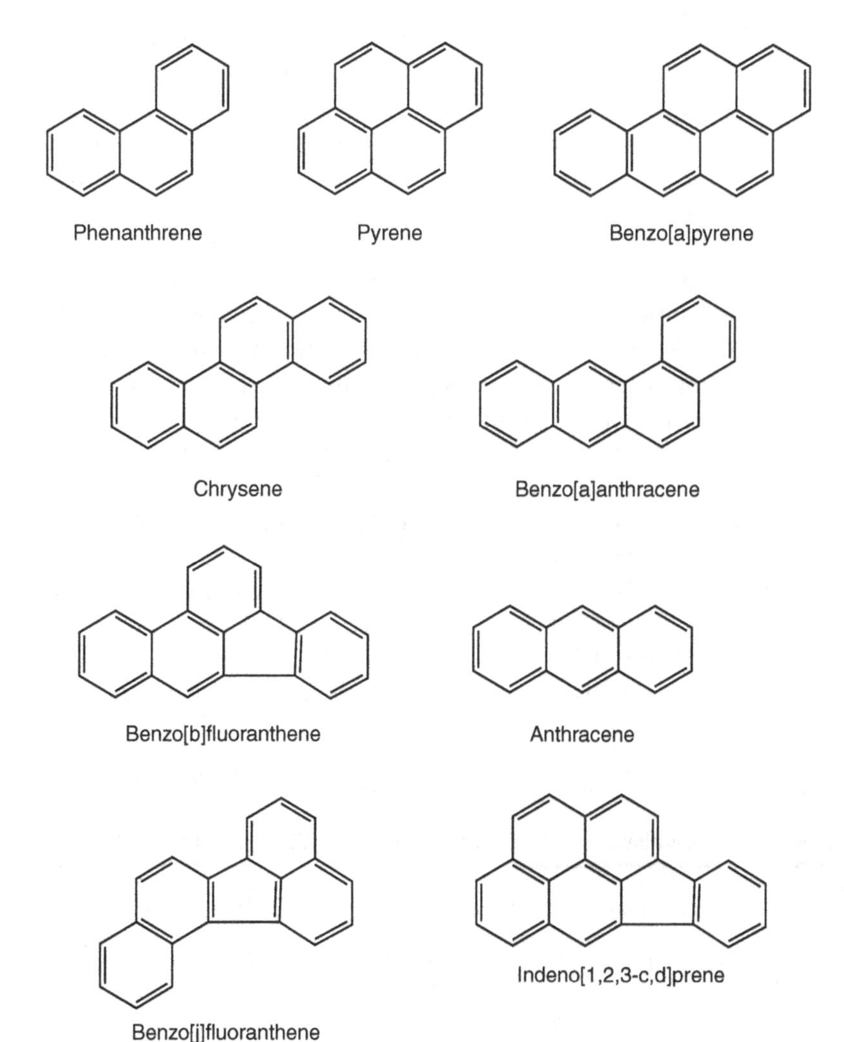

FIGURE 11.1
Structures of typical PAHs.

scrotal cancer in chimney sweeps and their continual contact with chimney soot. Research on the toxicity of PAHs, however, progressed somewhat slowly. In 1932, benzo[a]pyrene was isolated from coal tar and found to be highly carcinogenic in experimental animals.

Occurrence

One of the most abundant food sources of PAHs is vegetable oil. However, it is possible that the high levels of PAHs occurring in vegetable oils are due to endogenous production, with environmental contamination playing only a minor role. Some PAHs in vegetables are apparently due to environmental contamination because PAH levels decrease in vegetables cultivated farther from industrialized centers and freeways. The occurrence of PAHs in margarine and mayonnaise appears to be due to contamination of the oils used to make these products. Levels of PAHs in soil can also be quite high, even in areas distant from industrialized centers. Levels of PAHs of 100 to 200 ppm in the soil were found in some locations distant from human populations. It is thought that these levels result primarily as a residue from decaying vegetation. The significance of these relatively high levels of potentially carcinogenic substances in the soil is not fully understood.

Charcoal broiling or smoking of food also causes PAH contamination (Table 11.1). PAHs are formed mainly from carbohydrates cooked at high temperatures in the absence of oxygen. Broiling meat over hot ceramic or charcoal briquettes allows the melted fat to come into contact with a very hot surface. PAHs are produced in the ensuing reactions. These products rise with a resulting cooking fumes and are deposited on the meat. Similarly, the presence of PAHs in smoked meats is due to the presence of these substances in smoke. PAH levels in meat that is cooked at a greater distance from the coals are lower than in meat that is cooked close to the coals. Obviously, food processing produces certain levels of PAHs. It is of major

Table 11.1 Polycyclic Aromatic Hydrocarbons Found in Smoked Foods (ppb)

Food	Benzo[a]anthracene	Benzo[a]pyrene	Benzo[e]pyrene	Fluoranthene	Pyrene
Beef	0.4	—	—	0.6	0.5
Cheese				2.8	2.6
Herring				3.0	2.2
Dried herring	1.7	1.0	1.2	1.8	1.8
Salmon	0.5	—	0.4	3.2	2.0
Sturgeon	—	0.8	—	2.4	4.4
Frankfurters	—	—	—	6.4	3.8
Ham	2.8	3.2	1.2	14.0	11.2

importance to be aware of the presence of carcinogenic PAHs in our foods, and the overall public health hazard should be evaluated and controlled.

Benzo[a]pyrene (BP)

The most commonly known carcinogenic PAH is benzo[a]pyrene (BP), which is widely distributed in various foods (Table 11.2).

BP was reportedly formed at a level of 0.7 and 17 ppb at 370 to 390 and 650 °C, respectively, when starch was heated. Amino acids and fatty acids also produced BP upon high-temperature treatment (Table 11.3). Many cooking processes utilize the 370 to 390 °C range; for example, the surface temperature of baking bread may approach 400 °C and deep fat frying is 400 to 600 °C, suggesting that cooking produces some PAHs, including BP. The meat inspection division of the USDA and FDA analyzed 60 assorted foodstuffs and related materials for BP. Samples that contain relatively high levels of BP are shown in Table 11.1.

Table 11.2 Benzo[a]pyrene Found in Various Foods

Food	Concentration (ppb)
Fresh vegetables	2.85–24.5
Vegetable oil	0.41–1.4
Coffee	0.31–1.3
Tea	3.9
Cooked sausage	12.5–18.8
Smoked hot sausage	0.8
Smoked turkey fat	1.2
Charcoal-broiled steak	0.8
Barbecued ribs	10.5

Table 11.3 Amounts of PAHs Produced from Carbohydrates, Amino Acids, and Fatty Acids Heated at 500 and 700°C (μg/50 g)

	Starch		D-Glucose		L-Leucine		Stearic Acid	
PAH	500	700	500	700	500	700	500	700
Pyrene	41	965	23	1,680	—	1,200	0.7	18,700
Fluoranthene	13	790	19	1,200	—	320	—	6,590
Benzo[a]pyrene	7	179	6	345	—	58	—	4,440

Toxicity

BP is a reasonably potent contact carcinogen, and therefore has been subjected to extensive carcinogenic testing. Table 11.4 shows the relative carcinogenicity of BP and other PAHs. A diet containing 25 ppm of BP fed to mice for 140 days produced leukemia and lung adenomas in addition to stomach tumors. Skin tumors developed in over 60% of the rats treated topically with approximately 10 mg of benzo[a]pyrene three times per week. The incidence of skin tumors dropped to about 20% when treatment was about 3 mg × 3 per week. Above the 10 mg range, however, the incidence of skin tumors increased dramatically to nearly 100%.

BP is also carcinogenic when administered orally. In one experiment, weekly doses of greater than 10 mg administered for 10 weeks induced stomach cancers, although no stomach cancers were produced at the dose of 10 mg or less. At 100 mg doses, nearly 79% of the animals had developed stomach tumors by the completion of the experiment.

When 15 ppm of BP in feed was orally administered to mice, production of leukemia, lung adenomas, and stomach tumors were observed after 140 days. Table 11.5 shows the stomach tumor incidence (%) occurred in mice by BP.

Table 11.4 Relative Carcinogenicity of Typical Polycyclic Aromatic Hydrocarbons (PAH)

PAH	Relative Activity
Benzo[a]pyrene	+++
5-Methylchrysene	+++
Dibenzo[a,h]anthracene	++
Dibenzo[a,i]pyrene	++
Benzo[b]fluoranthene	++
Benzo[a]anthracene	+
Benzo[c]phenanthrene	+
Chrysene	+

+++, high; ++, moderate; +, weak.

Table 11.5 Stomach Tumor Incidence Caused by BP in Mice

Dose (oral, ppm)	Duration (days)	Incidence (%)
30	110	0
40–45	110	10
50–250	122–197	70
250	1	0
250	2–4	10
250	5–7	30–40
250	30	100

Mode of Toxic Action

BP is transported across the placenta and produces tumors in the offspring of animals treated during pregnancy. Skin and lung tumors appear to be the primary lesions in the offspring.

The biochemical mechanisms by which benzo[a]pyrene initiates cancer have been studied in some detail. Benzo[a]pyrene is not mutagenic and carcinogenic by itself, but first must be converted to active metabolites. The metabolic conversion initially involves a cytochrome P450-mediated oxidation, producing a 7,8-epoxide. The 7,8-epoxide, in turn, undergoes an epoxide hydrolase-mediated hydration, producing the 7,8-diol, which, upon further oxidation by cytochrome P450, produces the corresponding diolepoxide. This diolepoxide is highly mutagenic without metabolic activation and is also highly carcinogenic at the site of administration. The benzo[a]pyrene diolepoxide can react with various components in the cells, in which case it is possible that mutation will occur. This is thought to be the primary event in benzo[a]pyrene-induced carcinogenesis. The hypothesized chemical mechanisms of BP toxicity are shown in Figure 11.2.

The criterial factors for PAHs to be carcinogenic are (1) the entire polycyclic aromatic hydrocarbon must be coplanar; (2) a phenanthrene nucleus must be present together with some substituents, preferably at least one

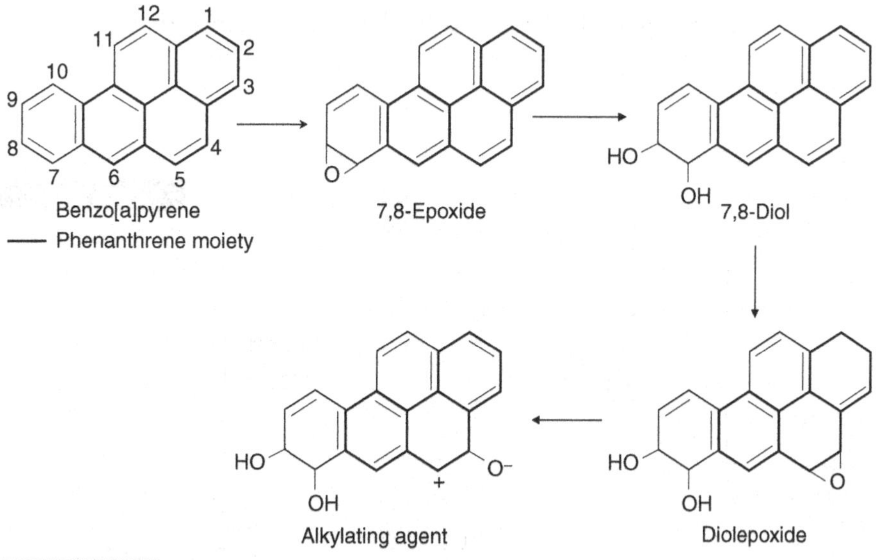

FIGURE 11.2 *Hypothesized chemical mechanisms of alkylating agent from benzo[a]pyrene.*

additional benzene ring; and (3) the covex edge (C_4 – C_5) of the phenanthrene moiety must be free of substituents.

MAILLARD REACTION PRODUCTS

In 1912, the French chemist L. C. Maillard hypothesized the reaction that accounts for the brown pigments and polymers produced from the reaction of the amino group of an amino acid and the carbonyl group of a sugar. Maillard also proposed that the reaction between amines and carbonyls was implicated in in vivo damage; in fact, the Maillard reaction later was proved to initiate certain types of damage in biological systems.

The Maillard reaction is also called nonenzymatic browning reaction. It occurs readily in any matrix containing amines and carbonyls under heat treatment. A summary of the Maillard reaction is shown in Figure 11.3.

Many chemicals formed from this reaction in addition to the brown pigments and polymers. Because of the large variety of constituents, a mixture obtained from a Maillard reaction shows many different chemical and biological properties: brown color, characteristic roasted or smoky odors, pro- and antioxidants, and mutagens and carcinogens, or perhaps antimutagens and anticarcinogens. It is common practice to use the so-called Maillard browning model system consisting of a single sugar and an amino acid to investigate complex, actual food systems. The results of much mutagenicity testing on the products of Maillard browning model systems have been reported. Some Maillard model systems that produced mutagenic materials are shown in Table 11.6.

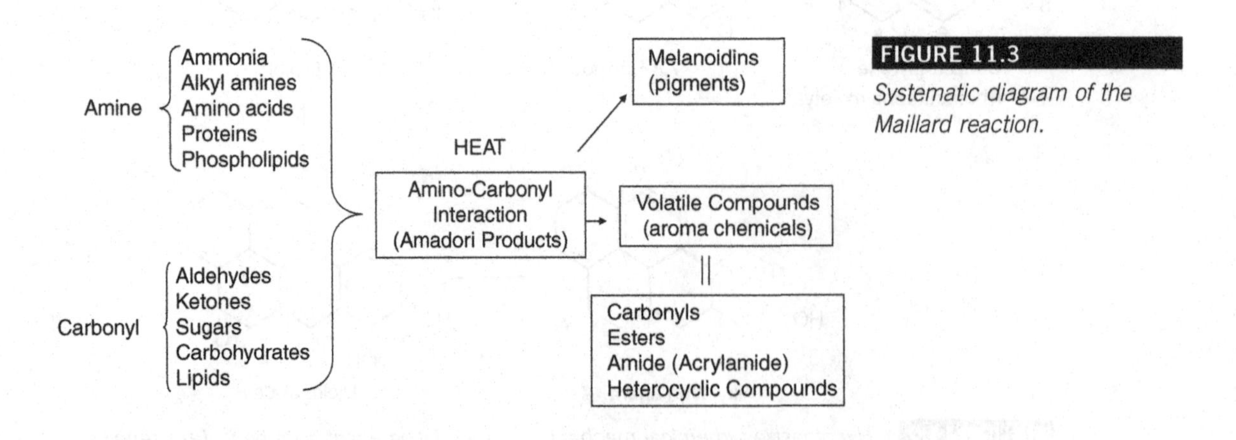

FIGURE 11.3
Systematic diagram of the Maillard reaction.

Table 11.6 Mutagenic Materials Produced from the Maillard Model System

Model System	*Salmonella typhimurium* Strains
D-Glucose/cysteamine	TA 100 without S9[a]
	TA 98 with S9
Cyclotene/NH_3	TA 98 without S9
	TA 1538 without S9
L-rhamnose/NH_3/H_2S	TA 98 with S9
Maltole/NH_3	TA 98 with S9
	TA 100 with S9
Starch/glycine	TA 98 with S9
Lactose/casein	TA 98 with S9
Potato starch/$(NH_4)_3CO_3$	TA 98 with S9
	TA 100 with S9
Diacetyl/NH_3	TA 98 with S9
	TA 100 with S9

[a]*Metabolic activation.*

POLYCYCLIC AROMATIC AMINES (PAA)

Occurrence

In the late 1970s, mutagenicity of pyrolysates obtained from various foods was reported. However, PAHs that could not be accounted for were those formed on the charred surface of certain foods such as broiled fish and beef. Table 11.7 shows the mutagenic activity of pyrolysates obtained from some foods.

Some classes of cooked protein-rich foods, such as beef, chicken, and dried seafood, tended to be more mutagenic than others and the extent of heating influenced the level of mutagenic activity. The most highly heated

Table 11.7 Mutagenicity of Pyrolysates Obtained from Select Foods Heated at Various Temperatures

	Revertants of TA98 + S-9 Mix/0.1 g Sample		
Food	**250°C**	**300°C**	**400°C**
Beef		178	11,400
Chicken		661	15,120
Egg, whole		121	4,750
Hairtail (or beltfish), raw		849	12,320
Eel, raw		309	6,540
Squid, dried	269	8,000	4,490
Skipjack tuna, dried	1,220	24,300	6,200
Sea weed, Nori		260	3,040

samples of milk, cheese, grains, and several varieties of beans, although heavily charred, were only weakly mutagenic or not mutagenic. Hamburger cooked at high temperatures was reported as mutagenic. The mutagenicity, however, was limited to the surface layers where most pyrolysates are found. On the other hand, no mutagenic activity was found in comparable samples of uncooked hamburger meat. The formation of these mutagenic substances seems to depend on temperature. For example, in heated beef stock, temperature dependent mutagen formation has been determined quantitatively.

The mutagenic principles of the tryptophan pyrolysates were later identified as nitrogen-containing heterocyclic compounds. A group of polycyclic aromatic amines is produced primarily during the cooking of protein-rich foods. Their structures are shown in Figure 11.4.

PAA are divided into two groups. Heating a mixture of creatine/creatinine, amino acids, and sugars produces the imidazoquinoline (IQ)-type. The IQ-type possesses an imidazole ring (i.e., a ring that is formed from creatine). The IQ-type PAAs are IQ, MeIQ, IQx (structure not shown), MeIQx, 4,8-DiMeIQx (structure not shown), and PhIP. The formation of IQ-type PAA is a result of heat treatment that causes cyclization of creatinine to form the imidazole moiety. The remaining moieties of the structure arise from pyridines and pyrazines formed through the Strecker degradation of amino acids and breakdown products of β-dicarbonyl products formed by the Maillard reaction. The other category of PAAs, the so-called non-IQ type, is composed of pyrolysis products formed from tryptophan: Trp-P-1, Trp-P-2, Glu-P-1, Glu-P-2, AαC, and MeAαC.

Toxicity

The acute toxicities of Trp-P-1 and Trp-P-2 are toxic to animals: LD_{50} (intragastric intubation) of Trp-P-1 are 200 mg/kg in mice, 380 mg/kg in Syrian golden hamsters, and 100 mg/kg in rats. Trp-P-2 is slightly more toxic than Trp-P-1. Animals that received over the LD_{50} usually died in convulsions within an hour. Trp-P-1 and Trp-P-2 induce a local inflammatory reaction when injected subcutaneously.

The early work on the isolation and production of these substances was based on their mutagenicity. They are also minor components of fried beef. Several other mutagens of this class are also present in cooked meat. Beef extracts, which contain IQ and MeIQx, are metabolically converted to active mutagens by liver tissue from several animal species and humans. Although these substances are highly potent mutagens, they are fairly weak carcinogens in rats. Following the mutagenicity studies on these pyrolysates,

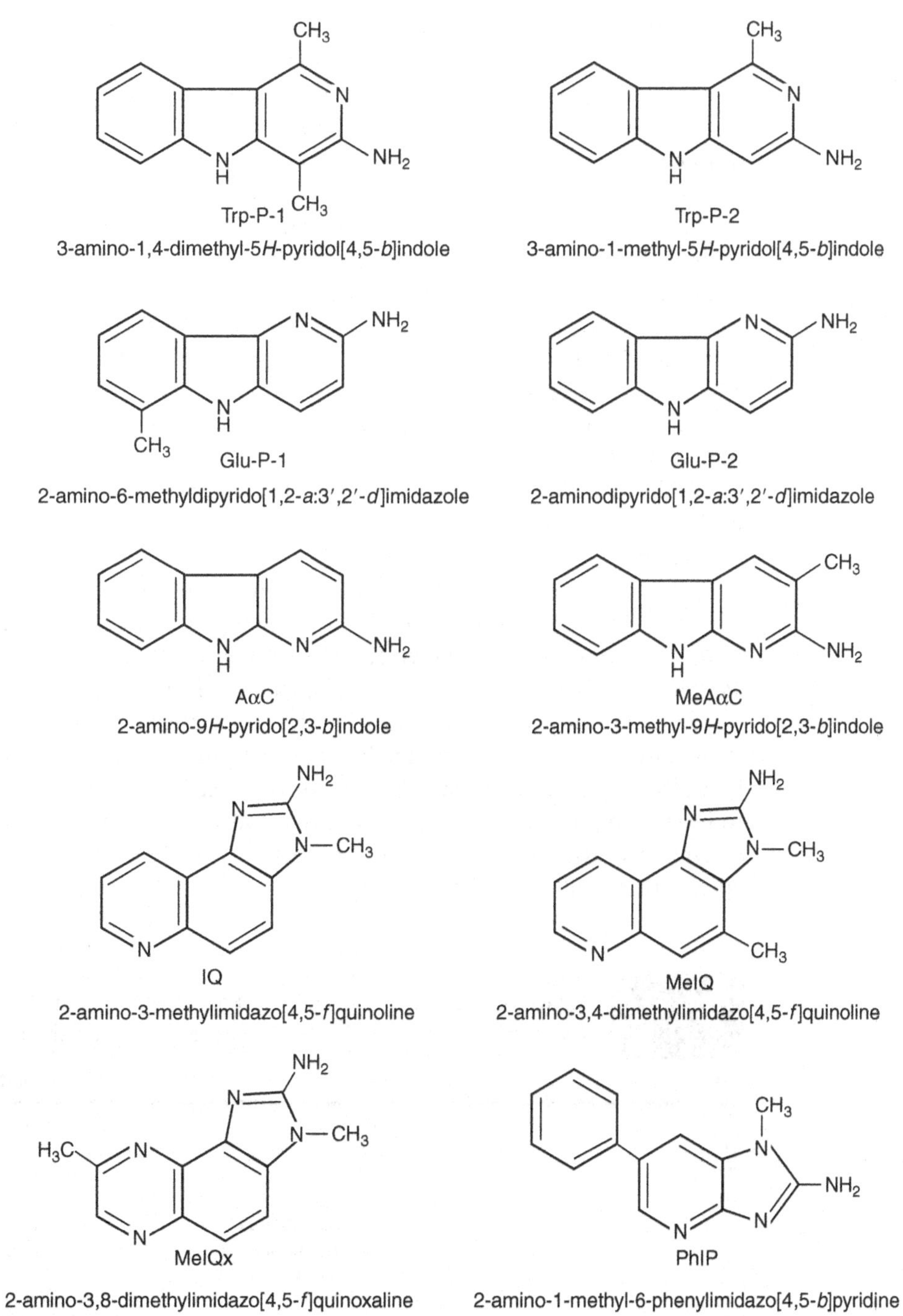

Trp-P-1
3-amino-1,4-dimethyl-5*H*-pyridol[4,5-*b*]indole

Trp-P-2
3-amino-1-methyl-5*H*-pyridol[4,5-*b*]indole

Glu-P-1
2-amino-6-methyldipyrido[1,2-*a*:3′,2′-*d*]imidazole

Glu-P-2
2-aminodipyrido[1,2-*a*:3′,2′-*d*]imidazole

AαC
2-amino-9*H*-pyrido[2,3-*b*]indole

MeAαC
2-amino-3-methyl-9*H*-pyrido[2,3-*b*]indole

IQ
2-amino-3-methylimidazo[4,5-*f*]quinoline

MeIQ
2-amino-3,4-dimethylimidazo[4,5-*f*]quinoline

MeIQx
2-amino-3,8-dimethylimidazo[4,5-*f*]quinoxaline

PhIP
2-amino-1-methyl-6-phenylimidazo[4,5-*b*]pyridine

FIGURE 11.4 *Structures of typical PAAs.*

the carcinogenicity of tryptophan (Trp-P-1 and Trp-P-2) and glutamine (Glu-P-1) was demonstrated using animals such as rats, hamsters, and mice. For example, a high percentage of tumor incidences were observed in mice fed a diet containing Trp-P-1 or Trp-P-2. High incidences of hepatoma were found in mice treated with Trp-P-1, Trp-P-2, Glu-P-1, and Glu-P-2. Female mice were more sensitive than males.

The various reports indicate that both amino acid and protein pyrolysates may act as carcinogens in the alimentary tracts of experimental animals. Extensive research is presently being conducted to determine whether PAAs produced during the cooking process are hazardous to humans. The identities of the mutagens produced under normal cooking conditions have been established in some cases. For example, the major mutagens in broiled fish are imidazoquinoline (IQ) and methylimidazoquinoline (MeIQx) (Figure 11.4).

Table 11.8 shows the mutagenicity of a typical PAA along with that of well-known carcinogens in *S. typhimurium* TA98 with S9 microsomal activation.

Some PAAs, such as IQ and MeIQ, exhibit strong mutagenic activity. Although Trp-P-1, Trp-P-2, and Glu-P-1 are highly mutagenic—being more mutagenic than the well-known carcinogen aflatoxin B_1, they are much less carcinogenic than aflatoxin B_1. This may be the result of their high initiating activities but low promoting activities. Carcinogenicity of these potent PAA mutagens is accounted for by hypothesizing that PAA can be metabolically activated by humans both through *N*-oxidation and *O*-acetylation, to produce highly reactive metabolites that form DNA adducts. The human enzyme activities for some substrates are comparable to those of the rat, a species that readily develops tumors when fed these PAA as part of the daily

Table 11.8 Mutagenic Activity of Typical PAA Found in Foods in *Salmonella typhimurium* Strain TA98 with S9 Microsomal Activation

PAA	Found in	Revertants/μg
MeIQ	Broiled sardine	661,000
IQ	Fried beef	433,000
MeIQx	Fried beef	145,000
Trp-P-1	Broiled sardine	104,000
Glu-P-1	Glutamic acid pyrolysate	49,000
Trp-P-1	Broiled sardine, broiled beef	39,000
Glu-P-2	Broiled, dried cuttlefish	1,900
Aflatoxin B_1	Corn	6,000
Benzo[a]pyrene	Broiled beef	320

diet. Therefore, these PAAs should be regarded as potential human carcinogens.

N-NITROSAMINES

Mixtures of inorganic salts, such as sodium chloride and sodium nitrite, have been used to cure meat and certain fish products for centuries. Some countries, not including the United States, also permit the addition of nitrate in the production of some varieties of cheese.

The nitrite ion plays at least three important roles in the curing of meat. First, it has an antimicrobial action. In particular, it inhibits the growth of the microorganisms that produce botulism toxin, *Clostridium botulinum*. It recently has been recognized that the bacterial reduction or curing action results from the nitrite ion. However, the mechanism and cofactors of this antimicrobial action are not clearly understood.

Nitrite also imparts an appealing red color to meats during curing. This arises from nitrosomyoglobin and nitrosylhemoglobin pigments. These pigments are formed when nitrite is reduced to nitric oxide, which then reacts with myoglobin and hemoglobin. If these pigments did not form, cured meats would have an unappetizing grayish color.

Finally, nitrite gives a desirable "cured" flavor to bacon, frankfurters, ham, and other meat products. In general, since cured meats often are stored under anaerobic conditions for extended periods, curing is very important in ensuring the safety of these foods.

Precursors

Nitrosation of secondary and tertiary amines produces stable nitrosamines. Unstable nitroso-compounds are produced with primary amines. The reaction rate is pH-dependent and peaks near pH 3.4. The nitrosation of weakly basic amines is more rapid than that of more strongly basic amines. Several anions, such as halogens and thiocyanate, promote the nitrosation process; on the other hand, antioxidants, such as ascorbate and vitamin E, inhibit the reaction of destroying nitrite. Diethylnitrosamine (DEN) and dimethylnitrosamine (DMN) occur in the gastric juice of experimental animals and humans fed diets containing amines and nitrite. The nitrosation reaction is also known to occur during the high-temperature heating of foods, such as bacon, which contains nitrite and certain amines. Figure 11.5 shows typical structures of N-nitrosamines.

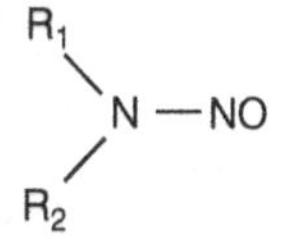

R_1
N—NO
R_2

e.g., $R_1 = CH_3$, $R_2 = CH_3$: Dimethylnitrosamine (DMN)

FIGURE 11.5 *Structure of dimethylnitrosamine.*

Occurrence in Various Foods

Nitrates are found in a wide variety of foods, both cured and uncured. In cured foods, nitrates vary in levels ranging from 10 to 200 ppm, depending on the country. Cured meats all have been shown to contain nitrosamines (Table 11.9), the higher levels appearing in cured meats that have been subjected to relatively high heating. It is important to note that the levels of nitrosamines detected in various foods are quite variable. The reasons for this variability are not clear but seem to be dependent on the type of food and conditions in the laboratory conducting the examination.

The levels of volatile nitrosamines in spice premixes, such as those used in sausage preparation, were found to be extraordinarily high. Premixes contained spices with secondary amines and curing mixtures included nitrite. Volatile nitrosamines formed spontaneously in these premixes during long periods of storage. The problem was solved simply by combining the spices and the curing mixture just prior to use.

In vegetables, nitrate is encountered often in relatively high levels (1000–3000 ppm). In produce such as cabbage, cauliflower, carrots, celery, and spinach, the nitrate levels are variable and the exact causes are uncertain. The dietary intake of nitrates and nitites for adult Americans has been estimated at 100 mg per day. Vegetables, especially leaf and root vegetables, account for over 85% of the total, and cured meats contribute about 9%. In certain areas, well water contains high levels of nitrate. Although exposure from meat products may have dropped in recent years, the use of nitrate in fertilizers means that vegetables continue to be significant sources of nitrates.

However, significant amounts of reduced nitrites are not found in most foods. An additional source of nitrites in humans results from the intestinal tract. Most ingested nitrite comes from saliva, which is estimated to contribute 8.6 mg of the total daily intake of 11.2 mg from the diet.

Table 11.9 Nitrosamine Content in Typical Cured Meats

Meat	Nitrosamine	Level (ppb)
Smoked sausage	Dimethylnitrosamine	< 6
	Diethylnitrosamine	< 6
Frankfurters	Dimethylnitrosamine	11–84
Salami	Dimethylnitrosamine	1–4
Fried bacon	Dimethylnitrosamine	1–40
	Nitrosoproline	1–40

Analyses of certain beers have also shown considerable variability in levels of nitrosamines. Although the mean concentration of volatile nitrosamines in both American and imported beer is generally quite low, the levels in certain samples can be as high as 70 ppb of DMN. It was soon found that beers produced from malt dried by direct fire rather than by air-drying had the highest levels of nitrosamine. The direct fire-drying process was shown to introduce nitrite into the malting mixture. Domestic beer manufacturers quickly converted to the air-drying process.

Heating of other nitrite-treated food products, such as animal feed, also has been shown to produce nitrosamines. In Norway in 1962, following an epidemic of food poisoning in sheep, extremely high levels of nitrosamines were detected in herring meal treated with nitrite as a preservative. The sheep suffered severe liver disease and many of them died. It was later shown that the rate of spontaneous formation of nitrosamine in nitrite-treated fish was dependent on the temperature of preparation following the addition of nitrite. Thus, refrigerated fish treated with nitrite had no more nitrosamine than fresh fish treated with nitrite, but heat treatment of fish increased the rate of nitrosamine formation following addition of nitrite. It was suggested that increased levels of nitrosamines in heated fish are due—at least in part—to increased concentrations of secondary amines resulting from protein degradation during the heating process.

Toxicity

The carcinogenic activity of many nitrosamines has been examined. In over 100 food substances assayed so far, approximately 80% were shown to be carcinogenic in at least one animal species. In fact, DEN is active in 20 animal species. As a result, DMN and DEN are two of the most potent carcinogens in this group. Administration of DMN at 50 ppm in the diet produces malignant liver tumors in rats in 26 to 40 weeks. Higher doses of DMN were shown to produce kidney tumors. In administration of DEN, when dosage is reduced below 0.5 mg/kg, a subsequent lag period between dosing and onset of tumors is shown to increase, with the total tumor yield remaining roughly the same. Alternatively, with a lower dose of 0.3 mg/kg, the lag time is 500 days, and finally, for a dose of 0.075 mg/kg, the lag time is increased to 830 days. No clear threshold dose for carcinogenicity of nitrosamines in the diet has been established yet.

Mode of Toxic Action

Nitrosamines, like other groups of chemical carcinogens, require metabolic activation to render them toxic. The activation process is mediated by

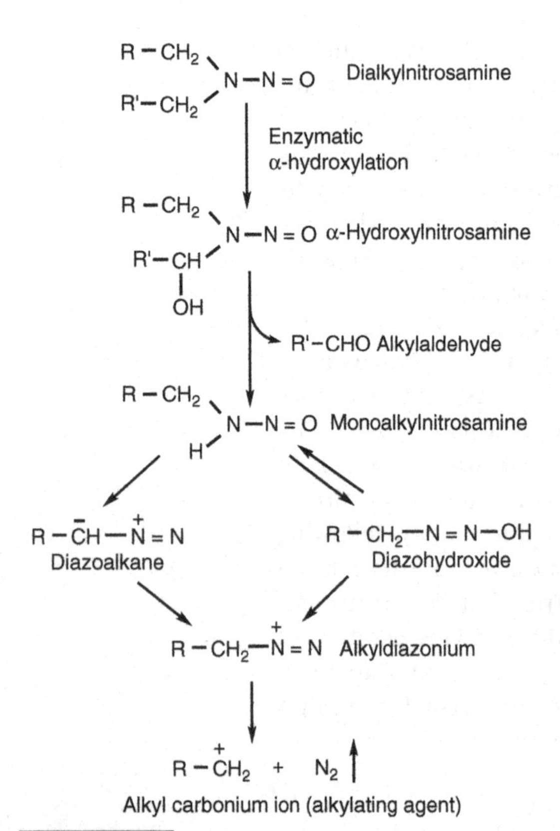

FIGURE 11.6 *Hypothesized formation mechanisms of an alkylating agent from nitrosamines.*

enzymes and involves, at least in some cases, hydroxylation of the α-carbon. Figure 11.6 shows the hypothesized formation mechanisms of an alkylating agent from nitrosamines.

The nitrosamines exhibit a good deal of organ specificity in their carcinogenic effect (Table 11.10). For example, DMN is an active liver carcinogen with some activity in the kidney, and benzylmethylnitrosamine is specific for the esophagus. This organ specificity is apparently due, at least in part, to site-specific metabolism.

Administration of certain nitrosamines to pregnant animals can result in cancer in the offspring. The time of administration seems to be critical. For example, in rats, administration of carcinogens must occur later than 10 days into gestation to produce cancer in the offspring, and the fetuses are most sensitive just prior to term. This development of sensitivity coincides with the development of the metabolic activation system of the fetuses. In addition, compared to the adults, the fetuses seem to be unusually sensitive to the carcinogenic effects of these substances. For example, at a maternal dose of only 2 mg/kg, which is 2% of the carcinogenic dose required for adults, N-nitrosoethylurea caused a carcinogenic response in the nervous system of offspring.

Under acidic pH, the nitrite ion can be protonated to form nitrous acid (HONO). The anhydrides of nitrous acid, N_2O_3, present in equilibrium with nitrous acid, can nitrosate a variety of compounds, especially secondary and tertiary amines. Halide and thiocyanate ions, present in foods and digestive fluids, can catalyze the formation of N-nitroso compounds.

General Considerations

Efforts to reduce nitrosamine formation in cured meats have been quite successful. Simply adding a reducing agent, such as erythsorbate or ascorbate, to the curing mix greatly reduced or eliminated nitrosamine formation in the final product. Now, domestic manufacturers of cured meat products generally add these reducing agents to the curing mixture along with the minimum amount of nitrite necessary to achieve the desired effect.

Table 11.10 Sites of Tumors Produced by N-Nitroso Compounds

Site	Compound
Skin	Methylnitrosourea
Nose	Diethylnitrosamine
Nasal sinus	Dimethylnitrosamine
Tongue	Nitrosohexamethyleneimine
Esophagus	Nitrosoheptamethyleneimine
Stomach	Ethylbutylnitrosamine
Duodenum	Methylnitrosourea
Colon	Cycasin
Lung	Diethylnitrosamine
Bronchi	Diethylnitrosamine
Liver	Dimethylnitrosamine
Pancreas	Nitrosomethylurethane
Kidney	Dimethylnitrosamine
Urinary bladder	Dibutylnitrosamine
Brain	Methylnitrosourea
Spinal cord	Nitrosotrimethylurea
Thymus	Nitrosobutylurea
Lymph nodes	Ethylnitrosourea
Blood vessels	Nitrosomorpholine

However, the nitrosamines found in foods are almost exclusively highly volatile. Very little is presently known about the concentrations of nonvolatile nitrosamines present in foods.

The risk to human health of dietary nitrite and nitrosamines is difficult to assess, especially since both catalysts and inhibitors of nitrosation may be present in a typical meal. As discussed in the preceding section, *in vivo* reduction of the ubiquitous nitrate ion to nitrite appears to be the major source of ingested nitrite, contributing more than three times the nitrite ingested with cured meats in the average American diet. In addition, there are significant nondietary sources of exposure to nitrosamines and nitrosatable compounds, including tobacco, some pharmaceuticals and cosmetics, and cutting oils used in industry. Isolating effects due to diet alone appears impossible. Nonetheless, it is prudent to minimize controllable exposures.

ACRYLAMIDE

Acrylamide ($CH_2 = CH\text{–}CONH_2$) has been an important industrial chemical; in particular, it has been used worldwide to synthesize polyacrylamide. Polyacrylamide has been used for various purposes, including removal of suspended solids from industrial wastewater, soil conditioner, grouting agents, surfactant for herbicide mixtures, a stationary phase for laboratory separations, and cosmetic formulations. Consequently, since acrylamide is formed from polyacrylamide upon degradation, it has been known to be present in drinking water for many years. Therefore, it may enter the food chain in the environment.

However, in April 2002, the discovery of significant levels of acrylamide in heat-processed starch-based foods, such as potato chips and French fries, triggered intensive studies confirming the presence and quantifying the amounts of acrylamide. Moreover, discovery of acrylamide in foods has prompted worldwide attention because it has been considered a probable human carcinogen, a neurotoxicant, and a genotoxicant. Table 11.11 shows amounts of acrylamide found in typical food items.

Formation Mechanisms of Acrylamide in Foods

Concerns of the hypothesized formation acrylamide in cooked foods, the Maillard browning reaction of a sugar and an amino acid have received much attention as the most likely mechanism. Significant amounts of acrylamide (221 mg/mol of amino acid) formation were reported in a browning model system consisting of an equimolar L-asparagine and glucose heated at 185 °C. When asparagine and glutamine, which have the same amide moiety at the end of their molecule, were heated alone at 180 °C for 30 min, 0.99 μg/g and 0.17 μg/g of acrylamide were formed, respectively. On the other hand, addition of glucose to asparagine increased acrylamide formation to 1200 μg/g. Also, acrylamide formation increased from 117 μg/g to 9270 μg/g with the addition of glucose in a system consisting

Table 11.11 Amounts of Acrylamide Found in Various Foods

Food	Amount (ppb)	No. of Items
Baby food	17–130	7
French fries	70–1036	17
Potato chips	117–2764	16
Protein foods	22–116	8
Breads and bakery products	10–354	10
Cereals	47–266	8
Snack foods	111–1168	6
Nuts and nut butters	28–457	8
Crackers	41–504	4
Chocolate products	45–909	7
Cookies	36–199	4
Coffee	275–351	5
Dried foods	11–1184	4

of asparagine, potato starch, and water. These results suggest that asparagine and carbonyl compounds—such as glucose, glyceraldehyde, and acrolein—play an important role in acrylamide formation in cooked foods. Figure 11.7 shows the hypothesized formation mechanism of acrylamide in foods.

The major route of acrylamide formation is from asparagine, which possesses an amide moiety. A carbonyl compound, such as glucose, seems to accelerate the acrylamide formation *via* Schiff's base formation as shown in scheme I. The other minor route of acrylamide formation is from acrolein produced from a lipid glyceride via a high temperature treatment as shown in scheme II. Acrolein is readily formed from glycerol produced from lipid glycerides. Once acrolein formed, it undergoes oxidation to form acrylic acid, which subsequently reacts with ammonia from an amino acid to yield acrylamide under high temperatures. Ammonia is formed from an amino acid upon Strecker degradation with the presence of a carbonyl compound. The radical form of acrolein may be formed by

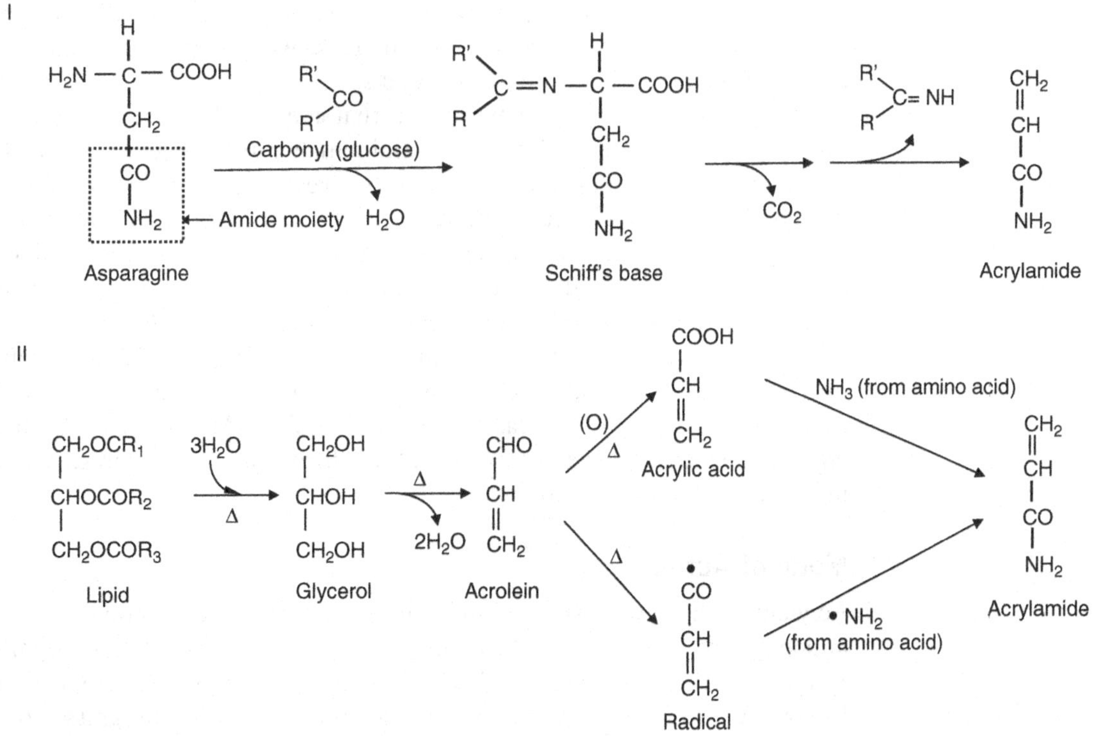

FIGURE 11.7 *Hypothesized formation mechanism of acrylamide in foods.*

a high temperature treatment. This radical then reacts with an amino radical, which also is formed from an amino acid under high temperature, to form acrylamide.

Toxicity

Until acrylamide was found in foods in 2002, human exposure to acrylamide was considered to come from dermal contact with solid monomers and inhalation of dust and vapor in the occupational setting. However, the major public concern about acrylamide exposure currently is centered on the ingestion of heat-treated starch-rich foods, such as potato chips and French fries.

The acute toxicity of acrylamide in oral LD_{50} in rats are in the range of 159 mg/kg to 300 mg/kg body weight and dermal LD_{50} in rabbits is 1680 μL/kg. As subchronic toxicity, repeated oral administration of acrylamide to rats at doses of 20 mg/kg bw/day and above produced peripheral neuropathy, atrophy of skeletal muscle, and decreased erythrocyte parameters. In monkeys, clinical signs of peripheral neuropathy occurred at doses of 10 mg/kg bw/day for up to 12 weeks. The genotoxicity study reported that acrylamide caused a high frequency of sister chromatid exchanges and breaks when it was fed to mice at a level of 500 ppm in the diet for three weeks.

Many laboratory studies have shown that acrylamide causes a variety of tumors in rats and mice. When male and female rats were given 3.0 mg/kg bw/day acrylamide in drinking water for two years, the incidence of tumors increased in the scrotum, adrenal, thyroid, mammary, oral cavity, and uterus. However, there is no definite evidence that acrylamide produces tumors in humans. Epidemiological studies with workers from acrylamide production plants did not result in conclusive evidence for human carcinogenicity. Acrylamide has been classified by the US EPA as a B2, a probable human carcinogen, by IARC (International Agency for Research on Cancer) as a 2B, a possible human carcinogen, and by ACGIH (American Conference of Industrial Hygienists) as an A3, confirmed animal carcinogen with unknown relevance to humans.

Mode of Action

Acrylamide has two reactive sites, the conjugated double bond and the amide group. The electrophilic double bond can participate in nucleophilic reactions with active-hydrogen-bearing functional groups both in vitro and in vivo. These include the SH of cysteine, homocysteine, and glutathione, α-NH_2 groups of free amino acids and N-terminal amino acid residues of proteins, the ε-NH_2 of lysine, and the ring NH group of histidine.

The major metabolite formed from acrylamide via the cytochrome P450 pathway is glycidamide in mice (Figure 11.8). Acrylamide may be conjugated by glutathione-S-transferase (GST) to N-acetyl-S-(3-oxopropyl)cysteine or it reacts with cytochrome P450 (CYY450) to produce glycidamide. Some metabolic studies on acrylamide show that liver, kidney, brain, and erythrocyte GST has significant binding capacity with acrylamide. Both acrylamide and glycidamide are electrophilic and can form adducts with sulfhydryl groups (–SH) on hemoglobin and other proteins. However, there are limited data regarding the potential for acrylamide to form DNA adducts. When isolated nucleosides were incubated with acrylamide *in vitro*, the adduct yield but the rate of formation was low.

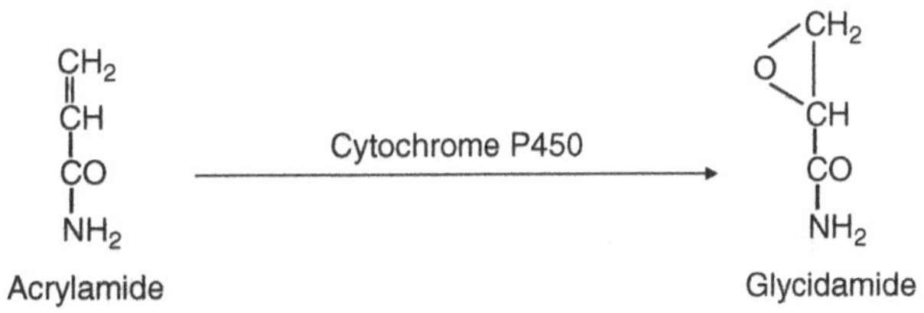

Acrylamide —Cytochrome P450→ Glycidamide

FIGURE 11.8 *Formation of glycidamide from acrylamide.*

General Considerations

Since the discovery of acrylamide in cooked foods, there have been numerous studies to reduce the amount of acrylamide formation. Some studies focused on how to reduce asparagine content in foods. Other studies have been searching for a cooking method to minimize the acrylamide formation. However, there has not yet been a satisfactory answer to this problem. The FAO (United Nations Food and Agriculture Organization) and WHO advise consumers that food should not be cooked excessively—for too long or at too high a temperature. If a high temperature treatment or a long cooking time is avoided, the formation of acrylamide in preferable cooked foods, such as roasted or toasted flavors, must be sacrificed. Figure 11.9 shows the relation between acrylamide and flavor chemical formation. If cooking was maintained at a temperature under 150°C, no preferable flavor chemicals are formed. Therefore, based on the current knowledge of acrylamide, the FDA has reemphasized its traditional advice to eat a balanced diet, choosing a variety of foods that are low in fat and rich in high-fiber grains, fruits, and vegetables. Additional recommendation will be made for cooking methods whenever more appropriate scientific information is advanced.

FOOD IRRADIATION

In the United States, commercial food processing is subject to regulation by the FDA and must meet specified standards of cleanliness and safety. In

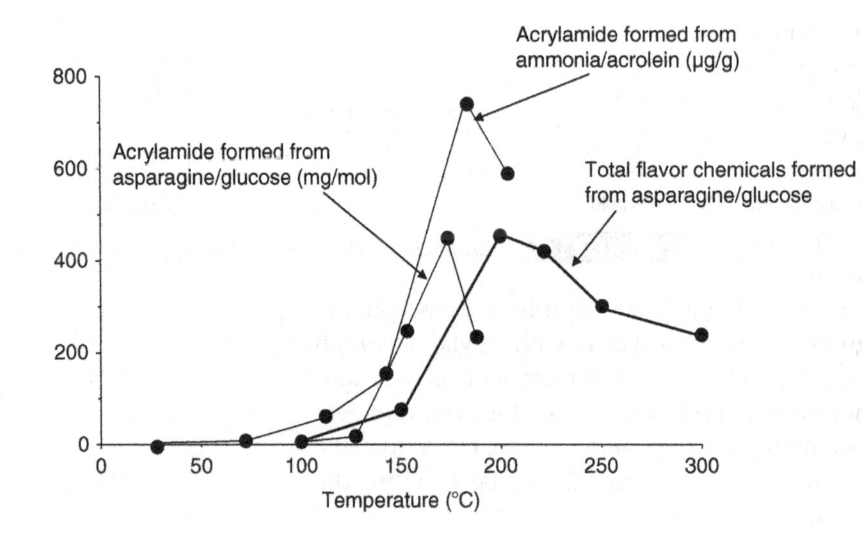

FIGURE 11.9
Formation of acrylamide and flavor chemicals in various temperatures.

certain situations, methods of food processing are considered under the category of food additives since they may intentionally alter the form or nature of food. The use of ionizing radiation to preserve food falls into this category.

Gamma radiation is used most often for food irradiation. Gamma rays are a form of electromagnetic radiation produced by such radioactive elements as Cobalt-60 and Cesium-137. Such sources emit radiation with energies of up to 10 million electron volts (MeV). This is sufficient to penetrate deep into foods, but is far below the range required to produce radioactivity in the target material. Since there is no direct contact between the source and the target, there is no mechanism that can produce radioactivity in irradiated foods.

Studies in the use of ionizing radiation to preserve food began shortly after World War II, in which a large number of potential applications have been identified. Ionizing radiation can sterilize foods, control microbial spoilage, control insect infestations, and inhibit undesired sprouting. Food irradiation has the potential to substantially reduce postharvest applications of pesticides to prevent spoilage due to insects and fungi. Irradiation can be used to destroy *Salmonella* in cases where heat treatment is not possible, for example, in frozen chicken.

Despite the potential of food irradiation as a preserving technique, it is widely misunderstood and controversial. Some opposition arises from apparent confusion between "irradiated" and "radioactive." Gamma irradiation of foods is in some ways analogous to sterilization of medical equipment with ultraviolet light. Both of these processes can kill a wide range of microorganisms by radiation.

Some other critics have raised questions about the toxicity of chemicals that may be produced during irradiation. The energies used are sufficient to produce free radicals, which can combine with each other or form new bonds to other compounds that may be present. However, it is important to remember that heat treatments commonly used in food processing are likely to produce a higher degree of chemical modification than is irradiation.

SUGGESTIONS FOR FURTHER READING

Claus, A., Carle, R., Schieber, A. (2008). Acrylamide in cereal products: A review. *J. Cereal Sci*. 47, 118-133.

Friedman, M. (2003). Chemistry, Biochemistry, and Safety of Acrylamide. A Review. *J. Agric. Food Chem*. 51, 4504-4526.

Ikan, R. (Ed.) (1996). The Maillard Reaction: Consequences for the Chemical and Life Sciences. John Wiley & Sons Ltd., New York.

Miller, E.C., Miller, J.A., Hirono, I., Sugimura, T., Takayama, S. (Eds.) (1979). Naturally Occurring Carcinogens-Mutagens and Modulators of Carcinogenesis. Japan Scientific Societies Press, Tokyo.

Sugimura, T., Kawachi, T., Nagao, M., Yahagi, T., Sano, Y., Okamoto, T., et al. (1977). Mutagenic principles in tryptophan and phenylalanine pyrolysis products, Proc. Japan Academy, Tokyo, 53B, 58-61.

Sugimura, T., Kondo, S., and Takebe, H. (Eds.) (1981). Environmental Mutagens and Carcinogens. University of Tokyo Press, Tokyo.

Turesky, R.J., Lang, N.P., Butler, M.A., Teitel, C.H., Kadlubar, F.F. (1991). Metabolic activation of carcinogenic heterocyclic aromatic amines by human liver and colon. *Carcinogenesis* 12, 1839-1845.

CHAPTER 12

Food Factors and Health

CHAPTER CONTENTS

The term functional foods was first introduced in Japan in the mid-1980s. Since then food factors, or functions of food, have begun to receive much attention as the third factor in the science of foods, after nutrients and flavors. The most common term for this type of concept is **functional food**. However, it has also been called by many different names, including designer foods, pharmafoods, nutraceuticals, medical foods, and a host of others, depending on the researcher's background and perspective.

Functional food sales in the United States were $16 billion in 1999, which represented about 3.3% of the total $747 billion food market. The sales of functional food are increasing by 8 to 9% a year and are estimated to reach $34 billion by 2010. This increase is accounted for by strong growth in novel food products such as probiotic yogurts, soya milks, and fiber-fortified breads. Functional foods comprise many products or supplements with active ingredients, including fiber-, mineral- or vitamin-fortified breakfast cereals, probiotic yogurts and yogurt drinks, and cholesterol-lowering margarines and spreads. In recent years, there are many so-called food supplement products, such as antioxidants, vitamins, minerals, and fiber that have come onto the market. The forces driving the growth of functional foods are:

- New research associating diet and the prevention of chronic disease
- The aging of the population in many developed countries and concerns about managing the health of this age group
- Growing pressure on public health spending
- Increased health consciousness among consumers
- Improvements in food science and technology

PROBIOTICS, PREBIOTICS, AND SYNBIOTICS

Probiotics

Probiotics are defined as live microbial food supplements that beneficially affect the host by improving its intestinal microbial balance. The Food and Agricultural Organization (FAO)/WHO defined probiotics as live microorganisms that, when administered in adequate amounts, confer a health benefit on the host.

The potential health benefits of probiotics include cholesterol-lowering, cancer chemopreventive, and immune-enhancing effects. The general functions of the probiotics, which are associated particularly with intestinal microflora, include energy salvage (lactose digestion and short chain fatty acid production), modulation of cell growth and differentiation, antagonism against pathogens, immune stimulation of the gut-associated lymphoid tissue, innate immunity against infections, production of vitamins, and reduction of blood lipids. For example lactic acid bacteria (*Lactobacillus and Bifidobacterium*) are the most common types of microbes used. They are able to convert sugars and other carbohydrates into lactic acid, which confers possible health benefits in preventing gastrointestinal infections. The functions of lactic acid bacteria include managing lactose intolerance, prevention of colon cancer, cholesterol lowering, lowering blood pressure, improving immune function and preventing infections associated with *Helicobacter pylori* and antibiotic-associated diarrhea, reducing inflammation, improving mineral absorption, preventing harmful bacterial growth under stress, and irritable bowel syndrome and colitis. Table 12.1 shows common strains used for probiotic fortified foods.

Prebiotics

Prebiotics are defined as nondigestible food ingredients that beneficially affect the host by selectively stimulating the growth and activity of one or

Table 12.1 Commonly Used Strains for Dairy Products and Probiotic Fortified Foods

Strain	Proven Effect in Humans
Bifidobacterium animalis DN 173 010	Stabilizes intestinal passage
Bifidobacterium animalis subsp. *lactis* BB-12	Immune stimulation, improves phagocytic activity, alleviates atopic eczema, prevents diarrhea in children and traveler's diarrhea
Bifidobacterium infantis 35624	Irritable bowel syndrome (IBS)
Bifidobacterium lactis HN019 (DR10)	Immune stimulation
Bifidobacterium longum BB536	Positive effects against allergies
Escherichia coli Nissle 1917	Immune stimulation
Lactobacillus acidophilus NCFM	Reduces symptoms of lactose intolerance, prevents bacterial overgrowth in small intestine
Lactobacillus casei DN114-001 (*Lactobacillus casei* Immunitas(s)/ Defensis)	Diarrhea and allergy reduction, immune stimulation, reduction of duration of winter infections, *H. pylori* eradication, antibiotic associated diarrhea, and *C. difficile* infections
Lactobacillus casei F19	Improves digestive health, immune stimulation, reduces antibiotic-associated diarrhea, induces satiety, metabolizes body fat, reduces weight gain
Lactobacillus casei Shirota	Maintenance of gut flora, immune modulation, bowel habits, and constipation
Lactobacillus johnsonii La1 (= Lactobacillus LC1)	Immune stimulation, active against *H. pylori*
Lactococcus lactis L1A	Immune stimulation, improves digestive health, reduces antibiotic-associated diarrhea
Lactobacillus reuteri ATTC 55730 (*Lactobacillus reuteri* SD2112)	Immune stimulation, against diarrhea
Lactobacillus rhamnosus ATCC 53013	Immune stimulation, alleviates atopic eczema, prevents diarrhea in children and many other types of diarrhea
Lactobacillus rhamnosus LB21	Immune stimulation, improves digestive health, reduces antibiotic-associated diarrhea
Lactobacillus salivarium UCC118	Positive effects with intestinal ulcers and inflammation
Saccharomyces cerevisiae (boulardii) lyo	Reduces antibiotic-associated diarrhea and *C. difficile* infections; to treat acute diarrhea in adults and children

a limited number of bacteria in the colon, and thus improving host health. Prebiotics are found naturally in many foods, and can also be isolated from plants (e.g., chicory root) or synthesized (e.g., enzymatically, from sucrose). They allow the selective growth of certain indigenous gut bacteria. To be an effective prebiotic, they must neither be hydrolyzed nor absorbed in the upper part of the gastrointestinal tract. They also should have a selective fermentation, such that the composition of the large intestinal microbiota is altered toward a healthier composition, and most importantly an ability to stimulate selectively the growth and activity of intestinal bacteria associated with health and well-being. Typical prebiotics and their functions are shown in Table 12.2.

The principal characteristic and effect of prebiotics in the diet is to promote the growth and proliferation of beneficial bacteria, such as

Table 12.2 Typical Prebiotics and Their Functions and Sources

Prebiotics	Function	Source and Nature
Fiber gums	Promote the production of short-chain fatty acids	Acacia, carrageenan, guar, locust bean, and xanthan
Fructo-oligosaccharides (FOS)	Stimulate the growth of *Bifidobacterium* and *Bactobacillus* strains, increase the absorption of calcium and magnesium, and decrease triglycerides, anticarcinogenic effects	Jerusalem artichokes, onions, leeks, grains, and honey
Inulins	Stimulate the growth of *Bifidobacterium* in large intestine, hold water, replace fat, and contribute minimal calories	Liliaceae, Amaryllidaceae, chicory, onions, leeks, garlic, bananas, asparagus, and artichokes
Isomalto-oligosaccharides isomaltotetratose,	Stimulate the growth of *Bifidobacterium* and *Lactobacillus* strains	Isomaltose, Panose, and other higher branched oligosaccharides
Lactitol	Releases constipation and hepatic encephalopathy	Disaccharide alcohol analogue of lactulose
Lactosucrose	Increases the growth of the *Bifidobacterium* species	Produced through enzyme action
Pyrodextrins	Promote the growth of *Bifidobacterium* in the large intestine	Mixture of glucose-containing Oligosaccharides derived from starch
Soy oligosaccharides	Stimulate the growth of *Bifidobacterium* species in the large intestine	Mainly in soybeans and peas
Transgalacto-oligosaccharides (TOS)	Stimulate the growth of *Bifidobacterium* in the large intestine	Produced from lactose via enzyme action
Xylo-linked oligosaccharide enzymes	Improve blood sugar levels and fat metabolism, restore normal intestinal flora, increase mineral absorption and vitamin B production, and reduce intestinal putrification	Oligosaccharides containing β-xylose residues, produced via action

Bifidobacterium and *Lactobacillus*, in the intestinal tract. Therefore, they potentially yield or enhance the effect of probiotic bacteria. Prebiotics also have increased the absorption of certain minerals including calcium and magnesium. They may also inhibit the growth of lesions, such as adenomas and carcinomas, in the gut, and thus reduce the risk factors associated with colorectal diseases. The immediate addition of substantial quantities of prebiotics to the diet may result in a temporary increase in gas, bloating, or bowel movement. It has been argued that chronically low consumption of prebiotic-containing foods in the typical Western diet may exaggerate this effect.

Synbiotics

Synbiotics are simply a marriage of the concepts of probiotics and prebiotics. Therefore, synbiotics generally consist of a live microbial food additive together with a prebiotic oligosaccharide. In other words, a synbiotic is a supplement that contains both a prebiotic and a probiotic that work together to improve the "friendly flora" of the human intestine. Products of fermented milks, such as yogurt and kefir, are considered to be true symbiotic products because they contain the live bacteria and the food they need to survive. The main reason for using a symbiotic is that a true probiotic, without its prebiotic food, does not survive well in the digestive system. Without the necessary food source for the probiotic, it will have a greater intolerance for oxygen, low pH, and temperature. Therefore, an appropriate supply of prebiotics to probiotics is essential to obtain effective function of probiotics.

Consuming a probiotic supplement that also includes the appropriate prebiotic has many beneficial effects. Most importantly, the combination has the ability to heal and regulate the intestinal flora, particularly after the destruction of microorganisms following antibiotic, chemotherapy, or radiation therapies. Without the beneficial organisms throughout the digestive system, proper digestion, absorption, and/or manufacture of nutrients cannot take place. A symbiotic will also suppress the development of putrefactive processes in the stomach and intestines, thus preventing the occurrence of a number of serious diseases: food allergies, ulcerous colitis, constipation, diarrhea, cancers, and gastrointestinal infections.

ANTIOXIDANTS

Among various food factors, cancer preventive substances, in particular antioxidants, present in foods and beverages have received much attention not only among scientists but also among consumers. There have been numerous reports on antioxidants associated with various diseases including atherosclerosis, cancer, diabetes, arthritis, immunodeficiency, Alzheimer's disease, and aging. Therefore, this chapter focuses on the role of antioxidants present in foods in disease prevention.

The Role of Oxygen in Living Organisms

Living matter is controlled by numerous chemical reactions. For example, oxygen trapped with hemoglobin in the blood is conveyed to the entire body and used for various reactions. The reactions involved with oxygen are called oxidation and reduction. Because we are living in atmosphere and constantly exposed to oxygen, reactions with oxygen necessarily play an important role in maintaining the world's living organisms. However, one problem is that there are many oxygen species, the so-called reactive oxygen species (ROS), which cause undesirable oxidative reactions. Typical ROS are shown in Table 12.3.

Table 12.3 Typical Reactive Oxygen Species (ROS)

ROS	Structure
Superoxide	$O_2^{\bullet}$
Singlet oxygen	$^1O_2$
Hydroxy radical	$\bullet OH$
Alkoxy radical	$RO\bullet$
Peroxy radical	$ROO\bullet$

As mentioned earlier, oxidative damage caused by reactive oxygen species is directly or indirectly associated with various diseases. Therefore, it is very important to supplement our systems with antioxidants to prevent diseases caused by oxidation.

In vivo Balance Between Oxidants and Antioxidants

Our body has certain enzymes to protect our living systems from oxidative damage. The most well known antioxidant enzyme is superoxide dismutase (SOD). There is an interesting correlation between the amount of SOD present and longevity of each species as shown in Figure 12.1. However, since the modern industrial revolution, people have been producing large amounts of ROS. Table 12.4 shows the typical sources of ROS in the modern world.

UV-light has irradiated the earth since the birth of the solar system. However, because of the creation of an ozone hole by anthropogenically produced alkyl halogens, the amount of UV-light that irradiates the surface of a certain area on the earth, such as Australia, is increasing significantly. Consequently, the amount of ROS produced by UV-light is increasing, and this may lead to an increase in the incidence of skin cancer. The various hazardous effects caused by cigarette smoke are well known. For example, lung cancer incidence among cigarette smokers is approximately six times that of nonsmokers. Cigarette

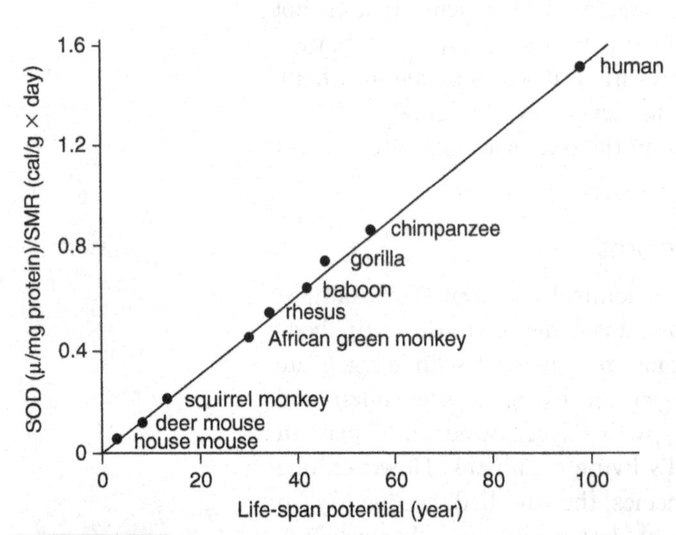

FIGURE 12.1 *Correlation between maximum life-span potential and amount of super oxide dismutase (SOD).*

Table 12.4 Typical Sources and Actions That Produce Reactive Oxygen Species (ROSs)

Source and Action	
Cigarette smoke	Metal ions
Automobile exhaust	Heat
Industrial wastes	Stress
Various incinerators	

smoke also causes various cardiovascular diseases. The presence of ROS in cigarette smoke must be responsible for some of these effects. In addition, ROS have been produced in large quantities by many man-made apparatuses, including automobiles, trains, airplanes, incinerators, and various modern industrial plants. Consequently, the balance between oxidants and anti-oxidants is being destroyed and currently the *in vivo* amount of oxidants exceeds that of antioxidants.

Lipid Peroxidation

Among the three major components of biological systems (proteins, carbohydrates, and lipids), lipids are the most susceptible toward oxidation. Once lipids are oxidized by ROS, many small molecular-weight carbonyl compounds—secondary oxidative products—are formed. This oxidative process is the so-called lipid peroxidation. Figure 12.2 shows proposed formation mechanisms of secondary oxidative products from a typical lipid, polyunsaturated fatty acid (PUFA) via exposure to ROS.

At the last stage of lipid peroxidation, many toxic reactive carbonyl compounds (RCCs), including formaldehyde, acetaldehyde, acrolein, glyoxal, methylglyoxal, and malonaldehyde, are formed. Table 12.5 shows typical RCCs formed by lipid peroxidation.

Heat treatments, including cooking and processes for food product preparation, cause oxidation of food components and, in particular, of lipids. The oxidation of lipids by heat may involve different mechanisms from those of autoxidation and photo-oxidation because the conditions of thermal oxidation are much more intense than those of autoxidation or photo-oxidation. Therefore, many secondary compounds have been identified from thermally treated lipids, including beef fat, cooking oils, pork fat, and dietary oils.

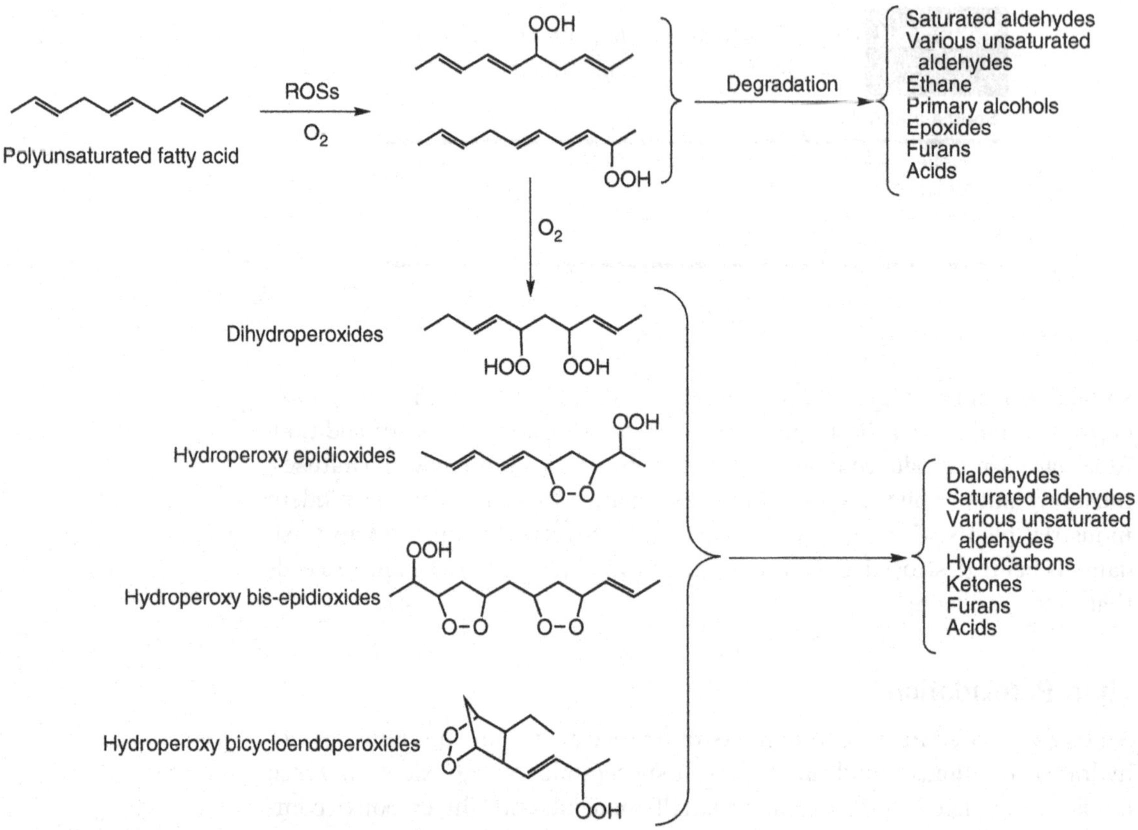

Polyunsaturated fatty acid
ROSs
O_2
OOH
OOH
Degradation
Saturated aldehydes
Various unsaturated aldehydes
Ethane
Primary alcohols
Epoxides
Furans
Acids
O_2
Dihydroperoxides
HOO OOH
Hydroperoxy epidioxides
OOH
O-O
Hydroperoxy bis-epidioxides
OOH
O-O O-O
Hydroperoxy bicycloendoperoxides
OOH
Dialdehydes
Saturated aldehydes
Various unsaturated aldehydes
Hydrocarbons
Ketones
Furans
Acids

FIGURE 12.2 *Proposed formation mechanisms of secondary oxidative products from a typical lipid, polyunsaturated fatty acid (PUFA) via exposure to ROS.*

Table 12.5 Reactive Carbonyl Compounds (RCCs) Formed by Lipid Peroxidation

RCC	Formula	B.P. (°C)	M.W.	Water Solubility (%)
Formaldehyde	HCHO	−19.5	30.03	55
Acetaldehyde	CH_3CHO	21	44.05	100
Acrolein	$CH_2{=}CHCHO$	52.5	56.0	67
4-Hydroxy-2-nonenal	$CH_3(CH_2)_4CH(OH)\text{-}CH_2{=}CH_2CHO$	275.6	156.22	<5
Malonaldehyde	$OHCCH_2CHO$	108.3	72.06	100
Glyoxal	OHCCHO	50.4	58.04	>90
Methyl glyoxal	CH_3COCHO	72.0	72.06	>90

Toxicity of RCCs

It is well known that oxidative damage, in particular, lipid peroxidation, is strongly associated with various diseases, as mentioned earlier. For example, oxidized methyl linoleate, containing 4-hydroxy-2-nonenal (4-HN) as the major component, caused lymphocyte necrosis in the thymus and Peyer's patches in mice. Palm oil oxidized by heat caused reduced rates of pregnancy (by 55%) in rats.

The toxicity of oxidized lipids is caused by the interaction of secondary products of RCCs rather than ROS directly, because ROSs are not readily absorbed by the intestines. Among the many products of lipid peroxidation, RCCs shown in Table 12.5 have received much attention as the chemicals implicated in various diseases. Formaldehyde has shown potential carcinogenicity in animal studies. Alterations in biological proteins were observed in the lungs of rats after they were exposed to gaseous formaldehyde at 32–37 mg/m^3 for 4 h/day for 15 days.

The chronic toxicities of acetaldehyde, such as carcinogenicity, have not been defined by appropriate long-term animal studies. An inhalation toxicity study on acetaldehyde resulted in 23 of the 59 mice dying by exposure to 10 mg/L for 2 h. Studies using cultured human cells indicate that mM concentration levels of acetaldehyde cause a wide range of cytopathic effects associated with multistep carcinogenesis.

Malonaldehyde (MA) may be the best-known lipid peroxidation product and the one that has been used most widely as a biomarker for various studies associated with lipid peroxidation. However, its toxicity has not been well established yet. The fact that MA reacts with DNA to form adducts to deoxyguanosine and deoxyadenosine subsequently implicates it in mutagenicity and carcinogenicity. When 500 μg MA/g body weight was administered to eight-week-old female Swiss mice, pancreatic lesions consisting primarily of atrophied exocrine cells with loss of zymogen granulation occurred.

A study using male outbreed Wistar rats indicated that glyoxal exerted tumor-promoting activity on rat glandular stomach carcinogenesis. Methyl glyoxal also is reported to have various biological implications. Methyl glyoxal inhibited protein, DNA, and RNA synthesis in villus and crypt cells as well as colonocytes. These reports clearly indicate that some RCCs produced from lipid peroxidation caused genotoxicity in experimental animals. Therefore, it is extremely important to find appropriate antioxidants to inhibit lipid peroxidation in order to prevent various diseases caused by oxidative damage.

FUNCTIONAL COMPONENTS FOUND IN FOOD FOR DISEASE PREVENTION

There have been numerous reports on possible prevention of diseases, in particular those caused by oxidative damage, using functional components found in foods. Table 12.6 shows typical functional components found in foods and their potential beneficial effects.

Vitamins have been known for many years to possess biological activities. Vitamin A is a metabolite or breakdown product of β-carotene. Therefore its activity is strongly associated with that of β-carotene. Vitamins C and E are known to scavenge ROS and consequently prevent diseases caused by various oxidative effects, including lipid peroxidation, cytotoxicity, DNA damage, and mutagenesis. For example, a study conducted using experimental mice suggested that Vitamin E had a fundamentally protective effect against breast cancer.

Many human prospective and retrospective studies strongly indicate that β-carotene protects against a variety of cancers. There are several possible mechanisms to account for the effects of β-carotene. β-carotene may alter carcinogen metabolism; it is an antioxidant and may enhance immune responses. Lycopene also belongs to the carotenoid group as well as β-carotene. It has been found in a ripe red tomato. Lycopene is the most abundant carotenoid in the prostate gland and reportedly reduces the risk of various cancers including prostate, breast, bladder, and skin.

Table 12.6 Typical Functional Components Found in Foods and Their Potential Beneficial Effect

Components	Major Sources	Effects
Vitamins A, C, E	Liver, dairy products, fish, bell peppers, citrus fruits, seed oils, mixed nuts, vegetable oils	Protecting cells ROSs, increasing immune and DNA repair functions
β-Carotene	Carrots, leafy green vegetables, corn, eggs, citrus fruits	Protecting cells from ROSs helping bolster cellular antioxidant defenses, increasing immune function
Lycopene	Tomatoes, watermelon	Helping to maintain healthy prostate
Anthocyanidins	Berries, cherries, red grapes	Helping bolster cellular antioxidant defenses, maintaining healthy brain function
Polyphenols, flavonoids	Green tea, cocoa, chocolate, apples, grapes, citrus fruits, onions, apples, tea broccoli, barley leaves	Protecting cells from ROSs, maintaining heart health, positive effect to heart and urinary tract health
Alkyl sulfides	Garlic, onions, leeks, scallions, chives	Enhancing detoxification of undesirable compounds, maintaining heart health and immune function
ω-3 Fatty acids	Fish oils, flaxseed oil	Reducing heart disease risk, maintaining mental and visual function

Flavonoids, found in a numerous variety of natural plants such as soybeans, grapes, and teas, have received much attention as a natural antioxidant in the last two decades. Flavonoids are polyphenolic C_{15} compounds composed of two phenolic nuclei connected by a three-carbon unit. They are categorized, according to chemical structure, into flavonols (quercetin, kaempherol), flavones (apigenin, luteolin), flavanols (catechin, epicatechin), flavanones (hesperetin, naringenin), isoflavones (genistein, daidzein), anthocyanidins (cyanidin, delphinidin), and chalcones. Natural flavonoids often are attached to sugars that affect their biological properties. Figure 12.3 shows the structure of typical flavonoids.

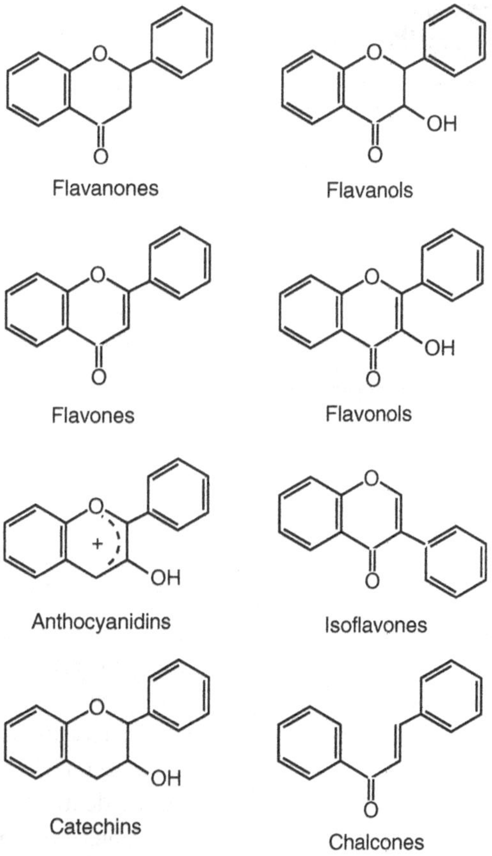

Flavanones

Flavanols

Flavones

Flavonols

Anthocyanidins

Isoflavones

Catechins

Chalcones

FIGURE 12.3 *Structures of typical flavonoids.*

Their potential health benefits include overall disease protection from enhanced cellular antioxidant defenses, roles in supporting heart and urinary tract health, osteoporosis, and maintenance of brain function as well as the prevention of neurodegenerative diseases and diabetes mellitus. There is increasing evidence for the cancer preventive effect of flavonoids emerging in studies of tea catechins, citrus flavonoids, and soy isoflavones.

Alkyl sulfides are present in allium herbs such as onions, garlic, leeks, chives, and scallions. Their potential health benefits include enhanced detoxification of undesirable compounds, cancer fighting, and maintenance of heart health and immune function. Epidemiological studies suggest that higher intake of allium vegetables is associated with a reduced risk of several types of cancers. The allium vegetables described earlier have a protective effect against both esophageal and stomach cancer as well as prostate cancer. The highest antioxidant activity in chives is observed in the leaves, which are also rich in flavonoids.

ω-3 fatty acids are a family of unsaturated fatty acids that have in common a carbon–carbon double bond in the n–3 position; that is, the third bond from the end of the fatty acid. Figure 12.4 shows a typical ω-3, α-linolenic acid (ALA).

$$CH_3CH_2CH=CHCH_2CH=CHCH_2CH=CH(CH_2)_6CH_2COOH$$

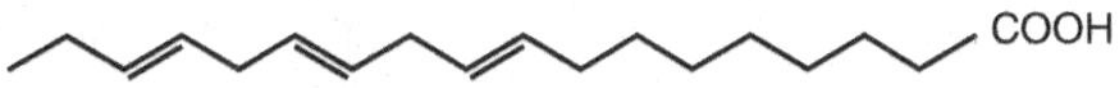

COOH

α-Linolenic acid: 9,12,15-octadecatrienoic acid

FIGURE 12.4 *Structure of α-linolenic acid (ALA).*

Table 12.7 Typical ω-3 Fatty Acids Found in Natural Plants and Animals

ω-3 Fatty Acid	Abr.	Lipid Name	Chemical Name[a]
Hexadecatrienoic acid	HTA	16:3 (*n*–3)	7,10,13-hexadecatrienoic acid
α-Linolenic acid	ALA	18:3 (*n*–3)	9,12,15-octadecatrienoic acid
Stearidonic acid	STD	18:4 (*n*–3)	6,9,12,15-octadecatrienoic acid
Eicosatetraenoic acid	ETE	20:3 (*n*–3)	11,14,17-eicosatrienoic acid
Eicosapentaenoic acid	ETA	20:4 (*n*–3)	8,11,14,17-eicosatetraenoic acid
Eicosapentaenoic acid	EPA	20:5 (*n*–3)	5,8,11,14,17-eicosapentaenoic acid
Docosapentaenoic acid	DPA	22:5 (*n*–3)	7,10,13,16,19-docosapentaenoic acid
Docosahexaenoic acid	DHA	22:6 (*n*–3)	4,7,10,13,16,19-docosahexaenoic acid
Tetracosapentaenoic acid	TPH	24:5 (*n*–3)	9,12,15,18,21-docosahexaenoic acid
Tetracosahexaenoic acid	THA	24:6 (*n*–3)	6,9,12,15,18,21-tetracosenoic acid

[a] *All in* cis *form.*

Table 12.7 shows the most common ω-3 fatty acids found naturally in plants and animals. The term *n*–3 (also called ω-3) signifies that the first double bond exists as the third carbon–carbon bond from the terminal ethyl end (*n*) of the carbon chain.

Some studies suggest that fish oil intake, such as cod liver oil, which contains high levels of ω-3 fatty acids such as EPA and DHA, reduces the risk of coronary heart disease. One study based on 11,324 patients with a history of myocardial infarction reported that treatment of 1 g/day of ω-3 fatty acids reduced the occurrence of death, cardiovascular death, and sudden cardiac death by 20, 30, and 45%, respectively. When 81 patients with unhealthy blood sugar levels were randomly assigned to receive 1800 mg daily EPA for two years, with the other half being a control group, those given the EPA had a statistically significant decrease in the thickness of the carotid arteries along with improvement in blood flow. Furthermore, a recent study indicated that DHA protects against the accumulation in the body of a protein believed to be linked to Alzheimer's disease. Therefore, on September 8, 2004, the USFDA gave "qualified health claim" status to the ω-3 fatty acids EPA and DHA.

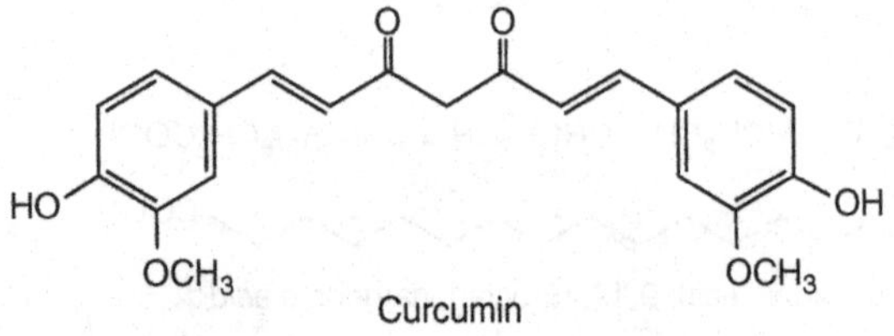

Curcumin (Figure 12.5) is a promising anticancer phytochemical that is present in the curie herb turmeric. Turmeric has been used in folk medicine since about 600 BCE for a wide range of maladies, including digestive disorders, arthritis, cardiovascular conditions, cancer, and bacterial infections. A large number of recent studies provide evidence that the major active component of turmeric, an unusual bisphenol called

FIGURE 12.5 *Structure of curcumin.*

curcumin, inhibits platelet aggregation, suppresses thrombosis and myocardial infarction, and suppresses symptoms associated with type II diabetes, rheumatoid arthritis, multiple sclerosis, and Alzheimer's disease. Curcumin also inhibits human immunodeficiency virus (HIV) replication, enhances wound healing, protects against liver injury, increases bile secretion, inhibits cataract formation, and protects against pulmonary toxicity and fibrosis. Among the many potential molecular targets that have been suggested for curcumin, it appears that inhibition of proinflammatory pathways and activation of the cellular antioxidant defenses are most promising as being essential to many of its biological effects.

Resveratrol (trans-3, 5, 4′-trihydroxystilbene) (Figure 12.6) is polyphenolic compound found in several plant sources, including nuts, berries, and grape skins. Studies in recent years have shown that this phytochemical exhibits chemopreventive activity in some cancer models, is cardioprotective, and can increase lifespans of a number of organisms, including small mammals. As with other potentially beneficial phytochemicals studied to date, no single molecular target seems to explain the many activities of resveratrol. Indeed, resveratrol has been shown to inhibit a plethora of enzymes belonging to different classes, including kinases, lipoxygenase, cyclooxygenases, sirtuins, and other proteins. In addition, this stilbene has been shown to bind to numerous cell-signaling molecules such as multidrug resistance protein, topoisomerase II, aromatase, DNA polymerase, estrogen receptors, tubulin, and mitochondrial F1F0-ATPase. Resveratrol also has been shown to activate various transcription factors, including NF-κB and HIF-1α; suppress the expression of antiapoptotic gene products such as survivin; inhibit important protein kinases such as Src, PI3K, and AKT; induce the cellular antioxidant response; suppress the expression of inflammatory biomarkers; inhibit the expression of angiogenic and metastatic gene products; and modulate cell cycle regulatory genes such as p53.

Sulforaphane (SFN) (Figure 12.6) is an isothiocyanate found in cruciferous vegetables and is present in especially large amounts in broccoli and broccoli sprouts. SFN has proved to be an effective chemoprotective agent in several relevant model systems including human tumor cell xenografts in rodents. The initial mechanistic studies focused on the induction of phase 2 GST conjugating enzymes by SFN, and the inhibition of phase 1 enzymes involved in carcinogen activation. Recent studies, however, suggest that SFN offers protection against tumor development during the postinitiation phases. Thus, SFN can suppress

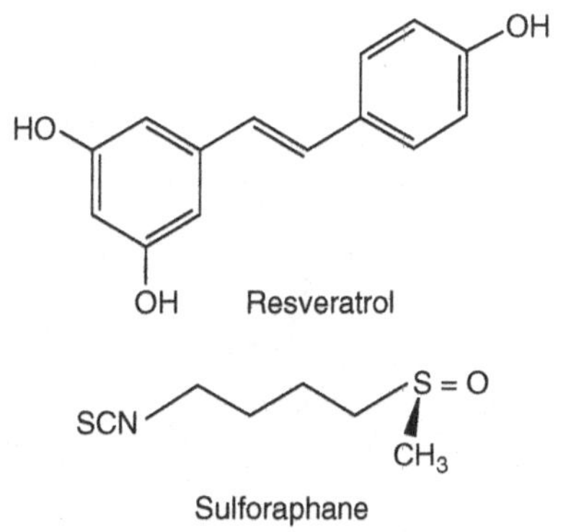

FIGURE 12.6 *Structure of resveratrol and sulforaphane (SFN).*

cancer development through molecular targets that are involved in controlling cell proliferation, differentiation, apoptosis, or cell cycle. It is interesting to note, however, that SFN can activate the cellular antioxidant defenses by a mechanism that involves the inhibition of mitochondrial electron transport and release of ROS.

Indole-3-carbinol (I3C) and its major gastric conversion product, 3,3′-diindolylmethane (DIM) (Figure 12.7), are produced from *Brassica* vegetables such as cabbage, broccoli, and Brussels sprouts. The glucosinolate precursor of I3C, glucobrassicin, occurs in some varieties of brussels sprouts at levels of up to 2,000 ppm based on the dry weight of the plant. Although hydrolysis of glucobrassicin by myrosinase in the plant is the major conversion pathway of glucobrassicin to DIM and other active indoles, activity also is produced following oral administration to rats of purified glucobrassicin. Results of several studies indicate that DIM exhibits promising cancer protective activities against cervical, prostatic, and esophageal neoplasia. Some of the most well-established cancer protective effects of I3C and DIM, however, are associated with their activities in mammary tissue. For example, dietary DIM reduced the incidence of DMBA-induced mammary tumors by as much as 95%, regardless of whether the DIM treatments were begun during the initiation or the postinitiation stages of carcinogenesis. Studies with I3C, however, suggest that this indole exhibits cancer-promoting activities in some tumor models.

Because of the well-documented cancer protective effects of I3C and DIM in rodents, their low toxicity, and their ready availability, I3C and DIM are undergoing several clinical trials as cancer chemotherapeutic and preventive agents against cancerous or precancerous lesions for several sites, including the uterine cervix and prostate gland. Furthermore, results of small studies with indole treatment of recurrent respiratory papillomatosis (RRP) are highly promising. DIM is currently the preferred treatment for this rare but debilitating disorder. Molecular targets of DIM include estrogen receptor β, mitochondrial F1F0-ATPase, topoisomerase IIα, the androgen receptor, and the Ah receptor. Furthermore, DIM can inhibit HIF-1 signaling and induce ROS release in hypoxic tumor cells, and potentiate the immune response to bacterial and viral antigens in rodents and cultured immune cells.

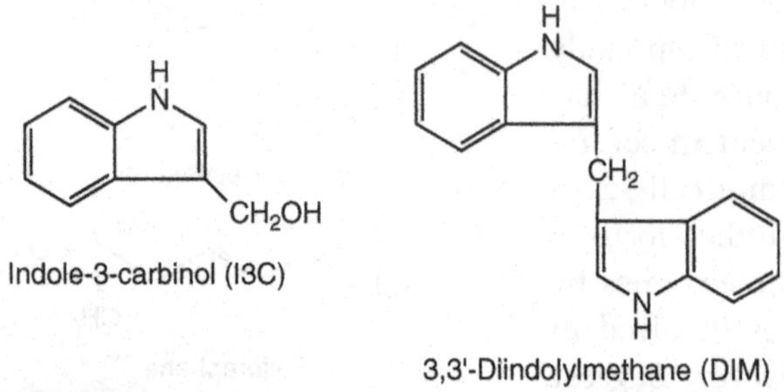

Indole-3-carbinol (I3C)

3,3′-Diindolylmethane (DIM)

FIGURE 12.7 *Structure of indole-3-carbinol (I3C) and 3,3′-diindolylmethane (DIM).*

As our understanding of the beneficial effects of phytochemicals has grown, it has become clear that several show considerable promise in the treatment or prevention of many important chronic diseases. Nevertheless, as a group they exhibit properties that are distinct from the properties of most established drugs. For example, most of the phytochemicals are active in the low micromolar concentration range and affect multiple molecular targets. In contrast, most established drugs are designed to be active in the nanomolar concentration range and target a single or small group of related molecular targets. Furthermore, most of the active phytochemicals are rapidly metabolized and show low toxicity, whereas most of the drugs are cleared more slowly from the body and can exhibit significant toxicities or side effects. Another potentially important difference in the two types of substances is that although the phytochemicals that are under investigation are chemical antioxidants, they can function as biological pro-oxidants, often by mechanisms that involve the inhibition of mitochondrial electron transport. In contrast, most drugs are effective at concentrations that do not induce the release of ROS and, thus, are not biological pro-oxidants. Continued studies of this important group of natural substances will determine their true clinical usefulness and safety, and will define the contributions of their effects on individual molecular targets or on general stress responses in their modes of action.

SUGGESTIONS FOR FURTHER READING

Calabrese, V., Cornelius, C., Mancuso, C., Pennisi, G., Calafato, S., Bellia, F., Bates, T.E., Giuffrida Stella, A.M., Schapira, T., Dinkova Kostova, A.T., Rizzarelli, E. (2008). Cellular stress response: A novel target for chemoprevention and nutritional neuroprotection in aging, neurodegenerative disorders and longevity. *Neurochem. Res.* 2008 Jul 16. [Epub ahead of print]

Clarke, J.D., Dashwood, R.H., Ho, E. (2008). Multi-targeted prevention of cancer by sulforaphane. *Cancer Lett.* 269:291-304.

Pari, L.,Tewas, D., Eckel, J. (2008). Role of curcumin in health and disease. *Arch. Physiol. Biochem.* 114:127-149.

Pirola, L., Fröjdö, S. (2008). Resveratrol: One molecule, many targets. *IUBMB Life.* 60:323-332.

Shibamoto, T., Terao, J., Osawa, T. (Eds.). (1998). Functional Foods for Disease Prevention I: Fruits, Vegetables, and Teas. ACS Symposium Series 701, ACS, Washington, DC.

Shibamoto, T., Terao, J., Osawa, T. (Eds.). (1998). Functional Foods for Disease Prevention II: Medicinal Plants and Other Foods. ACS Symposium Series 701, ACS, Washington, DC.

Shibamoto, T., Kanazawa, K., Shahidi, F., Ho, C.-T. (Eds.). (2008). Functional Food and Health, ACS Symposium Series 993,ACS Washington, DC.

Weng, J.R., Tsai, C.H., Kulp, S.K., Chen, C.S. (2008). Indole-3-carbinol as a chemopreventive and anti-cancer agent. *Cancer Lett.* 262:153-163.

Index

D

T

U

V

W

Food Science and Technology International Series

Maynard A. Amerine, Rose Marie Pangborn, and Edward B. Roessler, *Principles of Sensory Evaluation of Food*. 1965.
Martin Glicksman, *Gum Technology in the Food Industry*. 1970.
Maynard A. Joslyn, *Methods in Food Analysis*, second edition. 1970.
C. R. Stumbo, *Thermobacteriology in Food Processing*, second edition. 1973.
Aaron M. Altschul (ed.), *New Protein Foods*: Volume 1, *Technology, Part A*—1974. Volume 2, *Technology, Part B*—1976. Volume 3, *Animal Protein Supplies, Part A*—1978. Volume 4, *Animal Protein Supplies, Part B*—1981. Volume 5, *Seed Storage Proteins*—1985.
S. A. Goldblith, L. Rey, and W. W. Rothmayr, *Freeze Drying and Advanced Food Technology*. 1975.
R. B. Duckworth (ed.), *Water Relations of Food*. 1975.
John A. Troller and J. H. B. Christian, *Water Activity and Food*. 1978.
A. E. Bender, *Food Processing and Nutrition*. 1978.
D. R. Osborne and P. Voogt, *The Analysis of Nutrients in Foods*. 1978.
Marcel Loncin and R. L. Merson, *Food Engineering: Principles and Selected Applications*. 1979.
J. G. Vaughan (ed.), *Food Microscopy*. 1979.
J. R. A. Pollock (ed.), *Brewing Science*, Volume 1—1979. Volume 2—1980. Volume 3—1987.
J. Christopher Bauernfeind (ed.), *Carotenoids as Colorants and Vitamin A Precursors: Technological and Nutritional Applications*. 1981.
Pericles Markakis (ed.), *Anthocyanins as Food Colors*. 1982.
George F. Stewart and Maynard A. Amerine (eds), *Introduction to Food Science and Technology*, second edition. 1982.
Hector A. Iglesias and Jorge Chirife, *Handbook of Food Isotherms: Water Sorption Parameters for Food and Food Components*. 1982.
Colin Dennis (ed.), *Post-Harvest Pathology of Fruits and Vegetables*. 1983.
P. J. Barnes (ed.), *Lipids in Cereal Technology*. 1983.
David Pimentel and Carl W. Hall (eds), *Food and Energy Resources*. 1984.
Joe M. Regenstein and Carrie E. Regenstein, *Food Protein Chemistry: An Introduction for Food Scientists*. 1984.
Maximo C. Gacula, Jr. and Jagbir Singh, *Statistical Methods in Food and Consumer Research*. 1984.
Fergus M. Clydesdale and Kathryn L. Wiemer (eds), *Iron Fortification of Foods*. 1985.
Robert V. Decareau, *Microwaves in the Food Processing Industry*. 1985.
S. M. Herschdoerfer (ed.), *Quality Control in the Food Industry*, second edition. Volume 1—1985. Volume 2—1985. Volume 3—1986. Volume 4—1987.

F. E. Cunningham and N. A. Cox (eds), *Microbiology of Poultry Meat Products*. 1987.
Walter M. Urbain, *Food Irradiation*. 1986.
Peter J. Bechtel, *Muscle as Food*. 1986. H. W.-S. Chan, *Autoxidation of Unsaturated Lipids*. 1986.
Chester O. McCorkle, Jr., *Economics of Food Processing in the United States*. 1987.
Jethro Japtiani, Harvey T. Chan, Jr., and William S. Sakai, *Tropical Fruit Processing*. 1987.
J. Solms, D. A. Booth, R. M. Dangborn, and O. Raunhardt, *Food Acceptance and Nutrition*. 1987.
R. Macrae, *HPLC in Food Analysis*, second edition. 1988.
A. M. Pearson and R. B. Young, *Muscle and Meat Biochemistry*. 1989.
Marjorie P. Penfi eld and Ada Marie Campbell, *Experimental Food Science*, third edition. 1990.
Leroy C. Blankenship, *Colonization Control of Human Bacterial Enteropathogens in Poultry*. 1991.
Yeshajahu Pomeranz, *Functional Properties of Food Components*, second edition. 1991.
Reginald H. Walter, *The Chemistry and Technology of Pectin*. 1991.
Herbert Stone and Joel L. Sidel, *Sensory Evaluation Practices*, second edition. 1993.
Robert L. Shewfelt and Stanley E. Prussia, *Postharvest Handling: A Systems Approach*. 1993.
Tilak Nagodawithana and Gerald Reed, *Enzymes in Food Processing*, third edition. 1993.
Dallas G. Hoover and Larry R. Steenson, *Bacteriocins*. 1993.
Takayaki Shibamoto and Leonard Bjeldanes, *Introduction to Food Toxicology*. 1993.
John A. Troller, *Sanitation in Food Processing*, second edition. 1993.
Harold D. Hafs and Robert G. Zimbelman, *Low-fat Meats*. 1994.
Lance G. Phillips, Dana M. Whitehead, and John Kinsella, *Structure-Function Properties of Food Proteins*. 1994.
Robert G. Jensen, *Handbook of Milk Composition*. 1995.
Yrjö H. Roos, *Phase Transitions in Foods*. 1995.
Reginald H. Walter, *Polysaccharide Dispersions*. 1997.
Gustavo V. Barbosa-Cánovas, M. Marcela Góngora-Nieto, Usha R. Pothakamury, and Barry G. Swanson, *Preservation of Foods with Pulsed Electric Fields*. 1999.
Ronald S. Jackson, *Wine Tasting: A Professional Handbook*. 2002.
Malcolm C. Bourne, *Food Texture and Viscosity: Concept and Measurement*, second edition. 2002.
Benjamin Caballero and Barry M. Popkin (eds), *The Nutrition Transition: Diet and Disease in the Developing World*. 2002.
Dean O. Cliver and Hans P. Riemann (eds), *Foodborne Diseases*, second edition. 2002.
Martin Kohlmeier, *Nutrient Metabolism*, 2003.
Herbert Stone and Joel L. Sidel, *Sensory Evaluation Practices*, third edition. 2004.
Jung H. Han, *Innovations in Food Packaging*. 2005.

Da-Wen Sun, *Emerging Technologies for Food Processing*. 2005.
Hans Riemann and Dean Cliver (eds) *Foodborne Infections and Intoxications*, third edition. 2006.
Ioannis S. Arvanitoyannis, *Waste Management for the Food Industries*. 2008.
Ronald S. Jackson, *Wine Science: Principles and Applications*, third edition. 2008.
Da-Wen Sun, *Computer Vision Technology for Food Quality Evaluation*. 2008.
Kenneth David and Paul Thompson, *What Can Nanotechnology Learn From Biotechnology?* 2008.
Elke K. Arendt and Fabio Dal Bello, *Gluten-Free Cereal Products and Beverages*. 2008.
Debasis Bagchi, *Nutraceutical and Functional Food Regulations in the United States and Around the World*, 2008.
R. Paul Singh and Dennis R. Heldman, *Introduction to Food Engineering*, fourth edition. 2008.
Zeki Berk, *Food Process Engineering and Technology*. 2009.
Abby Thompson, Mike Boland and Harjinder Singh, *Milk Proteins: From Expression to Food.* 2009.
Wojciech J. Florkowski, Stanley E. Prussia, Robert L. Shewfelt and Bernhard Brueckner (eds) *Postharvest Handling*, second edition. 2009.

Printed in the United States
v Bookmasters